The AT Reader

Theory and practice in Appropriate Technology

The A̲T Reader

Theory and practice in Appropriate Technology

Edited by MARILYN CARR
with an Introduction by Frances Stewart

NEW YORK
INTERMEDIATE TECHNOLOGY DEVELOPMENT
GROUP OF NORTH AMERICA   1985

Published by Intermediate Technology Development Group
of North America, Publications Office, P.O. Box 337,
Croton-on-Hudson, NY 10520

© This selection Intermediate Technology Publications 1985

ISBN 0-942850-03-3

Printed in Great Britain by The Pitman Press, Bath, UK

Contents

3 Communication

Acknowledgements

It would be impossible to mention everyone who has contributed to the preparation of this book. Several people were kind enough to read through the preliminary contents list and to suggest additional extracts which would make the *AT Reader* more comprehensive. They include Andrew Barnett, Science Policy Research Unit; Stephen Biggs, University of East Anglia; Frances Stewart, Somerville College, Oxford; Terry Thomas, University of Warwick; Robert Waddell and Bill Lawson, University of New South Wales; and Don Mansell, University of Melbourne. Almost everybody at ITDG has been involved in one way or another, but special thanks are due to Patrick Mulvany, John Collett, Andrew Scott and Jeff Kenna for their comments on the case study chapters, and to George McRobie for his suggestions on current North American literature.

I would like to mention particularly Caroline O'Reilly and Sue Eckstein who worked with me as research assistants during the many months it took to compile the anthology. For their untiring help in the numerous tasks that are involved in preparing a publication of this nature, I owe them very special thanks.

Marilyn Carr,
London,
January 1985

ITDG would like to acknowledge the generous contribution of an anonymous donor which made this publication possible.

Introduction
by FRANCES STEWART

AN APPROPRIATE technology (AT) is a technology which is suited to the environment in which it is used. The idea that appropriate technology was often different from the technology actually being adopted was first conceived with reference to the Third World, but more recently it has been argued that in the advanced countries, too, much investment incorporates technologies which are inappropriate to the long-term needs of the community.

In the Third World, much of the technology is imported from the advanced countries; it is seen as inappropriate because it costs too much per workplace, creates too few jobs, involves excessive scale of production, leading to too rapid urbanization, while at the same time producing oversophisticated products which do not meet the basic human needs of the mass of poor people. Although advanced technology has permitted substantial increases in output, it has left the majority of people in the Third World impoverished and underemployed. In contrast, an *appropriate* technology would spread productive employment opportunities widely, and at less cost, and would produce products to meet basic needs. Appropriate technology would also make fuller use of local natural resources and would be consistent with the ecological needs of the environment.

Support for AT in advanced countries has similar roots — dissatisfaction with some of the consequences of the use of advanced technology. In the rich countries, too, modern technology has failed to create enough jobs; it has often had destructive effects on the environment; and has led to alienation among workers, as a result of the overspecialization and the employment relationships which are associated with it. AT for advanced countries would be job creating, producing more *satisfying* work; the ecological damage caused by the productive process would be less with AT, as a result of the greater use of renewable resources, so that a productive system based on AT would be sustainable over the long term.

Since the ideas of AT were introduced, the AT movement has gained a vast number of adherents, including some governments, such as those in Tanzania and India, and a good deal of the aid 'establishment' in advanced countries. Indeed, if AT is defined as described above, it is difficult to find any arguments against it. Yet despite the cogent arguments, and the prominent and numerous supporters, the success of the AT movement has so far been limited, where success is defined in terms of the proportion of investment resources incorporating AT. It is important to explore why this is so, because doing so points to the direction the AT movement will need to go, over the next decade, if AT is to become a reality for the majority, rather than a minority cause.

Obstacles which have impeded the adoption of AT may be summarized in three categories: problems of existence, of efficiency and of choice. To the extent that there is no appropriate technology in existence, inappropriate technology is unavoidable. But existence alone is not sufficient. If a small-scale, low-cost alternative technology already exists and has a high unit cost but is very inefficient (and leads to a much lower level of output than a large-scale, high-cost alternative) then in most circumstances the latter will be chosen. Hence AT needs to be 'efficient', offering comparable levels of productivity to the alternatives. Finally, there is the question of choice; even where an efficient AT exists, inappropriate technologies are often selected for a variety of reasons. Issues arising in connection with these three problems — existence, efficiency and choice — form the substance of many of the extracts in this volume. Here I will consider some of the issues very briefly.

The question of *existence* is closely connected to that of choice of product. Product choice largely determines technology choice. If the final product is a modern one, embodying advanced technology, for example an airplane, a recent model of car, or a computer, then the technology necessary to produce it will generally also be a recently developed technology with inappropriate characteristics, in terms of capital cost, scale etc. Even among older products, if the product is required to have the characteristics generally available in the mass markets of advanced countries — being standardized, of uniform quality, highly packaged etc. — the method of production will tend to be capital-intensive and large in scale. This effect has been demonstrated in a number of empirical studies — for example of maize milling, sugar refining, bread manufacture. For some products, then, there is a very limited range of technolgies for the 'core' process, although a wider choice is available for ancillary activities, such as preparation of materials, packaging and transport. For others, there is more choice but only if some variation in product characteristics is accepted. Hence AT has to encompass appropriate product choice, not only because appropriate products are a vital element in appropriate technology, but also because unless the product choice is widened, the technology choice will be very limited.

The range of technology in existence at any time depends on the historical development of technology and the current direction of research and development (R and D) and technological change. The main reason why many modern products are associated with inappropriate technologies is that they have been developed to suit high income economies, where there is a much greater abundance of capital, labour costs are much higher, and firms are to a great extent organized on a large scale. The direction of technological change also affects the *efficiency* of different techniques. Since most R and D has been devoted to large-scale techniques, their efficiency has been increased relative to ATs. The AT movement has produced some change here, but of a limited magnitude — indeed, the efforts have been very important as an example of what might be achieved, but insufficient to change the broad direction of technological change, except in a very few cases. Most R and D is still conducted by large firms in advanced countries which in most cases leads to a continual increase in the efficiency of inappropriate technologies. A vital element for the future viability of AT is therefore the *continuous, systematic* and *substantial* R and D (and dissemination) on appropriate techniques and appropriate products. Without this, the efficiency of AT will decline relative to inappropriate technologies, making it less attractive. This need for research applies to AT both in the Third World and in advanced countries: pioneering efforts, such as those of the late M. K. Garg in India have shown how effective such research can be, but such examples are unfortunately rare. That this is so is partly due to the whole system of production and incentives in both advanced countries and developing countries which, by leading to choice of inappropriate technologies, tends to discourage R and D on AT. In many situations, even where ATs exist and are efficient, inappropriate technologies are nonetheless chosen. This arises from numerous distortions in the selection mechanisms which determine the choice of technology. Price signals are often distorted, with cheap capital and expensive labour. Credit allocation in many countries is biased towards the large-scale formal sector, while machinery suppliers from advanced countries also provide credit for expensive equipment, often at subsidized rates. Machinery producers in developing countries, too, sometimes produce inappropriate machines which they promote vigorously. Despite the nominal commitment to AT among donors, many aid projects use high-cost imported equipment. Highly unequal income distribution creates markets for advanced products among the élites, which require inappropriate technologies to produce, while impoverishment among the masses means they lack the purchasing power for the consumption of appropriate products. In agriculture, unequal land distribution, together with biases in the credit system and imbalance in research, often leads to the adoption of unnecessarily capital-intensive technology.

There are a host of reasons why distorted selection mechanisms lead to inappropriate choices. Similar mechanisms apply in the advanced countries, where most selection mechanisms (e.g. credit allocation, information systems, price signals) tend to favour the large-scale sector. If the AT movement is to be successful in terms of affecting a substantial proportion of investment decisions, it will be necessary to change these selection mechanisms. So far the AT movement has been effective in demonstrating the need for AT, and, in some areas, its existence and efficiency, as the extracts in the earlier chapters of this book show. The next stage is to secure a change in selection mechanisms in such a way that the widespread adoption of AT ensues. But to secure such a change is extremely difficult because it often involves a challenge to the existing vested interests which have gained from the present inappropriate set of choices. At this stage AT necessarily becomes political because the changes necessary require major shifts in economic power. This is not to argue that any single type of political economy is uniquely associated with AT. Both capitalist and socialist systems are capable of generating appropriate technological choices, while both have been associated with inappropriate choices. The change in political economy required is one which increases opportunities of the low-income groups both in terms of investment resources and consumption. As is clear from the extracts that follow, the fundamental objective of those who advocate AT is to enhance the economic position of the impoverished, and those who have not fully participated in the orthodox growth process, based as it is on inappropriate technology. The first generation literature showed the need for this, and the possibility of achieving it. It is the task of the second generation literature (of which much less is available) to explore the conditions which will make this possibility into a reality.

The issue of AT is not an all-or-nothing one. It is possible to move part of the way, taking more appropriate choices in some areas, and this can be seen by looking at the world around us. Some countries have clearly taken more appropriate decisions in both industry and agriculture than others: contrast, for example, Japan, China, India and Taiwan with Mexico, the Philippines or Nigeria. The first group of countries has adopted a much more appropriate set of technologies than the second, even though there are many examples of inappropriate decisions within the first group. But, as Mao said, it is possible to 'walk on two legs': and given the limited availability of AT in some areas, this is a posture which may be necessary at a particular stage of technological development — with small-scale labour-intensive technologies in some sectors, and more capital-intensive technologies in others.

Similar possibilities apply to advanced countries: the role of AT may be gradually extended. While the actual technologies that constitute the appropriate technology are likely to differ as between Third World and advanced countries, the sort of changes required to secure AT are very similar, while both share the same ultimate objective — a sustainable system of production and consumption in which every member of society participates.

One special contribution of AT thinking has been to emphasize the role of technology as an independent force determining the pattern of employment and consumption. But technology is not itself autonomous: it is created by social forces. In the past, these forces have favoured the creation of inappropriate technology. Just as appropriate choice of technology has to establish its own routine procedures, so the direction of technological change has to be redirected, so that R and D leads to improvements in ATs and to new ATs, in the normal process of scientific and technological activity, and without exceptional interventions. In the long run, changing the direction of technological change may be more important to the success of AT than changing the choice made today out of the technological alternatives now available. However, the two go together because changing the current choice is a necessary aspect of changing the direction of technological change, while without the latter, future technological choice will be confined to inappropriate technologies.

The AT movement of the past has suffered from two weaknesses. First, there has tended to be too much attention on the 'hardware' and too little attention on the 'software' aspects of introducing and disseminating improved technologies. As recorded in many of the extracts in this book, experience shows that for a technology to be really appropriate and to reach the masses of people who could benefit from it, a great deal of effort has to be put into encouraging economic change, providing institutional back-up and understanding interactions between technological and social change. Second, as will be evident from the extracts which follow — outsiders have performed an important role in Appropriate Technology — as, for example, thinkers and as financiers. The endogenizaton that is necessary for the spread of Appropriate Technology implies of course a reduced role for outsiders and a much greater role for the 'insiders' and the people themselves — upon whom the success of AT will rest.

Overleaf: A photovoltaic array, which can convert sunlight directly to electricity.

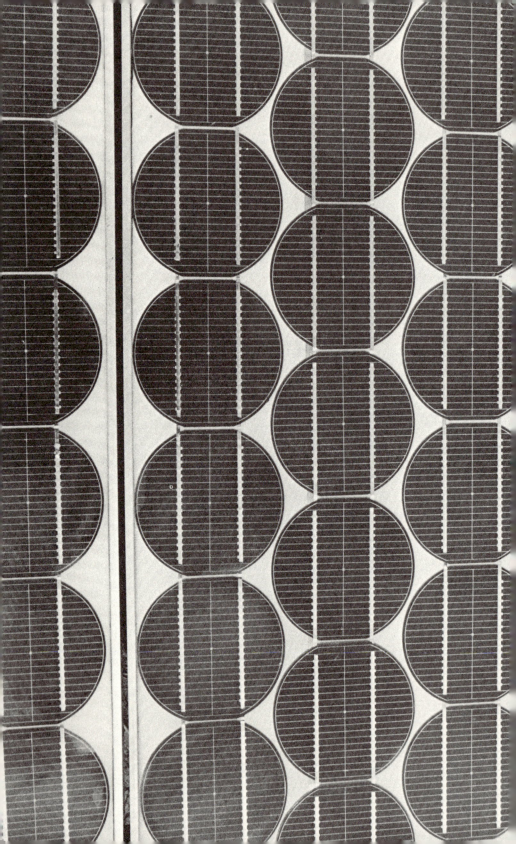

I

Appropriate Technology

History, Concepts and Evolution

IN THE past quarter of a century the developing countries have made great progress in terms of increasing the per capita incomes of their rapidly growing populations. Progress has also been made in the non-income fields such as health, education and infrastructure, but despite this growth record, poverty has not been appreciably lessened. Levels of unemployment and income inequality are increasing, as are the numbers of people living at or below subsistence level. According to the World Bank, there were by the early 1980s over 800 million people (almost half of the total population of the developing world) living in 'absolute poverty'. Such people are denied the basic components of a decent human life — adequate food supply, adequate clothing and housing, and access to basic services such as clean water, energy, schools and health facilities — which are taken for granted by most people in developed countries.

The pattern is identical in almost every developing country. Approximately 80 per cent of the population lives in rural areas where poor returns from agricultural work and lack of non-farm employment opportunities offer little in the way of a brighter future. This results in a massive migration of people (especially younger people) from the countryside, and in a growth of urban population with which urban industry has simply been unable to keep pace. The amenities which attracted people to the city in the first place are only available to the relatively few who can find steady well-paid employment. Others scrape a living by such jobs as shoe-shining, selling matches, selling lottery tickets and so on — jobs which circulate money, but do not create wealth. Meanwhile, from the point of view of food, the cities are parasites demanding bigger and bigger surpluses from the countryside — surpluses which the countryside is increasingly unable to provide.

This situation is one which E.F. Schumacher has appropriately described as a 'process of mutual poisoning', whereby industrial development in the cities destroys the economic structure of the hinterland and the hinterland takes its revenge by mass migration into the cities, poisoning them and making them utterly unmanageable.[1]

The solution to any problem must lie in its cause. Undeniably, a major

---

1 E.F. Schumacher, *Small is Beautiful: A Study of Economics As If People Mattered* (Blond and Briggs, 1973).

1

cause of the most pressing problems of the Third World has been the transfer and use of technologies which are totally inappropriate to prevailing conditions. A famous case study from one African country illustrates this point perfectly. Two plastic-injection moulding machines costing US$100,000 each were imported to produce plastic shoes and sandals. Working three shifts and with a total labour force of only forty workers, the machines produced 1.5 million pairs of shoes and sandals a year. At US$2 per pair, these were better value and had a longer life than cheap leather footwear at the same price. Thus, 5,000 artisan shoemakers lost their livelihood which, in turn, reduced the markets for the supplies and makers of leather, hand tools, cotton thread, tacks, glues, wax and polish, fabric linings, laces, wooden lasts and carton boxes — none of which were required for plastic footwear. As all the machinery and the material (PVC) for the plastic footwear had to be imported, while the leather footwear was based largely on indigenous materials and industries, the net result was a decline in both employment and real income within the country.[2]

Strategies such as this, even if they have produced the desired rate of economic growth, have very obviously contributed to an inability to create full employment, to a loss of employment opportunities in the traditional sector and to the increasing rate of migration from rural areas to the cities. The problem has been compounded by a relative lack of policy measures aimed at increasing agricultural productivity, at generating rural off-farm employment opportunities and at generally improving living standards in the rural areas. Agricultural development plans, for instance, have often emphasized the increased use of imported tractors and other types of expensive engine-driven agricultural implements, which are now to be found, in various states of disrepair, throughout the Third World. These machines may be appropriate in Europe and North America, where capital and skills are plentiful and where labour is scarce, but in developing countries, where the opposite holds true, they are (except in a few circumstances) far from being appropriate. The imposition of imported technologies on Third World farming systems is unlikely to achieve the major objective of increasing food output when the skills necessary to maintain and repair complicated machines, and often the fuel and spare parts necessary to keep them running, are lacking.

Similarly, in the area of the provision of basic services such as water and heath facilities, there have been attempts to emulate the standards of the West by providing, for instance, modern hospitals and individual water connections in the town. These, however, simply increase the standard of living of an already better-off minority while the majority of the people in greatest need receive no benefits at all. For example, it costs between US$4,000 and US$25,000 to provide one bed in a new modern hospital, while operational costs vary between US$1,000 and US$4,000 per bed year. For the same expenditure, a small rural health centre could be built, equipped and maintained. Such a centre could provide at least some medical assistance for up to

2 Keith Marsden, 'Progressive Technologies for Developing Countries', *International Labour Review*, Vol. 101 (May 1970).

6,000 people. Of course, modern hospitals are still needed, but millions of rural people require medical assistance and this is something they will never get if planners think only in terms of modern hospitals.

The tasks which have to be faced lie in helping millions of small farmers to be more productive; in creating millions of new jobs — in farming, but even more in non-farm work — in the rural areas; and in providing millions of families with at least reasonably adequate basic services.

Under the circumstances, planners and development economists have had to take a new look at the meaning of development, and arising out of this has come a trend towards development strategies which emphasize the maximization of employment, the reduction of income inequalities, and the provision of basic human needs, rather than the maximization of output. An integral part of this change in emphasis has been an increased interest in the development and dissemination of technologies which are appropriate to the needs and means of the masses of people in developing countries. In general, such technologies need to be small, simple and cheap enough to harmonize with local human and material resources and lend themselves to widespread reproduction with the minimum of outside help. They need to provide new and improved workplaces in the rural areas where people live — workplaces which can be created at low cost without making impossible demands on savings and imports. They should be the basis of production directed mainly at meeting local basic needs and using indigenous raw materials and local skills. They should not be as complex as those transferred from the West with such disastrous effect, but should enable higher levels of labour productivity and surplus than the technologies traditionally used in developing countries.

These characteristics are those which E.F. Schumacher was describing when, in 1973, he first brought the concept of 'intermediate' technology to the attention of the public in his best-selling book, *Small is Beautiful*. Schumacher was of course, not the first or the only person to identify the need for intermediate technologies in the Third World. Such eminent figures as Gandhi and Julius Nyerere have also talked about, and campaigned for, such technologies. Schumacher did, however, encapsulate society's concern about the state of the world at the very time when the problems of the developing countries were becoming increasingly serious, and when the developed countries themselves were facing increasingly severe economic and environmental problems.

Schumacher's concept of 'smallness' offered a potential solution to the problems of both the developing and the developed world. In the case of the latter, his analysis of the problems and his suggested solution mirrored those of many other writers and philosophers from Morris and Ruskin to Huxley, Rachel Carson and John Kenneth Galbraith. All warned of the dangers of environmentally careless growth, and argued for a more humane, ecologically sound and human-scale society. The idea that both the First World and the Third should move towards technologies appropriate to a sustainable, balanced economy began to gain acceptability.

From being a relatively unknown concept in the 1960s, Appropriate Technology has come to be a commonly used term in the literature of develop-

3

ment. With the change in emphasis in development strategies in the Third World, and the increasing difficulty experienced by the First World countries in sustaining their present pattern of development, the concept is also one which has gained widespread acceptance from governments, academics, international agencies and the business world. When Schumacher first started his campaign for more appropriate technologies, his ideas were not well received. A quarter of a century later, the wisdom of his words and the dangers involved in ignoring them are becoming only too obvious.

Despite, or perhaps because of, its increased popularity confusion still exists as to the meaning of Appropriate Technology and the difference between Appropriate and Intermediate Technology. Numerous writers over the years have addressed themselves to this issue and in the first section of this chapter extracts have been selected which attempt to define and explain the AT or IT concept in various ways. Particularly important is Schumacher's explanation of what he himself meant by 'intermediate' technology. A useful expansion is provided by James Robertson who elaborates on the idea of a 'One World Technology'. Also very relevant are Julius Nyerere's view of intermediate technology, Gandhi's definition of Swadeshi (self-reliance) and Schumacher's, Morris's and Robertson's views on work — essentially something which is enjoyed rather than being a burden.

The AT concept and movement have not been without their critics. In Section 2, some of the more common criticisms, including the accusation that AT promotes 'second best' technology and that it cannot solve problems in the absence of political change, are presented — along with rejoinders. One of the more recent challenges to the movement is to be found in Aghiri Emmanuel's book *Appropriate or Underdeveloped Technology?* in which he argues that a technology 'appropriate' to the underdeveloped countries would be an underdeveloped technology — one which freezes and perpetuates underdevelopment. The critical response to this book by Frances Stewart is well worth reading. Particularly good is the extract by Hans Singer in which he argues 'to declare the gap between rich and poor countries to be a technological one, and then try to apply the same technology to both, will widen rather than narrow the real economic gap. On the other hand, different technologies serve to reduce the economic gap and hence, ultimately to eliminate the need for different technologies'. Also useful is George McRobie's response to those who argue for revolutionary changes and see appropriate technology as useless without these: he argues simply that rules can be changed within existing structures, so as to give some help to the poor.

There can be little doubt that some of the new development strategies emerging in recent years — particularly those of the basic needs family — go hand in hand with the concepts of appropriate technology. Section 3 outlines the main features of the 'alternative' development strategies (concentrating heavily on human basic needs) and discusses how they call for and are in turn supported by the kind of appropriate technology described by Schumacher. The extract by Reginald Green gives a useful account of how the basic needs strategies came into being, while that by Diwan and Livingstone compares the

4

characteristics of basic needs-type strategies with development strategies based on the maximization of output, and gives some examples of the former. While the advent of the basic needs-type strategy has undoubtedly lent academic respectability to Appropriate Technology, it also has its critics. To end this section, an opposing view is given in an extract by Deepak Lal in which he accuses basic needs-type strategies of being paternalistic and entailing a vast increase in state control and in the influence of bureaucracy. This is accompanied by a critical response by Keith Griffin.

Although the AT concept was first made popular by Schumacher in *Small is Beautiful*, it has been part of the philosophy and writings of the world's great 'thinkers' throughout the Ages. The extracts in Section 4 show how the thoughts and fears for society of writers as diverse as John Stuart Mill, Aldous Huxley and George Orwell have mirrored those of Schumacher. To complement Schumacher's *Small is Beautiful* are Kirkpatrick Sale's 'Steady-State Economy'; Kenneth Boulding's 'Spaceship Earth'; Jeremy Rifkin's 'Low-Entropy Economy'; John Kenneth Galbraith's 'Affluent Society'; and Jaques Ellul's 'Technological Society'. All of these maintain that it is madness to pursue limitless growth and warn of the dangers of technology controlling society, rather than man controlling technology. The extract by Ellul is particularly sobering, reflecting as it does the way in which technology can become an end in itself, and the way in which our lives have come to be dominated by 'the artificial necessity of the technical society'. We are facing a future, he claims, that even Huxley never dreamed of.

But what of Schumacher himself? His way of describing the metaphysical errors of the modern age was to talk in terms of 'giantism', 'complexity', and 'violence'. The extract in which he elaborates on these problems is followed by extracts and examples which illustrate each point. Aldous Huxley and Kirkpatrick Sale talk of the need for decentralization. The extracts by Stephen Salter and John Davis examine the meaning of simplicity. The issue of environmental violence is brilliantly illustrated in Rachel Carson's 'Fable for Tomorrow', from *Silent Spring*, and is backed up by recent accounts of acid rain pollution.

To end this chapter, it is appropriate to review the growth and transformation of the AT movement since the publication of *Small is Beautiful*. One clear trend is the extent to which interest and involvement in the AT concept has grown. An OECD survey has estimated that by the early 1980s over 1,000 groups claimed to be working on Appropriate Technology and that a new one is being established somewhere in the world at the rate of one a week. But more important is the way in which existing institutions — governments, UN agencies, and big businesses — have embraced the concept. Unlike many other development 'fads' Appropriate Technology has not withered away after a brief period of popularity. Rather, it has moved from the periphery into the mainstream of development and has become, as Nicolas Jéquier puts it, an important political phenomenon. It is on this aspect that the extracts in Section 5 of the chapter concentrate. Particularly interesting are those by Jéquier, who reviews Schumacher's ideas in the world of the 1980s, and

George McRobie, who analyses the process by which Appropriate Technology can and has been moving from an 'alternative' to a normal part of administrative, business and community activity.

1 DEFINITIONS AND CONCEPTS

Schumacher on Intermediate Technology

If we define the level of technology in terms of 'equipment cost per workplace', we can call the indigenous technology of a typical developing country — symbolically speaking — a £1-technology, while that of the developed countries could be called a £1,000-technology. The gap between these two technologies is so enormous that a transition from the one to the other is simply impossible. In fact, the current attempt of the developing countries to infiltrate the £1,000-technology into their economies inevitably kills off the £1-technology at an alarming rate, destroying traditional workplaces much faster than modern workplaces can be created, and thus leaves the poor in a more desperate and helpless position than ever before. If effective help is to be brought to those who need it most, a technology is required which would range in some intermediate position between the £1-technology and the £1,000-technology. Let us call it — again symbolically speaking — a £100-technology.

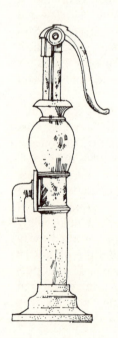

Such an intermediate technology would be immensely more productive than the indigenous technology (which is often in a condition of decay), but it would also be immensely cheaper than the sophisticated, highly capital-intensive technology of modern industry. At such a level of capitalization, very large numbers of workplaces could be created within a fairly short time; and the creation of such workplaces would be 'within reach' for the more enterprising minority within the district, not only in financial terms but also in terms of their education, aptitude, organizing skill, and so forth.

This last point may perhaps be elucidated as follows:

The average annual income per worker and the average capital per workplace in the developed countries appear at present to stand in a relationship of roughly 1:1. This implies, in general terms, that it takes one man-year to create one workplace, or that a man would have to save

6

Intermediate Technology

Simplest Technology
Indonesian farmers clearing land —
the simplest tools cost little to buy
and nothing to operate, but the work
is hard and slow and produces the
least of any technology.

Intermediate Technology
A Jordanian peasant tilling land using a wooden-shared plough — the tool
makes the work easier, costs little and can be made locally, but a plough
drawn by animals is not as productive as mechanised equipment.

Advanced Technology
In Mexico, one farmhand learns from another how to operate a modern tractor —
the machinery is quick and efficient, but is expensive to buy and maintain,
may deprive people of work and be ecologically harmful.

one month's earnings a year for twelve years to be able to own a workplace. If the relationship were 1:10, it would require ten man-years to create one workplace, and a man would have to save a month's earnings a year for 120 years before he could make himself owner of a workplace. This, of course, is an impossibility, and it follows that the £1,000-technology transplanted into a district which is stuck on the level of a £1-technology simply cannot spread by any process of normal growth. It cannot have a positive 'demonstration effect'; on the contrary, as can be observed all over the world, its 'demonstration effect' is wholly negative. The people, to whom the £1,000-technology is inaccessible, simply 'give up' and often cease doing even those things which they had done previously.

The intermediate technology would also fit much more smoothly into the relatively unsophisticated environment in which it is to be utilized. The equipment would be fairly simple and therefore understandable, suitable for maintenance and repair on the spot. Simple equipment is normally far less dependent on raw materials of great purity or exact specifications and much more adaptable to market fluctuations than highly sophisticated equipment. Men are more easily trained; supervision, control, and organization are simpler; and there is far less vulnerability to unforeseen difficulties.

E.F. Schumacher,
Small is Beautiful,
1973.

What are 'Appropriate Technologies'?

Appropriate technologies:
1   are low in capital costs;
2   use local materials whenever possible;
3   create jobs, employing local skills and labour;
4   are small enough in scale to be affordable by a small group of farmers;
5   can be understood, controlled and maintained by villagers wherever possible, without a high level of Western-style education;
6   can be produced out of a small metal-working shop, if not in a village itself;
7   suppose that people can and will work together to collectively bring improvements to their communities, recognizing that in most of the world important decisions are made by groups rather than by individuals;

8

8   involve decentralized renewable energy sources, such as wind power, solar energy, water power, methane gas, animal power and pedal-power (such as in that highly efficient machine, the bicycle);

9   make technology understandable to the people who are using it and thus suggest ideas that could be used in further innovations;

10 are flexible so that they can continue to be used or adapted to fit changing circumstances;

11 do not involve patents, royalties, consultant fees, import duties, shopping charges, or financial wizards; practical plans can be obtained free or at low cost and no further payment is involved.

Ken Darrow and Rick Pam, *Appropriate Technology Sourcebook*, 1978.

A Few Definitions of Technology

Alternative technology is the term used to describe new types of equipment or new organizational forms which represent a viable alternative to the existing 'main-stream' technologies of today. Examples: 'self-help' housing schemes instead of conventional urban development programmes, or small-scale organic farming instead of large-scale energy-intensive cultivation techniques.

Appropriate technology (AT) is now recognized as the generic term for a wide range of technologies characterized by any one or several of the following features: low investment cost per work-place, low capital investment per unit of output, organizational simplicity, high adaptability to a particular social or cultural environment, sparing use of natural resources, low cost of final product or high potential for employment.

Capital-saving technology (CST) or **light-capital technology** (LCT), a concept pioneered by Congressman Clarence D. Long of the US House of Representatives and now widely used by the US Agency for International Development, is a technology characterized primarily by its low cost in capital and the small size of the investment needed to create a job. Building roads with efficient labour-intensive methods embodies light-capital technologies; building them with bulldozers and scrapers does not.

Community technology (CT), a term widely used in the American counterculture and by such writers as Karl Hess, is a small-scale technology which does not require a complex infrastructure, which is specifically tailored to the needs and capabilities of small urban or rural communities, and which seeks to foster community participation in the decision-making processes. Examples: small-scale co-operative industrial activities or decentralized water supply and waste disposal systems.

Environmentally sound and appropriate technology (ESAT), a concept developed by the United Nations Environment Programme and Amulya K. Reddy of India, is an appropriate technology which is particularly well adapted to the local social and economic environment, and which uses renewable rather than non-renewable resources. Examples of ESATs in the energy field: biogas plants or biomass conversion systems.

Hardware, a term borrowed from the computer industry and now widely used by the AT community, is the physical embodiment of technology: tools, implements, machines, devices and equipments.

Intermediate technology (IT) is a technology which stands halfway between traditional and modern technology. Intermediateness is a relative notion: in Black Africa, the ox-drawn plough is an intermediate technology (more sophisticated than the traditional hoe, but less complex than the tractor) but in South-East Asia, it can be considered as a traditional technology. The concept of intermediate technology was developed by E. F. Schumacher, author of the best-selling book *Small is Beautiful.*

Low-cost technology (LCT) is a technology whose main feature is the low cost of the final product or service, or the low cost of the investment required to provide this product or service. Example: stabilization ponds for sewage treatment.

Socially appropriate technology (SAT) is a technology which is likely to have beneficial effects on income distribution, employment, work satisfaction, health and social relations. Example: a vaccine against malaria or schistosomiasis.

Soft technology (ST) is a technology which is well adapted to the local cultural and social environment, which uses renewable rather than non-renewable resources, and which does as little damage as possible to the surrounding eco-system. Examples: windmills, small hydro-power plants.

Software is the non-material dimensions of technology, e.g. knowledge, experience, organizational forms, managerial tools, institutional structures, legal provisions and financial incentives.

Village technology is small-scale technology aimed primarily at meeting the basic needs of rural dwellers in the developing countries. This concept was pioneered among others by the United Nations Children's Fund (UNICEF). Examples: small scale on-farm storage systems for food, low-cost dryers.

Nicolas Jéquier & Gerard Blanc, *The World of Appropriate Technology,* 1983.

One World Technology

An important feature of an equilibrium economy will be a shift of emphasis from big technology, as in today's industrial and hyper-industrial economies, to the development of advanced technologies of an appropriate form and scale. More effort will go into designing and producing machines and systems for individuals and small communities to use. These will be specifically aimed at helping people to meet more of their own household or local needs in spheres such as food and agriculture, building, repairs and maintenance of all kinds, leisure and entertainment, and also energy and transport. More generally, the idea that technology can be appropriate or inappropriate will have a much greater influence than it does today — the idea being that technology ought to be good to work with, sparing its use of resources, produce a good end product, and be kind to the environment.

The idea that capital-intensive technology is appropriate for the advanced countries and intermediate technology for the developing countries will not apply in an equilibrium world economy.

James Robertson,
The Sane Alternative, 1978.

What *is* Appropriate Technology? (*Larry*)

Except for some large sisal estates and a very small number of commercial coffee, tea and mixed farms, Tanzanian agriculture has always been based on technologically very backward peasant production. The farm implements were — and indeed to a large extent still are — the hand hoe, the panga and the small axe. The farming system was that of shifting agriculture — i.e. cultivating for a year or two and then abandoning that plot to the bush when its fertility declined. The system and the tools were thus inherently inefficient, and viable even at subsistence level only while the population was living in scattered homesteads and kept low by a high mortality rate.

If there was to be any prospect for economic and social development, this structure had to be changed. The policy of villagization therefore brought our people together in communities which could be served by a primary school, clean water and a basic health service centre such as a dispensary. In practice, when the villages were being established, too much emphasis was sometimes given to the size, which was best for the provision of these social services, and not enough to the demands of maximizing agricultural production; where it is necessary we are trying to correct these mistakes so as to reduce the distance between the shamba farm and the home.

But also in practice (and contrary to myth) farming was not 'collectivized' in the new villages. The predominant mode of production has continued to be the private peasant holding, although almost all villages also have a 'village farm'. This is either run by the village government or owned co-operatively by a voluntary group of the peasants who get the proceeds from it. The form of this village farm, and its size, is democratically determined by each village; it varies very greatly from one place to another. Our party does seek to encourage each village to have a village farm of at least 100 hectares, the proceeds from which will help to pay for village activities and development, but only a minority of village farms are yet as large as this.

Inevitably such re-locating has its short-term costs, but taking the country as a whole these were only significant for quantity of output in relation to the production of cashewnuts. The real constraint on increased output commensurate with the growing population and urban economy has come from our failure to move fast enough from the use of hand implements to animal-drawn or other intermediate power tools, and to better seeds and/or cultivation practices.

Production incentives have to be considered in the context of the society in which they are being applied, and even in the context of the world economy. The prices of Tanzania's agricultural exports have gone down in relation to the costs of its imports — the capital goods and raw materials on which local manufacturing and even modernized agricultural production depend. Yet it is

12

goods which the farmers want, not money. And about 85 per cent of Tanzania's population live in the rural areas; they cannot be completely protected from the fall in our national income which has occurred during the last three or four years, any more than they can be defended against the effects of drought, flooding, and other natural or political disasters (such as the Amin war) which have hit Tanzania since 1977. The government, through its pricing and tax policies, and through the pattern of its expenditures, has tried to reduce the burden on the rural areas as much as possible, and certainly the welfare of working people in the urban areas has worsened relative to that of rural dwellers. But in a period of world recession, foreign exchange shortage and domestic inflation, it is not possible for an agriculturally dependent country to give positive 'incentives' to farmers, however much the government might wish to do so. Tanzania does try; the annual fixed prices set for the most important food and export crops have now for a number of years been increased at, or in most cases above, the estimated level of inflation.

Tanzania has always produced the bulk of its own food, but its dependence on primitive tools and rainfed agriculture, and the growth in population, have meant that the increased food output is not proportionate to the rise in the population. The government's endeavours to improve agricultural methods and the availability of better seeds and tools are expected to have an effect on all output. The recent policy statement will help to ensure that it is everywhere worth the peasant's while to invest in

improved fertility in the land he is working; government differential pricing policy is designed to encourage food crops most appropriate to different ecological areas; reforms in the marketing system should reduce distribution and credit rigidities while safeguarding the development of a national food strategic reserve; and many other reforms are being made. It is, however, too early to judge the changes made in the very recent reassessment of national policies in relation to agriculture.

Traditionally our villages, and even our extended families, lived and worked on a basis of self-reliance. During the independence struggle the nationalist movement aroused the people to a wider consciousness of what they could themselves achieve by their own activities. Thus it was the people's efforts, led by TANU, which led to Tanganyika's independence in December 1961 (and Zanzibar's independence two years later). It was also the TANU Youth Leage and the Tanganyika Parents' Association (set up by TANU) which began self-help adult literacy classes and 'bush-schools' for the young children etc.

After independence there was a tremendous upsurge of these 'self-help' activities. Schools, dispensaries, community centres and other buildings were erected by the people themselves all over the country; so much so that it became necessary to try to introduce greater planning into these spontaneous efforts, because teachers, medical personnel and drugs, etc., could not be supplied to provide service in the new buildings. This kind of work, however, still continues. Even if the government now often helps by providing the tin roof, or the cement, or the craftsman for a village school, etc., it is very rarely that the villagers do not themselves provide the bulk of the labour and a proportion of the costs. Indeed it is only because of this continued 'self-reliance activity' that universal primary education has become possible, that 75 per cent of our villages have a shop, 40 per cent of the rural people have an easily accessible clean water supply, and that most of them have a basic health facility available within five kilometres.

Yet our ambitions have grown much larger; many of the services needed for a decent life in our villages, and for modernized agriculture, better roads, etc., require a level of technical expertise and an input of capital which is not commonly available within the villages, and is often not available — or available in sufficient quantities — within the country.

This is a fact of life, of our under-development. And with this fact has come the tendency of leadership to down-play the importance and the relevance of what we can do for ourselves as they compare the result with the sophistication of other countries' manufactures offered for sale — or on credit. It becomes a case of 'the best' (judged in international terms) being the enemy of 'the good'. Those leaders who advocate carts or animal-drawn ploughs then find themselves accused of trying to 'hold our country back'. And the villager who wants one of these articles finds them so difficult to obtain that he is often forced to continue carrying goods on his head and cultivating with the hand-hoe.

When I said that the most serious internal problem was the failure at all

Where farms are small, machinery should not be big. (*UNITED NATIONS*)

14

levels to understand the practice and principles of self-reliance, I was referring to the widespread results of this leadership attitude. But the first need, if a movement is to be reversed, is to recognize it! This is happening. Over the last three to four years, the ever-increasing shortage of foreign exchange and all the goods which depend on it, and the shortage even of public credit, is now having some effect — at all levels. Government and party officials are becoming more realistic in drawing up national plans and in the type of things they are encouraging the villages to plan and work for. The villagers and the urban workers are themselves more confident and more vocal in pressing their demands for goods which are relevant to their needs.

Thus we are at last taking seriously what are often called 'intermediate technology' methods, for example, by encouraging and helping with the provision of simple animal-powered agricultural tools, the local making of soap, the development of local pottery, the use of bricks and tiles for building, etc. We have much more to do; not everyone is yet converted to a realization that for us the 'best' will often be quite different technically from what it is for a developed society. And we are now handicapped in promoting this change because there is, generally speaking, no mechanical tradition in our villages, and because even the simplest tools usually have some small import content even if it is only scrap iron — and the amount of anything which we can buy from abroad decreases year by year. However, we are and we will be moving from now on in the right direction.

Julius Nyerere, *The Courier*, 1983.

Gandhi on the Swadeshi Movement

After much thinking, I have arrived at a definition of 'Swadeshi' that perhaps best illustrates my meaning. Swadeshi is that spirit in us which restricts us to the use and service of our immediate surroundings to the exclusion of the more remote. Thus, as for religion, in order to satisfy the requirements of the definition, I must restrict myself to my ancestral religion. That is the use of my immediate religious surrounding. If I find it defective, I should serve it by purging it of its defects. In the domain of politics I should make use of the indigenous institutions and serve them by curing them of their proved defects.

In that of economics I should use only things that are produced by my immediate neighbours and serve those industries by making them efficient and complete where they might be found wanting. It is suggested that such Swadeshi, if reduced to practice, will lead to the millennium. And as we do not abandon our pursuits after the millennium, because we do not expect to reach it within our times, so may we not abandon Swadeshi, even though it may not be fully attained for generations to come.

Let us briefly examine the three branches of Swadeshi as sketched above. Hinduism has become a conservative religion and therefore a mighty force, because of the Swadeshi spirit underlying it. It is the most tolerant because it is non-proselytizing, and it is as capable of expansion today as it has been found to be in the past. It has succeeded, not in driving out, as I think it has been erroneously held, but in absorbing Buddhism. By reason of the Swadeshi spirit, a Hindu refuses to change his religion, not necessarily because he considers it to be the best, but because he knows that he can complement it by introducing reforms. And what I have said about Hinduism is, I suppose, true of other great faiths of the world, only it is held that it is specially so in the case of Hinduism. Following out the Swadeshi spirit, I observe the indigenous institutions and the village 'panchayat' (council) hold me. India is really a republican country, and it is because it is that that it has survived every shock hitherto delivered. Princes and potentates, whether they were Indian born or foreigners have hardly touched the vast masses except for collecting revenue. The latter in their turn seem to have rendered unto Caesar what was Caesar's, and for the rest have done much as they have liked. The vast organization of caste answered not on the religious wants of the community, but it answered to its political needs. The villagers managed their internal affairs through the caste system, and through it they dealt with any oppression from the ruling power or powers. It is not possible to deny of a nation that was capable of producing the caste system its wonderful power of organiz-ation. One had but to attend the great Kumbha Mela at Hardwar last year to know how skilful that organization must have been, which without any seeming effort was able effectively to cater for more than a million pilgrims. Yet it is the fashion to say that we lack organizing ability. This is true, I fear, to a certain extent, of those who have been nurtured in the new traditions. We have laboured under a terrible handicap owing to an almost fatal departure from the Swadeshi spirit. We, the educated classes, have received our education through a foreign tongue. We have therefore not reacted upon the masses. We want to represent the masses, but we fail. They recognize us not much more than they recognize the English officers. Their hearts are an open book to neither. Their aspirations are not ours. Hence there is a break. And you witness, not in reality, failure to organize but want of correspondence between the representatives and represented. If during the last fifty years had we been educated through the vernaculars, our elders and our servants and our neighbours would have partaken of our knowledge; the discoveries of a Bose or a Ray would have been household treasures as are the Ramayan and Mahabharat. As it is, so far as the masses are concerned, those great discoveries might as well have been made by foreigners. Had instruction in all the branches of learning been given through the vernaculars, I make bold to say that they would have been enriched wonderfully. The question of village sanitation, etc., would have been solved long ago. The village panchayats would be now a living force in a special way, and India would almost be enjoying self-government suited to its requirements and would have been spared the humiliating spectacle of organized assassination on its sacred soil. It is not too late to mend. And you can help if you will, as no other body or bodies can.

And now for the last division of Swadeshi. Much of the deep poverty of the masses is due to the ruinous departure from Swadeshi in the economic and industrial life. If not an article of commerce had been brought from outside India, she would be today a land flowing with milk and honey. But that was not to be. We were greedy and so was England. The connection between England and India was based clearly upon an error. But she (England) does not remain in India in error. It is her declared policy that India is to be held in trust for her people. If this be true, Lancashire must stand aside. And if the Swadeshi doctrine is a sound doctrine, Lancashire can stand aside without hurt, though it may sustain a shock for the time being. I think of Swadeshi not as a boycott movement undertaken by way of revenge. I conceive it as a religious principle to be followed by all. I am no economist, but I have read some treatises which show that England could easily become a self-sustained country, growing all the produce she needs. This may be an utterly ridiculous proposition, and perhaps the best proof that it cannot be true is that England is one of the largest importers in the world. But India cannot live for Lancashire or any other country before she is able to live for herself. And she can live for herself only if she produces and is

helped to produce everything for her requirements, within her own borders. She need not be, she ought not to be drawn into the vortex of mad and ruinous competition which breeds fratricide, jealousy and many other evils. But who is to stop her great millionaires from entering in to the world competition? Certainly not legislation. Force of public opinion, proper education, however, can do a great deal in the desired direction. The handloom industry is in a dying condition. I took special care during my wanderings last year to see as many weavers as possible, and my heart ached to find how much they had lost, how families had retired from this once flourishing and honourable occupation. If we follow the Swadeshi doctrine, it would be your duty and mine to find out neighbours who can supply our wants and teach them to supply them where they do not know how to proceed, assuming that there are neighbours who are in want of healthy occupation. Then every village of India will almost be a self-supporting and self-contained unit, exchanging only such necessary commodities with other villages where they are not locally producible. This may all sound nonsensical. Well, India is a country of nonsense. It is nonsensical to parch one's throat with thirst when a kindly Mohammedan is ready to offer pure water to drink. And yet thousands of Hindus would rather die of thirst than drink water from a Mohammedan household. These nonsensical men can also, once they are convinced that their religion demands that they should wear garments manufactured in India only and eat food only grown in India, decline to

17

wear any other clothing or eat any other food. Lord Curzon set the fashion of tea-drinking. And that pernicious drug now bids fair to overwhelm the nation. It has already undermined the digestive apparatus of hundreds of thousands of men and women and constitutes an additional tax upon their slender purses. Lord Hardinge can set the fashion for Swadeshi, and almost the whole of India will forswear foreign goods. There is a verse in the Bhagavad Gita, which freely rendered, means masses follow the classes. It is easy to undo the evil of the thinking portion if the community were to take the Swadeshi vow, even though it may for a time cause considerable inconvenience. I hate legislative interference in any department of life. At best, it is the lesser evil. But I would tolerate, welcome, indeed plead for a stiff protective duty upon foreign goods. Natal, a British Colony, protected its sugar by taking the sugar that came from another British colony, Mauritius. England has sinned against India by forcing free trade upon her. It may have been food for her, but it has been poison for this country.

M.K. Gandhi, *Speeches and Writings of M.K. Gandhi*, 1919.

Mass Production or Production by the Masses?

As Gandhi said, the poor of the world cannot be helped by mass production, only by production by the masses. The system of mass production, based on sophisticated, highly capital-intensive, high energy-input dependent, and human labour-saving technology, presupposes that you are already rich, for a great deal of capital investment is needed to establish one single workplace. The system of production by the masses mobilizes the priceless resources which are possessed by all human beings, their clever brains and skilful hands, and supports them with first-class tools. The technology of mass production is inherently violent, ecologically damaging, self-defeating in terms of non-renewable resources, and stultifying for the human person. The technology of production by the masses, making use of the best of modern knowledge and experience, is conducive to decentralization, compatible with the laws of ecology, gentle in its use of scarce resources, and designed to serve the human person instead of making him the servant of machines. I have named it intermediate technology to signify that it is vastly superior to the primitive technology of bygone ages but at the same time much simpler, cheaper and freer than the super-technology of the rich. One can also call it self-help technology, or

democratic or people's technology — a technology to which everybody can gain admittance and which is not reserved to those already rich and powerful.

E.F. Schumacher, *Small is Beautiful,* 1973.

The New Paradigm of Work

The dominant paradigm of work in Western Europe and North America in the nineteenth and twentieth centuries emerged from the individualist, puritan ethic which was generated by the Reformation and confirmed by the industrial revolution. As that period of history comes to an end, will we go back to pre-industrial or pre-Reformation paradigms of work? I do not think so, at least not exactly.

The new paradigm of work will, I believe, owe quite a lot to what Schumacher has called Buddhist economics. He says that the Buddhist point of view takes the function of work to be threefold: to give a man a chance to utilize and develop his faculties; to enable him to overcome his ego-centredness by joining with other people in a common task; and to bring forth the goods and services needed for a becoming existence. This is not unlike William Morris's view that 'a man at work, making something which he feels will exist because he is working at it and wills it, is exercising the energies of his mind and soul as well as of his body....If we work thus we will be men, and our days will be happy and eventful'. But I believe that both Schumacher and William Morris underestimated the significance of increasing economic equality between men and women, and therefore also the centrally important change that will probably take place in our relative valuation of paid and unpaid work.

I suggest that the new paradigm of work which is now emerging will see work as something which every human being should be able to take satisfaction in doing. Work, whether paid or unpaid, will signify those activities which are undertaken to satisfy human needs — one's own and other people's; and those needs will be assumed to include the higher level needs for love, esteem and personal growth as well as the basic needs for food, clothing, shelter and safety. Work will no longer be regarded as a chore — as something to be endured by the less fortunate (like the slaves in ancient Greece), to be shirked whenever possible, and ultimately to be abolished by automation. It will no longer be regarded as a job, to be created and pre-

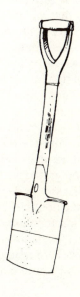

19

James Robertson,
*The Sane
Alternative*, 1978. served, counted and recorded, for its own sake. In a sane, humane, ecological society work will be necessary and desirable activity which confirms people in the knowledge of their own worth, which confirms the meaning of their relationships with other people, and which confirms their unity with the natural environment in which they live.

 # Useful Work Versus Useless Toil

The above title may strike some of my readers as strange. It is assumed by most people nowadays that all work is useful, and by most well-to-do people that all work is desirable. Most people, well-to-do or not, believe that, even when a man is doing work which appears to be useless, he is earning his livelihood by it — he is 'employed', as the phrase goes; and most of those who are well-to-do cheer on the happy worker with congratulations and praises, if he is only 'industrious' enough and deprives himself of all pleasure and holidays in the sacred cause of labour. In short, it has become an article of the creed of modern morality that all labour is good in itself — a convenient belief to those who live on the labour of others. But as to those on whom they live, I recommend them not to take it on trust, but to look into the matter a little deeper.

Let us grant, first, that the race of man must either labour or perish. Nature does not give us our livelihood gratis; we must win it by toil of some sort or degree. Let us see, then, if she does not give us some compensation for this compulsion to labour, since certainly in other matters she takes care to make the acts necessary to the continuance of life in the indi-

vidual and the race not only endurable, but even pleasurable.

You may be sure that she does so, that it is of the nature of man, when he is not diseased, to take pleasure in his work under certain conditions. And, yet, we must say in the teeth of the hypocritical praise of all labour, whatsoever it may be, of which I have made mention, that there is some labour which is so far from being a blessing that it is a curse; that it would be better for the community and for the worker if the latter were to fold his hands and refuse to work, and either die or let us pack him off to the workhouse or prison — which you will.

Here, you see, are two kinds of work — one good, the other bad; one not far removed from a blessing, a lightening of life; the other a mere curse, a burden to life.

What is the difference between them, then? This: one has hope in it, the other has not. It is manly to do the one kind of work, and manly also to refuse to do the other.

What is the nature of the hope which, when it is present in work, makes it worth doing?

It is threefold, I think — hope of rest, hope of product, hope of pleasure in the work itself; and hope of these also in some abundance and of

20

good quality; rest enough and good enough to be worth having; a product worth having by one who is neither a fool nor an ascetic; pleasure enough for all of us to be conscious of it while we are at work; not a mere habit, the loss of which we shall feel as a fidgety man feels the loss of the bit of string he fidgets with.

I have put the hope of rest first because it is the simplest and most natural part of our hope. Whatever pleasure there is in some work, there is certainly some pain in all work, the beast-like pain of stirring up our slumbering energies to action, the beast-like dread of change when things are pretty well with us; and the compensation for this animal pain in animal rest. We must feel while we are working that the time will come when we shall not have to work. Also the rest, when it comes, must be long enough to allow us to enjoy it; it must be longer than is merely necessary for us to recover the strength we have expended in working, and it must be animal rest also in this, that it must not be disturbed by anxiety, else we shall not be able to enjoy it. If we have this amount and kind of rest we shall, so far, be no worse off than the beasts.

As to the hope of product, I have said that Nature compels us to work for that. It remains for us to look to it that we do really produce something, and not nothing, or at least nothing that we want or are allowed to use. If we look to this and use our wills we shall, so far, be better than machines.

The hope of pleasure in the work itself: how strange that hope must seem to some of my readers — to most of them! Yet I think that to all living things there is a pleasure in the exercise of their energies, and that even beasts rejoice in being lithe and swift and strong. But a man at work, making something which he feels will exist because he is working at it and wills it, is exercising the energies of his mind and soul as well as of his body. Memory and imagination help him as he works. Not only his own thoughts, but the thoughts of the men of past ages to guide his hands; and, as part of the human race, he creates. If we work thus we shall be men, and our days will be happy and eventful.

Thus worthy work carries with it the hope of pleasure in rest, the hope of the pleasure in our using what it makes, and the hope of pleasure in our daily creative skill.

All other work but this is worthless; it is slaves' work — mere toiling to live, that we may live to toil.

Therefore, since we have, as it were, a pair of scales in which to weigh the work now done in the world, let us use them. Let us estimate the worthiness of the work we do, after so many thousand years of toil, so many promises of hope deferred, such boundless exultation over the progress of civilization and the gain of liberty.

William Morris, *Useful Work Versus Useless Toil*, 1885.

No race can prosper till it learns there is as much dignity in tilling a field as in writing a poem. — Booker T. Washington

21

Entertaining Objections

Since the idea of intermediate technology was first put forward, a number of objections have been raised. The most immediate objections are psychological: 'You are trying to withhold the best and make us put up with something inferior and outdated'. This is the voice of those who are not in need, who can help themselves and want to be assisted in reaching a higher standard of living at once. It is not the voice of those with whom we are here concerned, the poverty-stricken multitudes who lack any real basis of existence, whether in rural or in urban areas, who have neither 'the best' nor 'the second best' but go short of even the most essential means of subsistence. One sometimes wonders how many 'development economists' have any real comprehension of the condition of the poor.

There are economists and econometricians who believe that development policy can be derived from certain allegedly fixed ratios, such as the capital/output ratio. Their argument runs as follows: The amount of available capital is given. Now, you may concentrate it on a small number of highly capitalized workplaces, or you may spread it thinly over a large number of cheap workplaces. If you do the latter, you obtain less total output than if you do the former; you therefore fail to achieve the quickest possible rate of economic growth. Dr. Kaldor, for instance, claims that 'research has shown that the most modern machinery produces much more output per unit of capital invested than less sophisticated machinery which employs more people'. Not only 'capital' but also 'wages goods' are held to be a given quantity, and this quantity determines 'the limits on wages employment in any country at any given time'.

If we can employ only a limited number of people in wage labour, then let us employ them in the most productive way, so that they make the biggest possible contribution to the national output, because that will also give the quickest rate of economic growth. You should not go deliberately out of your way to reduce productivity in order to reduce the amount of capital per worker. This seems to me nonsense because you may find that by increasing capital per worker tenfold you increase the output per worker twentyfold. There is no question from every point of view of the superiority of the latest and more capitalistic technologies.

The first thing that might be said about these arguments is that they are evidently static in character and fail to take account of the dynamics of development. To do justice to the real situation it is necessary to consider the reactions and capabilities of people, and not confine oneself to machinery or abstract concepts. As we have seen before, it is wrong to assume that the most sophisticated

22

equipment, transplanted into an unsophisticated environment, will be regularly worked at full capacity, and if capacity utilization is low, then the capital/output ratio is also low. It is therefore fallacious to treat capital/output ratios as technological facts, when they are so largely dependent on quite other factors.

The question must be asked, moreover, whether there is such a law, as Dr. Kaldor asserts, that the capital/output ratio grows if capital is concentrated on fewer workplaces. No one with the slightest industrial experience would ever claim to have noticed the existence of such a 'law', nor is there any foundation for it in any science. Mechanization and automation are introduced to increase the productivity of labour, i.e. the worker/output ratio, and their effect on the capital/output ratio may just as well be negative as it may be positive. Countless examples can be quoted where advances in technology eliminate workplaces at the cost of an additional input of capital without affecting the volume of output. It is therefore quite untrue to assert that a given amount of capital invariably and necessarily produces the biggest total output when it is concentrated on the smallest number of workplaces.

The greatest weakness of the argument, however, lies in taking 'capital' — and even 'wages goods' — as 'given quantities' in an under-employed economy. Here again, the static outlook inevitably leads to erroneous conclusions. The central concern of development policy must be the creation of work opportunities for those who, being unemployed, are consumers — on however miser-able a level — without contributing anything to the fund of either 'wages goods' or 'capital'. Employment is the very precondition of everything else. The output of an idle man is nil, whereas the output of even a poorly equipped man can be a positive contribution, and this contribution can be to 'capital' as well as to 'wages goods'. The distinction between those two is by no means as definite as the econometricians are inclined to think, because the definition of 'capital' itself depends decisively on the level of technology employed.

Let us consider a very simple example. Some earth-moving job has to be done in an area of high unemployment. There is a wide choice of technologies, ranging from the most modern earth-moving equipment to purely manual work without tools of any kind. The 'output' is fixed by the nature of the job, and it is quite clear that the capital/output ratio will be highest, if the input of 'capital' is kept lowest. If the job were done without any tools, the capital/output ratio would be infinitely large, but the productivity per man would be exceedingly low. If the job were done at the highest level of modern technology, the capital/output ratio would be low and the productivity per man very high. Neither of these extremes is desirable, and a middle way has to be found. Assume some of the unemployed men were first set to work to make a variety of tools, including wheel-barrows and the like, while others were made to produce various 'wages goods'. Each of these lines of production in turn could be based on a wide range of different technologies, from the simplest to the most sophisticated. The task in every case would be to find an intermediate technology which obtains a fair level of productivity without having to resort to the purchase of expensive and sophisticated equipment. The outcome of the whole venture would be an economic development going far beyond the completion of the initial earth-moving project. With a total input of 'capital' from outside which might be much smaller than would have been involved in the acquisition of the most modern earth-moving equipment, and an input of (previously unemployed) labour much greater than the 'modern' method would have demanded, not only a given project would have been completed, but a whole community would have been set on the path of development.

I say, therefore, that the dynamic approach to development, which treats the choice of appropriate, intermediate technologies as the central issue, opens up avenues of constructive action, which the static, econometric approach totally fails to recognize. This leads to the next objection which has been raised against the idea of intermediate technology. It is argued that all this might be quite promising if it were not for a notorious shortage of entrepreneurial ability in the under-developed countries. This scarce resource should therefore be utilized in the most concentrated way, in places where it has the best chances of success, and should be endowed with the finest capital equipment the world can offer. Industry, it is thus argued, should be established in or near the big cities, in large integrated units, and on the highest possible level of capitalization per workplace.

The argument hinges on the assumption that 'entrepreneurial ability' is a fixed and given quantity, and thus again betrays a purely static point of view. It is, of course, neither fixed nor given, being largely a function of the technology to be employed. Men quite incapable of acting as entrepreneurs on the level of modern technology may nonetheless be fully capable of making a success of a small-scale enterprise set up on the basis of intermediate technology — for reasons already explained above. In fact, it seems to me that the apparent shortage of entrepreneurs in many developing countries today is precisely the result of the 'negative demonstration effect' of a sophisticated technology infiltrated into an unsophisticated environment. The introduction of an appropriate, intermediate technology would not be likely to founder on any shortage of entrepreneurial ability. Nor would it diminish the supply of entrepreneurs for enterprises in the modern sector; on the contrary, by spreading familiarity with systematic, technical modes of production over the entire population, it would undoubtedly help to increase the supply of the required talent.

Two further arguments have been advanced against the idea of intermediate technology — that its products would require protection within the country and would be unsuitable for export. Both arguments are based on mere surmise. In fact a considerable number of design studies and costings, made for specific products in specific districts, have universally demonstrated that the products of an intelligently chosen intermediate technology could actually be cheaper than those of modern factories in the nearest big city. Whether or not such products could be exported is an open question; the unemployed are not contributing to exports now, and the primary task is to put them to work so that they will produce useful goods from local materials for local use.

E.F. Schumacher, *Small is Beautiful*, 1973.

Appropriate Technology, Technological Dependence and the Technological Gap

The idea of appropriate technology sometimes meets with the criticism that it carries a thinly veiled neo-colonial implication — an implication that the existing, modern, efficient technology is not right for the developing countries, which should be satisfied with something inferior, second-best, less efficient. This is a serious reaction which merits careful study.

To begin with, one must recognize that there is an element of truth in this objection. If the situation is such that all technological power is concentrated in the rich countries, and that therefore the only efficient technology developed and existing is the capital-intensive and sophisticated technology, then it is quite true that the use of this technology is inevitable, even though it may not be ideally suited to the needs and requirements of developing countries. Its superior efficiency would outweigh its inappropriateness. In that case, it is very proper that all efforts be concentrated on achieving a transfer of this modern technology under the best possible conditions, free of undue restrictions and involving the least possible drain of other resources. That certainly is the situation in certain sectors (for example,

oil refineries or atomic energy installations) where safety and precision and other related product requirements are supreme. Bot those who advocate appropriate technology, while quite ready to admit that such areas exist, would argue; (a) that this is not the situation in all or most sectors of the economy; and (b) that even where it does exist it is only a second-best solution for the developing countries. The best solution would be to develop a technology which is at once efficient and modern and yet better geared to the resources and requirements of developing countries.

Those who are suspicious of the idea of an appropriate technology sometimes also argue as follows: 'The industrialized countries are rich; they use a modern, sophisticated, capital-intensive technology; *ergo*, they must be rich because they use this particular technology'. If the idea is put in this form, the error in the reasoning will spring to the eye. The industrialized countries are not rich because they use a capital-intensive technology. On the contrary, one could argue that if in the early stages of their industrial revolutions the industrialized countries had tried

25

to use the technology which they rightly use today they would never have become as rich as they are now (although this is a hypothetical argument that cannot be proved one way or the other). The industrialized countries are not rich because they use the sophisticated, capital-intensive technology; the line of causation is the other way round. They use the capital-intensive technology — and rightly so — because they are rich.

The suspicion of appropriate technology is increased when the proposal takes the form of suggesting an intermediate technology. In fact, the appropriate technology in a number of important respects, such as capital intensity, will be intermediate between the traditional technology now prevailing over the activities of most people in developing countries and the modern, capital-intensive technology widely prevailing in the industrialized countries. But the idea of 'intermediate technology' can be misinterpreted as suggesting a technology which is intermediate in efficiency between the low efficiency of present traditional technologies and the high efficiency of modern technology. This is not in the mind of those who use the term 'intermediate technology', they rightly emphasize that the intermediate technology they advocate is in fact the most efficient for the circumstances of most developing countries. All the same, it is preferable to speak of 'appropriate technology' rather than 'intermediate technology'.

The criticisms of appropriate technology may be further allayed by emphasizing that it is a transitional policy and not a permanent policy. As developing countries succeed in achieving development, their factor proportions will become more similar to those of the industrialized countries, and the difference in the technologies appropriate for the two groups of countries will diminish and perhaps finally disappear. One can say that the purpose of those who emphasize that the technology appropriate for developing countries is different from the technology appropriate for rich industrialized countries is precisely to make their own statement redundant. As development is successfully achieved by means of appropriate technology, the need for a different appropriate technology for developing countries will gradually disappear.

Again, the critics of the concept sometimes argue that it in effect establishes two different standards and will therefore create and perpetuate a 'technological gap'. This is a misconception. The gap exists in the fact that some countries are poor while other countries are rich. The task is to reduce or eliminate this gap — the economic gap. Different technologies will serve to reduce the economic gap and hence, ultimately, to eliminate the need for different technologies, as just argued. If we misdefine the problem by declaring that the gap is a technological gap, and then try (disregarding the economic gap) to apply exactly the same technology to the two groups of countries, the real economic gap will widen further instead of narrowing. Thus, to say that the application of appropriate technology perpetuates the gaps between rich and poor countries is a travesty of the true position.

H. Singer, *Technologies for Basic Needs*, 1977.

Underdeveloped Technology?

There are two themes running through Emmanuel's book [Arghiri Emmanuel: *Appropriate or Underdeveloped Technology*, 1982]: the first concerns of the multinationals in economic development; the second, an assessment of appropriate technology. In general Emmanuel believes that the most advanced and capital-intensive technology is also the technology which maximizes output, and consequently is in the best interests of the developing countries. He also believes that the multinational companies are the most effective mechanism for transferring advanced country technology and therefore should be welcomed as a medium of development, not criticized as a cause of underdevelopment. Advocates of intermediate technology — Schumacher, Latham-Koenig and McRobie — are criticized for a short-sighted view, putting jobs before output: 'the goal is not to put people to work, cost what it may, but to give them something to eat'.

Emmanuel's book contains a considerable number of logical errors, factual mistakes and misleading comparisons. Moreover, there are no arguments which have not been heard before, either with respect to multinational companies (MNCs) or appropriate technology (AT). Nonetheless, the book must be taken seriously. Emmanuel is a distinguished writer and thinker, who is likely to have considerable influence. He writes with a misleading confidence — one would say dogmatism — and apparent clarity, which makes his arguments persuasive. In my view, these arguments are seriously flawed and his conclusions incorrect with respect to both the major themes.

Accepting that developing countries need advanced-country technology, it does not follow that MNCs are the most effective vehicle of technology transfer as Emmanuel claims. Countries can and do acquire advanced-country technology in a variety of ways, many of which break down the package and separate acquisition of technology from finance and ownership of resources. Such arm's-length acquisition of technology has much to recommend it, because it permits countries to avoid many of the adverse effects of MNC activities, for example, the inhibiting effects on local learning, the close influence of MNCs on government policies, the poor terms the developing country tends to receive, the tax allowance, and the often devastating effects on consumption patterns. Emmanuel ignores each of these effects. He cites the cases of Japan, of South Korea and of Taiwan as examples of successful use of advanced-country technology, but ignores the fact that these countries have, to a large extent, acquired their technology at arm's length rather than directly through MNCs, and have therefore managed to get the best out of the technology, without suffering its worst effects.

Appropriate technology is dismissed by Emmanuel for putting jobs before output and income. The attack is misconceived because the appropriate technologies advocated do not do this: they maximize employment and output. Lack of research in the area of ATs has undoubtedly meant that in many areas efficient appropriate techniques are non-existent. This is precisely why groups such as ITDG were established: their activities consist in developing and promoting efficient ATs which increase output as well as jobs. Many examples could be cited from cement production to sugar processing, egg tray manufacture to windpower. The argument for appropriate technology is not that jobs should be put before output, but that techniques can be developed which promote both. The need for such development arises from the distorting effects of advanced-country technology in many countries, concentrating investment resources on a minority, leaving the great majority in conditions of unemployment or underemployment, with minimal incomes. AT is intended to raise productivity and incomes outside the advanced technology sector and so to extend the benefits of development throughout the population. In many countries, there is no effective redistribution of income, so that while capital-intensive techniques may raise incomes, they do not give the many poor 'something to eat'. The only effective way of raising

the incomes of the poor is to raise their ability to earn income: appropriate technology is designed to help secure precisely that. The objective of AT is to raise the incomes of the poor; the technologies and jobs are vehicles to achieve this, not ultimate objectives.

Again, Emmanuel cites the cases of South Korea and Taiwan to show that advanced technology works. He is right to consider them as generally successful, but what they show is rather different. In these countries, more labour intensive technologies were selected, and many advanced-country techniques were adapted to the special conditions of the countries, while rural technologies were also widely introduced. In these countries, then, the benefits of growth were widely spread, to a considerable extent because of the careful selection of more appropriate technologies. The Latin American countries provide examples of near 'pure' use of advanced-country technology, using the MNC as the major vehicle of transfer (the combination advocated by Emmanuel). Among these countries, output generally grew quite fast, but its benefits were mainly confined to the upper income groups; the poor remained poor, unemployed or underemployed, and their numbers increased. Emmanuel's chosen strategy did not succeed in giving them sufficient to eat.

Frances Stewart *Appropriate Technology* 1983.

There should be no place for machines that concentrate power in a few hands and turn the masses into mere machine minders. — Mahatma Gandhi

28

New Wine in Old Wineskins

Appropriate technology is the latest, or one of the latest, strategies devised to soothe the social conscience of the affluent countries, which do not cease to protest against the enormous human suffering that economic underdevelopment perpetuates — suffering that cohabits in this world with exaggerated and, so to speak, insulting luxury.

What has occurred for decades can be likened to new wine being poured into old wineskins: new technological strategies being poured into the same old, dependent, autocratic social structures of most underdeveloped countries. Unless deep-rooted changes in the social structures of the developing countries take place and are accompanied by fundamental changes in the attitudes of the developed ones, all partial strategies — basic needs, mass education, fertility control, 'sites and services', 'the green revolution', and others — are doomed to have only limited success. Let us face the facts. All these strategies have been tried before, in one form or another, in the past twenty-five years. In fact, we have hardly any new solutions for the 1980s. Not even the wine is new. What in most cases is missing is the determination to act. Both local and international powers and decision centres (such as the Pentagon, ITT and the like) are unwilling to change radically a social structure that is cumulatively beneficial to a handful of people but that inexorably damages the vital chances of the majority and is at all times compliant with the economic interests of the developed countries. It is, however, extremely difficult to generate this determination to change in a world that is still·strained by the fear of communism and still owes its prosperity to the present international division of labour.

Luis de Sebastián in J. Ramesh and C. Weiss, *Mobilizing Technology for World Development*, 1979.

Political and Technological Change

The main conclusion that we must draw is that, to the extent that technology becomes not only a function of but in fact an integral part of the dominant political interests, it is impossible to think in terms of technological change unless we are simultaneously prepared to consider the need for social and political change. Indeed, it is only through political change, and in particular through achieving liberation from the economic and political shackles of a dominant class, that the possibility of

29

significant technological change can emerge. The category of 'inappropriateness' in technology, as that of underdevelopment in general, is the inevitable function of a world trade system that reinforces the supremacy of the united interests of foreign capital and an indigenous elite: technology becomes one of the prime means by which this supremacy is maintained.

Within this perspective, we see clearly why intermediate technology, although in principle admirably suited to the conditions prevailing in the traditional sector of the economy of many underdeveloped countries, is unlikely to be successfully applied in practice unless accompanied by the necessary structural political changes. These changes are not merely of a cosmetic and superficial nature, but reach to the heart of the political systems of the underdeveloped countries. They therefore relate as much to changes in the modern sector as to those demanded by the traditional sector, if a situation in which the two sectors work in co-operation is to be substituted for the present one in which the one dominates and suppresses the other.

To take a specific example, supporters of intermediate technology often point to China as an example of a relatively underdeveloped country in which a pattern of dual development embracing both modern and traditional sectors of the economy has been achieved. There are many instances of ways in which the Chinese have been able to use human labour to carry out tasks that would have been mechanized in the West, and Chinese technology is frequently referred to as a prototype of genuine

intermediate technology. It is important to realize, however, that this has only been achieved since the People's Revolution in 1948, and in particular since the Great Leap Forward that began at the end of the 1950s, during which the Chinese broke with their Soviet technical advisers and Mao Tse Tung first put forward his 'walking on two legs' strategy for industrial and technological development. This strategy refers to the technique of combining agricultural with industrial development, new with traditional production techniques, and small-scale labour-intensive local industry with large-scale capital-intensive modern industry. Much of China's advanced technology was originally imported from the Soviet Union, Japan and Europe, and it has been estimated that, next to the US and the USSR, China must soon be considered as the third most important technological power in the world.

How relevant is China's experience to the situation of other underdeveloped countries, faced with a similar set of initial conditions? It is important to stress that science and technology in China cannot be examined apart from the social, economic, political and ideological setting in which they are pursued. It has been pointed out that Chinese planners have set about solving the type of problems now faced by many developing countries by imposing social controls which policy-makers in other societies may not have access to. These range from controls on patterns of consumption and the level of real wages to the redistribution of income on egalitarian grounds. They are closely tied to the structure of

30

political power and the social objectives of those responsible for policy-making. It appears to be the non-competitive structure of the Chinese economy which enables the sharing of technology and technological know-how between enterprises, and the protection of the markets of small-scale industries.

Other countries in which an intermediate technology has already become an important part of official development strategy are Cuba and North Vietnam. Achievements of North Vietnamese scientists and technologists, for example, working under war conditions and on severely limited resources, range from the development of an efficient transport system based mainly on bicycles, to the breeding of various water plants for use in flooded fields as an alternative to chemical methods of nitrogen fixation. The distinctive feature of these countries is that they have created, often only after bitter struggle, an economic system based on a planned, rather than a market, economy. This would appear to substantiate the suggestion that any alternative to the advanced technology that characterizes the industrialized countries, presents a political as much as a technological challenge to the existing economic system.

David Dickson, *Alternative Technology*, 1974.

Not Revolution.....but Changing the Rules

We have had cases where people in big industry tried to oppose AT. But they were half-hearted. We haven't met great opposition — but that may be a function of the fact that the movement is still very small. Even so, our experience is that a number of governments have done fantastic work without great upheavals. Look at what India has done. She has reserved a whole range of products for the small-scale sector to produce, and made credit much easier to get. Nobody has raised great political screams about this. The Indians have demonstrated that you can begin to change the rules of the game within the existing structure. These are steps towards my ideal, everybody's ideal, of nonviolent revolutions. Where you have particular types of structure, as used to be the case in Ethiopia and Iran, where the structure itself was violent toward the people, there is no chance for nonviolent change. But where you've got a relatively open political structure, there are lots of things that can be done within that to make it easier for the poor to help themselves.

George McRobie, *Ceres*, 1983.

Development does not start with goods: it starts with people and their education, organization and development. — E. F. Schumacher

Appropriate Technology and Basic-Needs Strategy

It is now becoming increasingly clear that conventional development strategies which emphasize the growth of gross national product *per se*, without at the same time inquiring into the pattern of growth which determines its fruits, do not, in most developing countries, alleviate mass poverty and unemployment. What has happened in many cases is that through conventional growth strategies the fruits of growth have been concentrated in the hands of a small privileged minority and have not reached the bulk of the population. . . . One of the factors which seems to have contributed to the perpetuation of poverty is that rapid growth has occurred in the small modern sector of the economy using most advanced imported technology. This growth has not spilled over into the rural traditional and urban informal sectors. In fact, quite often growth in the modern sector has occurred at the expense of these sectors. Technological progress in the former often does not lead to the raising of technological levels in the latter.

If past patterns of development have not yielded the desired results, there is clearly a need for current development strategies to be reoriented towards the elimination of poverty and the unemployment and the fulfilment of basic needs. These three elements are all inter-related. Both unemployment and underemployment prevent the majority of the population in developing countries from having access to minimum personal consumption needs such as adequate food and shelter, and to minimum social services such as water, education, sanitation, medical facilities and transport. Thus the technology required for a basic-needs strategy in a developing country must concentrate more than in the past on meeting the requirements of the small farmer, small-scale rural industry and the informal sector producer. Such a strategy calls for, and is in turn supported by, a special kind of appropriate technology: a technology which differs from that developed in the industrialized countries by the industrialized countries and for the industrialized countries even more than the difference in factor proportions would require. This is so because under a basic-needs strategy technology must bear the double burden of adapting existing or imported new technology to the general situation of the developing country *and* of underpinning the redistribution of incomes which goes with a basic-needs strategy. For this reason it might be called a 'doubly appropriate' technology. It can safely be assumed that a 'doubly appropriate' technology must contain a greater element of technological innovation (although possibly based on pre-existing knowledge nor currently selected for use or development by the industrialized countries) than a 'simple appropriate' technology, which can more often use the instruments of selective choice and adaptation applied to existing technologies, usually developed in the industrialized countries.

Hans Singer, *Technologies for Basic Needs*, 1977.

Basic Human Needs

As an organizing concept for a development strategy, Basic Human Needs — as the name implies — is concerned with the primary needs of communities and individuals. It rejects the sacrifice of a minimum decent (socially determined) standard of life for workers and peasants, either to provide the 'incentive' for capitalist accumulation or the means to socialist reconstruction for the putative benefits of rather vaguely identified future generations. In rejecting the maximization of the rate of growth of productive forces, it also denies the primacy of accumulation.

BHN as a strategy has five broad targets. It seeks to provide:

— basic consumer goods — food, clothing, housing, basic furnishings and other socially defined necessities (including, in China, a decent burial);
— universal access to basic services, e.g. primary and adult education, pure water, preventative and curative health programmes, habitat (environmental sanitation, urban and rural community infrastructure) and communications;
— The right to productive employment (including self-employment) yielding both high enough productivity and equitable enough remuneration to allow each household to meet its basic personal consumption out of its own income;
— an infrastructure capable of: producing the goods and services required (whether directly via domestic production or indirectly through foreign trade); generating a surplus to finance basic communal services; providing investment sufficient to sustain the increases in productive forces needed to advance towards the fulfilment of BHN;
— mass participation in decision-taking and the implementation of projects.

The BHN strategy envisaged here is production-oriented — transfer payments in the sense of secondary redistribution of consumption power are not central. Its emphasis is on primary redistribution — of income, assets, power — because it views the separation of production and distribution as theoretically unsound and practically impossible. The need for productive employment is therefore both an end and a means and accumulation, while denied the status of an end, is seen as critical.

But BHN is not concerned with absolute poverty or minimum needs

33

alone. Its stress on the social determination of needs implies that, as initial targets are approached, new ones will succeed them, and in this sense it is process-oriented. At least immediate BHN goals, therefore, differ sharply from state to state depending on historic, political, social and economic circumstances. Meeting initial BHN targets is infinitely more difficult technically, and *a fortiori* politically, with high degrees of inequality. In part however, the stress on equality depends on value

Basic Needs strategies try to leave no-one out. (*UNITED NATIONS*)

judgements - radical inequality of results (not just of opportunity) is perceived as an evil.

Participation is a vital means to achieving these goods as well as an end in itself. The strategy does not propose marginal tinkering, but a form of liberation much closer to revolution - non-violent or otherwise. Consensus and mutual interest models of the state which deny class differences and the integral nature of struggle are rejected: there is no way the strategy can work without mass participation and control.

The BHN strategy taken as a whole is not materialistic: priority emphasis on a number of the basic goods and services mentioned are hard to justify on market force or socialist production grounds. Some (especially the material) components of a BHN strategy can be listed as global, universal needs: others are specific to time, community and place.

The main influences behind the emergence of explicit BHN and BN (Basic Needs) include:

— The attempt to articulate a socialist economic and pricing calculus more relevant to a socialist society's aims. This is associated with the work of Kalecki and I. Sachs, and involved what Minhas has termed, in a slightly different context, the rejection of the Benthamite calculus (which is basically marginalist economics turned into a general social model).
— The 'mass needs' debate (particularly in its Mahgrebin-Egyptian forms), particularly the

examination of the limits of socio-economic reconstruction under Nasser and those imposed by the initial (Bettelheim) heavy industry centred Algerian strategy.
— Certain Latin American thinking arising from the limitations and failures as well as insights of ECLA's 'gapmanship' model (Cardoso). The disaggregation of dependence models in order to study in detail their impact on exploited and excluded groups (Stavenhagen, Furtado) was also an important Latin American element.
— Interaction between the debate on the New International Economic Order (NIEO) and that on Self-Reliance. Especially relevant was the recognition that changes at international trade levels were meaningless without parallel national strategic changes being made both on the periphery and by stronger trade partners at the centre. Otherwise, the excluded, exploited and oppressed in the periphery would be unlikely to benefit from so-called gains achieved within the framework of NIEO.
— Reactions agains the arguments in *Limits to Growth* that world resource constraints required continue inequality. Particularly important here was the work of the Bariloche Foundation on a Latin American model which sought to demonstrate the feasibility of meeting basic material needs in a reasonable time.
—The attempt by the United Nations Environmental Programme (UNEP) (and particu-

larly by Maurice Strong) to develop an 'inner limit' of minimum human needs as a co-constraint with the ecological 'outer limit' in the development of environmental policy.

— The World Bank's (IBRD) (and particularly Robert McNamara's) growing concern from 1969 onwards that the old development model excluded at least 40 per cent of the world's population from its benefits. This concern had previously led to 'absolute poverty eradication' and 'redistribution with growth'.

— The International Labour Organization's World Employment Programme (ILO-WEP) and the conversion of those most concerned from strategies concerned with narrowly defined wage employment to those stressing full productive employment.

— A general revolt against intellectual over-centralism: both a 'revolt of the periphery' against Eurocentric intellectual paradigms and a questioning of the 'top down' analyses made by central decision takers.

— The experience of several nations which did pursue strategies markedly unlike that of the dominant paradigm: China and Tanzania (BHN); Taiwan, South Korea (BN). Sri Lanka's BN approach was a perplexing influence as it was basically non-participatory, only tenuously linked to primary (as opposed to secondary fiscal and subsidy) redistribution, and neither economically nor socially self-sustaining.

Of these influences the last is probably the most important. However, the UNEP-IBRD-ILO strands contributed much of the analysis leading up to the current BHN debate. Three influences, often asserted to have been critical, almost certainly were not; indeed, they were rejected by a majority of those involved at an early stage. These were:

— the European conceived African and Asian 'community development' movement of the 1950-60s, seen as offending both against freedom (paternalism and Eurocentrism) and necessity (inadequate attention to the basic need of poor people to produce more).

— the social statistics movement - including 'social cost/benefit analysis - seen as economistic, always in danger of 'black boxing' experts' values as truth and ignoring needs as perceived by workers and peasants, and likely to serve as a substitute or excuse for not acting in respect of visible needs. This can be seen in part as a Third World revolt against Western intellectual hegemony, and in part as a practitioner reaction against arid formalism.

— the most austere 'alternative life-style', 'zero growth' forms of First World environmentalism, seen as relating to totally different objective conditions and as embodying values (e.g. austerity for its own sake) the Third World did not share.

R.H. Green, *IDS Bulletin*, 1978.

Conventional versus Alternative Development Strategies

Conventional Development Strategies (CDS) have been rationalized on the grounds that poor people cannot, and do not, save; and that large-scale production — in factories and on farms — is more efficient. In view of these two assumptions, there is little reason for the bulk of poor people to participate in the development process because the bulk of the population in developing countries is poor and the major part of their productive activities is carried on in small-scale production. They have small land holdings, small farms, and small businesses. Their participation, whenever it is possible, therefore involves inefficiency which by definition reduces growth. People, accordingly, are a problem and their participation leads to inefficiencies.

In Alternative Development Strategies (ADS) this logic and the implied assumptions do not apply. The assumptions instead are that poor people can, and do save; and that small-scale production is more efficient, particularly from the point of view of resource cost and the long run. This is the basic meaning of the phrase 'small is beautiful'. There is now evidence on both these propositions and accordingly, ADS seeks participation by the people...

A Comparison between the Characteristics of CDS and ADS

General Characteristics	CDS	ADS
I. Objective	Maximum GNP per Capita or or Welfare of Rich	Development of a Human Being or Welfare of Poor
Indicator	Level of GNP	Level and Composition of GNP
II. Technology	Imported	Indigenous
Modes of production	Centralized	Decentralized
Local institutions	Unimportant	Crucial
People participation in decision-making	Unnecessary	Fundamental
Local Solutions	Uniform	Diverse
Social change for for people's benefit	Unnecessary	Necessary
Role of people vs. experts	People are the problem, experts are the solution	People are solutions, experts are advisors
III. Role of theoretical model	Standard theory is fundamental, answers come from theory	There is no standard theory. Experimental

The need for ADS is becoming more and more urgent. The ideas on ADS, on the other hand, are still in the initial stages of development, and there is no one ADS (nor could there be). There are, instead, various forms of ADS depending upon the emphasis on different objectives. Accordingly, we present below some of the variants of ADS.

When China gained independence, the strategy of development in socialist economies followed the practice of the Soviet Union. Soviet economic development placed heavy emphasis on investment, so that consumption was postponed year after year; and on centralized modes of production in terms of heavy industry, collective farms, big cities, and capital-intensive technologies. However, Mao did not follow this strategy in China. There are different opinions among intellectuals of various persuasions regarding the nature, character, and success of development strategy in China. It is our contention that the development strategy in China, up to the death of Mao, has been a variant of ADS. It contains all the elements (of ADS) namely, redistribution and production of goods in favour of the poor in rural areas by decentralized means encouraging local initiative.

The elements of the Chinese development strategy may be listed as follows.

— Production is concentrated in a small number of material goods considered essential for all the population.
— Major productive activities are very labour intensive.
— A large part of production is agricultural and is carried on in communes where the production decisions are highly decentralized.
— The development strategy has not encouraged industrialization as it has in other developing countries. If some industries have been needed, these have been set up and promoted.
— Specialization is minimal.
— The emphasis has been on the use of education, production and technology to serve the people.
— There has been little growth in, or of, cities.

As a result, this strategy has been able to eliminate both poverty and unemployment. It has also resulted in sharply reducing, if not completely eliminating, consumption differences and inequalities. This has been accomplished without any help from foreign aid since the Soviet withdrawal.

Professor John Gurly summarizes these achievements and places them in a larger development context.

The truth is that China over the past two decades has made very remarkable economic advances (though not steadily) on almost all fronts. The basic overriding fact about China is that for twenty years it has fed, clothed and housed everyone, has kept them healthy, and has educated most. Millions have not starved; sidewalks and streets have not been covered with multitudes of sleeping, begging, hungry and illiterate human beings; millions are not disease ridden. To find such deplorable conditions, one does not look to China these days but, rather, to India, Pakistan and almost

38

Community involvement is essential for Development.

everywhere in the underdeveloped world. These facts are so basic, so fundamentally important that they completely dominate China's economic picture, even if one grants all of the erratic and irrational policies alleged by its numerous critics.

Part of the success of this strategy has been due to the fact that the economic and social institutions were fundamentally changed so that the new institutions, such as communes, were favourable to the objectives of the ADS. Soviet Russia had also changed the economic and social institutions drastically. However, the new institutions they created were not responsive to ADS goals; these were more suitable to other goals; basically those of CDS. If we compare the Soviet Union's and China's development strategies we find a number of differences and the following table outlines some of them. The table is self-explanatory.

Comparison Between Soviet and Chinese
Development Strategies

Categories/Emphasis	Soviet	Mao
1. Modes of Production	Centralized	Decentralized
2. Production	Capital Goods	Goods for Basic Needs
3. Techniques	Capital-Intensive	Labour-Intensive
4. Education/ Technology	Specialization	Non-specialization
5. Rewards	Material	Nonmaterial
6. Location	Urban	Rural

Our conclusion from this comparison is that Soviet economic development is a variant of CDS while Maoist development is a variant of ADS.

39

The philosophy and strategy of development in Tanzania was laid out in 1956 in the Arusha declaration. It emphasized the development in rural areas and de-emphasized the growth of cities and urban areas. The most important institution through which this process was to be encouraged is a 'Ujamaa Village'. The objectives of development and the nature of the Ujamaa Village have been analyzed and revised on the basis of experience. Most recently, President Nyerere defined the objective of development as the development of people. In his own words, 'For the truth is that development means the development of people. Roads, buildings, the increase of crop output, and other things of this nature are not development; they are only tools of development'. However, people cannot be developed. They develop themselves. This is fully recognized. Accordingly, the concept of Ujamaa Village has also undergone a revision. An 'Ujamaa Village' is a voluntary association of people who decide of their own free will to live together and work together for the common good'. Hence the policy of Ujamaa Village is not intended to be merely a revival of the old settlement schemes under another name.

Obviously, an Ujamaa Village has all the elements of ADS. The production and distribution are based on local resources and local initiatives. The process of decision-making is decentralized and participatory. 'Village' implies an emphasis on the rural and the needs of the poor.

It is difficult to gauge the success of the development experience in Tanzania. The indicators of success are different from the standard GNP, production, and price statistics. The information on other variables is not easily available.

Guinea-Bissau gained its independence from Portuguese rule after a fifteen-year struggle, in 1974. Dennis Goulet has studied the development strategy of Guinea-Bissau. He defines it as one of the variants of ADS 'in which distribution of benefits is more important than mere economic growth, and in which serious efforts are made to involve local communities in vital decisions affecting them'. There are three priorities in this development: agricultural development, improving human resources through education, and improvements in health and nutrition. The emphasis is on participation by the people. A new theory and practice of education is being developed which places more emphasis on 'political' instead of 'linguistic' literacy.

In the above sections we have listed some of the countries which have followed variants of ADS at a national level. However, virtually all countries aim to achieve some of the objectives of ADS, either nationally or regionally. For example, full employment is now a major goal of economic policy in virtually all developed countries. Japan, in the process of its own development, has relied heavily on small farms and tri-centre areas. In Taiwan, production in agriculture has been encouraged by institutions which are in many respects similar to the communes of the People's Republic of China. One can find some examples in virtually every country.

Ramesh K. Diwan and Dennis Livingston,
Alternative Development Strategies and Appropriate Technology, 1979.

An Opposing View

Influenced by the continuing absolute poverty of millions in the Third World, many observers have despaired of being able to deal with it through what the ILO calls 'conventional high growth' policies. Instead, they would seek to meet the basic needs of the people by 'the production and delivery to the intended groups of the BN basket through 'supply-management' and a 'delivery system'. The 'BN basket' consists of:

First, ... various items of private consumption: food, shelter, clothing. Secondly, various public services such as drinking water, sanitation, public transport, health and educational facilities.

Thus a number of components of BN are publicly-provided services (though not necessarily public goods). The aim of the BN approach is to expand the supply of these basic services as well as to convert the bulk of private consumption into publicly-provided goods and services. The 'strategy' is fundamentally paternalistic, entailing a vast increase in state

41

control and bureaucratic discretion.

The following kind of contrast is a persistent theme in the writings on BN. First, even with efficient growth, poverty has not been alleviated to the extent possible because of various market imperfections. Secondly, a perfectly-functioning bureaucratic system of allocation can achieve the optimal degree of growth to alleviate poverty. The fallacy in this is obvious, as the last Section sought to show.

Yet various dirigiste regimes, such as Mrs. Bandaranaike's Sri Lanka and Nyerere's Tanzania, which are cited as having achieved improvements in various indicators of the quality of life (such as longevity and literacy) even with little growth, are held up as the examples to be emulated. On these indices of welfare improvement, the post-war period has been one of the most beneficial for all strata of Third World populations, compared with both their own historical experience and that of today's developed countries. It is argued, however, that the differences in growth performance among developing countries are not necessarily related to changes in these social indicators. Hence, income growth is not necessarily associated with the alleviation of poverty.

Amartya Sen has classified countries according to their respective performance in longevity and literacy. He concludes that, for longevity, the communist countries have performed best. But the best performers in terms of both indicators are Taiwan, South Korea, Hong Kong and Singapore. However, low-income countries like Sri Lanka (longevity) and Tanzania (literacy) are also relative success stories. Sen proceeds to argue that there are two ways of removing poverty:

Ultimately, poverty removal must come to grips with the issue of entitlement guarantees. The two strategies differ in the means of achieving this guarantee. While one relies on the successfully fostered growth and the dynamism of the encouraged labour market, the other gives the government a more direct role as the provider of provisions.

This approach, however, glosses over the differences in the nature of the guarantee provided by each of these supposed ways of removing poverty. Not all guarantees are equally iron-clad! Despite stagnant economies, countries such as Sri Lanka and Tanzania have hitherto managed to operate social welfare programmes to meet so-called basic needs. The security of such politically-determined entitlements of the poor is, however, coming increasingly into question as the inexorable increase in their cost confronts a fixed economic pie from which to finance them.

By contrast, the poverty which, in the East Asian countries, has been alleviated through rising incomes cannot be so easily reversed by political fiat, since entitlements have been earned and are underwritten by the wealth created in the growth process. Moreover, there is good reason to believe that, by promoting efficiency, Sri Lanka and Tanzania could have achieved both growth and the alleviation of poverty — as Korea and Taiwan have done.

Growth, or its quality, has been

inadequate in many developing countries partly because their policies have been based on simplistic and one-dimensional notions of the dominant constraint, or bottleneck on development. Many of the same people who are now wringing their hands at the insufficient alleviation of poverty by past growth were the proponents of various mechanistic models based on developmental gaps — such as skills, savings, and foreign exchange — which their particular 'strategy' was proposed to fill.

This intellectual framework has not changed. The new gap is between the different goods and services actually consumed by the Third World's poor and those deemed by technocrats to be necessary to meet basic needs. Filling that gap is considered to be a matter of social engineering, which the bureaucracies of the Third World can readily perform. Further support is thereby lent to their dirigiste impulses which, in attempting to supplant the price mechanism, have done so much indirect damage to the prospects of the Third World's poor. By not emphasizing enough the inherent limitations of an imperfect bureaucracy at the same time as they castigate imperfect markets, those seeking to supplant the price mechanism in the provision of basic needs may yet again divert attention from the most important lesson of the varied development performance of the Third World in the last three decades, namely, that efficient growth which raises the demand for unskilled labour by 'getting the prices right' is probably the single most important means of alleviating poverty.

Deepak Lal, *The Poverty of 'Development Economies'*, 1983.

On Misreading Development Economics

The Poverty of 'Development Economics' is a polemic: it is not a serious work of scholarship. The book's title is a pun on a volume by Thomas Balogh and its major conclusion is an old saw about 'getting prices right'. . . .

The major charge against development economics is that it provides intellectual arguments in support of government intervention to accelerate growth and reduce poverty. This is branded by Mr Lal as 'the *dirigiste* dogma'. Nowhere is the dogma defined — it is really a straw man — although its 'essential elements' are said to be a belief that the market economy should be supplanted by direct controls, that resource allocation is unimportant, that the arguments for free trade are invalid and that governments should intervene to improve the distribution of income and wealth. Against this dogma Mr Lal upholds the 'economic principle' — the nearly universal assumption that people will take advantage of the opportunities presented to them — and asserts without evidence or argument that denying the economic principle is 'the hall-mark of much development economics'.

43

It is perfectly possible to believe the prices matter and that government intervention often is necessary to achieve specified objectives. Indeed, the great majority of economists believe precisely this. To believe, as Mr Lal apparently does, that prices are the only things that matter one must assume (i) that short-run elasticities are high and that once disturbed, the movement of market prices to a new equilibrium is rapid; (ii) that income distribution does not matter, or that governments should not influence the present distribution of income, whatever that happens to be; and (iii) that if governments do intervene they are more likely to make things worse than better.

There is abundant evidence that the first assumption is false. Free exchange rates adjust only slowly to their long-run equilibrium values; the composition of a country's output responds only slowly to changes in its international terms of trade, as can readily be seen today in many Third World countries; and the labour market in industrial countries corrects an unemployment disequilibrium painfully slowly, as the massive unemployment in the West associated with the current prolonged recession indicates. The list of additional examples is endless.

As for the distribution of income, it is simply a political fact of life that governments do care about inequality, some favouring it in the name of incentives and others deploring it in the name of social justice. Whether one likes it or not, and Mr Lal clearly does not, governments will continue to intervene in market processes to alter the distribution of income and wealth. To say blandly that 'we lack a consensus about the ethical system for judging the desirability of a particular distribution of income' does about as much good as spitting into the wind. One doesn't have to be a Marxist, or even a development economist, to recognize that much of modern politics consists of a struggle among classes, groups and coalitions to alter the distribution of income. Moreover, once this point is recognized it follows that the set of 'efficient' relative prices is itself a function of the distribution of income, since a different distribution will result in a different pattern of final and intermediate demand and hence in different market clearing prices.

Even Mr Lal acknowledges that there is a theoretical case for government intervention, but he argues that in practice the case is weak. The reasons for this are partly the transactions costs of intervention and the consequent 'bureaucratic failure', but more important to his argument is the 'theory of the second best'. The theory of the second best demonstrates that when there are multiple distortions in an economy, the removal of only one distortion may actually reduce overall efficiency. For example, if both gas and electricity are subsidized and gas is the lower cost source of energy, the removal of the subsidy on gas would result in lower efficiency in the use of fuel, because it would raise gas prices and shift demand to the higher cost electricity. Mr Lal then uses this theory to argue that it may be, and by implication often is, 'second-best' to do nothing. This, of course, begs the question as to whether governments should not intervene to remove both (or all) distortions. The theory of the second best can be used to justify pervasive intervention just as easily, and perhaps more easily, than it can be used to support Mr Lal's policy of non-intervention.

Mr Lal argues, quite correctly, that 'there are few, if any, instruments of government policy which are non-distortionary, in the sense of not inducing economic agents to behave less efficiently in some respects'. It follows from this neither that governments should refrain from interven-tion nor that they should attempt to adopt the utopian prescription of imposing lump-sum taxes and sub-sidies. Indeed 'distorting' interven-tions, e.g. food rationing, may well be preferable whenever specific redistributive measures would meet with widespread support and cooper-ation, whereas general redistributive measures would not. Mumbo-jumbo about the theory of the second best and lump-sum tax/subsidy systems really doesn't get us very far in the real world.

Keith Griffin, *Third World Quarterly*, 1984

4 SOCIETY AND ENVIRONMENT

Appropriate Technology through the Ages

Evidence — or at least tightly rea-soned argument — in support of the Schumacherian thesis is forthcoming from a variety of sources, both recent and ancient. As early as 322 BC, Aris-totle said:

> Most persons think that a state in order to be happy ought to be large; but even if they are right, they have no idea of what is a large and what is a small state...........
> To the size of states there is a limit, as there is to other things, plants, animals, implements; for none of these retain their natural power when they are too large or too small, but they either wholly lose their nature, or are spoiled.

In this statement, Aristotle appears to anticipate some contem-porary concepts relating to the need of limiting economic growth and con-trolling the drain on the world's resources.

The potential harmony of man with nature, and especially the rela-tively modest absolute material needs he has to lead a productive life, is no better illustrated than by Thor-eau's idyllic repose at Walden Pond. Thoreau's descriptions were graphic, simple, and nostalgically appealing — not presenting a highly structured metaphysical argument, but reaching out the latent quality in all of us for seeking simplicity and satisfaction in a direct basic relationship to nature. Maintaining this primordial relation-ship is one of the objectives of the appropriate technology concept as it is envisioned by many.

The problem, however, concerns man's tendency never to be satisfied with his material status, to be caught in a self-perpetuating cycle of ever-increasing wants, and to be progressively more isolated from fulfilling innate associations with his natural surroundings.

Jacques Ellul of the faculty of law at Bordeaux University produced in 1964 a reasoned, complex and brilliant discourse on the effect of technology on man. Ellul describes how *la technique* has so engulfed man, and so unconsciously, as to completely subvert his nature and his option for self-determination. All of 'civilized' humanity is pre-empted from achieving a rapport with nature that will finally bring modern man into a stable, acceptable relationship.

The aims of technology which were clear enough a century and a half ago, have gradually disappeared from view. Humanity seems to have forgotten the wherefore of all its travail, as though its goals had been translated into an abstraction or had become implicit; or as though its ends rested in an unforeseeable future of undetermined date, as in the case of Communist society. Everything today seems to happen as though ends disappear, as a result of the magnitude of the very means at our disposal.

None of our wise men ever pose the question of the end of all their marvels. The 'wherefore' is resolutely passed by. The response which would occur to our contemporaries is: for the sake of happiness. Unfortunately, there is no longer any question of that. One of our best-known specialists in diseases of the nervous system writes: 'We will be able to modify man's emotions, desires, and thoughts, as we have already done in a rudimentary way with tranquilizers.' It will be possible he says to produce a conviction or an impression of happiness without any real basis for it. Our man of the golden age, therefore, will be capable of 'happiness' amid the worst privations. Why then promise us extraordinary comforts, hygiene, knowledge, and nourishment if, by simply manipulating our nervous systems, we can be happy without them? The last meagre motive we could possibly ascribe to the technical adventure thus vanishes into thin air through the very existence of technique itself.

But what good is it to post questions of motives? Of Why? All that must be the work of some miserable intellectual who balks at technical progress. The attitude of the scientists, at any rate, is clear. Technique exists because it is technique. The golden age will be because it will be. Any other answer is superfluous.

His arguments are reminiscent of Huxley's *Brave New World Revisited*. Huxley writes about a future in which man is 'happy', pleasantly anaesthetized and stress-free, but with complete loss of volition in matters where the political masters wish that there be none. Huxley's world is governed through behaviour reinforcement abetted by chemical means — the Orwellian '1984' by coercive, terrorist methods — but they both have in common the extinction of any true 'freedom' that mankind may feel it possesses. This reference to the possible illusion of man's self-determination reflects some of the thesis of B.F. Skinner, the noted behavioural psychologist, who states:

What we need is a technology of behaviour. We could solve our problems quickly enough if we could adjust the growth of the world's population as precisely as we adjust the course of a spaceship, or improve agriculture and industry with some of the confidence with which we accelerate high-energy particles, or move toward a peaceful world with something like the steady progress with which physics has approached absolute zero (even though both remain presumably out of reach). But a behavioural technology comparable in power and precision to physical and biological technology is lacking, and those who do not find the very possibility ridiculous are more likely to be frightened by it than reassured. That is how far we are from 'understanding human issues' in the sense in which physics and biology understand their fields, and how far we are from preventing the catastrophe toward which the world seems to be moving.

He speaks of a 'technology of operant behaviour', which suggests (at the risk of simplistically analyzing Skinner's thesis) the imminent possibility of conditioning human behaviour through a technology based on demonstrable scientific principles that can lead mankind toward some predetermined condition of interpersonal relationships.

What the foregoing references to behaviouristic aspects and possibilities of human direction have to do with appropriate technology is that activity conducive to economic and social development may be possible to induce and, in the process, may become, by definition 'appropriate technology'. Of course, the Ellulian-Huxleyan-Orwellian aspects of this are obvious. What is suggested here, however, is that the imperfectly understood process of development would almost certainly benefit from a more profound examination of the motivational aspects.

Arguments regarding the concept of appropriate technology relate to the utility of technology and to the feasibility of its application. Ralph Nader in the contemporary US scene has had a great effect on US attitudes toward environmental impacts of private business. Nader views the activities of MNCs (Multinational Corporations) as being commercially exploitive in the international scene. Such enterprise often engages in activity that does not work to the best advantage of the developing country. The implication is one of immediate, capitalistic gain at the possible expense of the best interests of the involved country.

Barry Commoner of the United States strongly emphasizes the use of natural products to offset the negative ecological impacts of man's use of synthetics. Although not against 'growth', he supports the use of technology that involves such natural products and processes. In terms of the precepts of appropriate technology, Commoner is considered an effective thought leader.

John Kenneth Galbraith has written several books relevant to the application of technology to the problems of economic development. In his 1958 work *The Affluent Society*, he developed an argument supporting the idea that, especially in the United States, man has become so materialistic and removed from rational allocation of resources that the country is immensely over-balanced in favour of high-consumption consumer goods production. Public programmes that have great potential for improving the 'quality of life' are consistently and progressively deprived of support. One of his formulae for offsetting this situation is emphasizing a humanistic education process (another element of appropriate technology for development).

One branch of conventional wisdom clings nostalgically to the conviction that brilliant, isolated and intuitive inventions are still a principal instrument of technological progress and can occur anywhere and to anyone. Benjamin Franklin is the sacred archetype of the American genius and nothing may be done to disturb his position. But in the unromantic fact, innovation has become a highly organized enterprise. The extent of the result is predictably related to the quality and quantity of the resources being applied to it. These resources are men and women. Their quality and quantity depends on the extent of the investment in their education, training, and opportunity. They are the source of technological change. Without them investment in material capital will still bring growth, but it will be the inefficient growth that is combined with technological stagnation.

Although he was writing with specific reference to the US scene, the view is equally applicable to the developing countries. Galbraith further states: 'Finally, with better social balance, investment in human resources will be kept more nearly abreast of that in material capital. This, we have seen, is the touchstone for technological advance. As such, it is a most important and possibly the most important factor in economic growth'.

Donald D. Evans in Donald D. Evans and Laurie Nogg Adler (ed), *Appropriate Technology for Development: A Discussion and Case Histories*, 1979.

The Steady-State Economy

It was in 1848, that notably fruitful year, that the idea of the steady-state economy was first put forth, by the political economist John Stuart Mill, in his path-breaking *Principles of Economics*. Growth, as he saw, was inherent in capitalism, particularly the industrial capitalism then beginning to stamp itself upon the United Kingdom, and yet it was by definition a finite process: 'the increase of wealth is not boundless'. Capital will eventually cease to produce any sensible return for the capitalist, he posited, when the cost of extraction, manufacture, and disposal become great

48

enough and when the redistributive tax systems of the state become extensive enough. At that point there will no longer be any point in being a capitalist: the breed will vanish. In place of the 'progressive state', the 'stationary state':

I am inclined to believe that it would be, on the whole, a very considerable improvement on our present condition. I confess I am not charmed with the ideal of life held out by these who think that the normal state of human beings is that of struggling to get on; that the trampling, crushing, elbowing, and treading on each other's heels, which form the existing type of social life, are the most desirable lot of human kind, or anything but the disagreeable symptoms of one of the phases of industrial progress...

The best state for human nature is that in which, while no one is poor, no one desires to be richer, nor has any reason to fear being thrust back by the efforts of others to push themselves forward...

It is scarcely necessary to remark that a stationary condition of capital and population implies no stationary state of human improvement. There would be as much scope as ever for all kinds of mental, cultural, and moral and social progress; as much room for improving the Art of Living, and much more likelihood of its being improved, when minds ceased to be engrossed by the art of getting on.

Like many ideas, the measure of its worth varied indirectly with the time it took to become recognized by others: for more than 150 years it was virtually ignored, and the few who noticed it — professionals, mostly — understood it little. With the revival of interest in ecology in the 1960s, however, and then with the energy crisis of the 1970s, a number of modern economists began to re-examine the notion of a stationary economy and reshape it for modern conditions.

Economist Kenneth Boulding in 1966 introduced the idea of 'spaceship earth', by which he meant to suggest that we live within a basically closed system that must ultimately, like the spaceship, depend on its use of recycled materials and renewable energy. British economist Ezra Mishan in 1967 offered the idea of 'growthmania' in his elegant and influential *Costs of Economic Growth* and suggested the rejection of 'economic growth as a prior aim of policy in favour of a policy seeking to apply more selective criteria of welfare...to direct our national resources and our ingenuity to recreating an environment that will gratify and inspire men'. In 1971 Rumanian emigre Nicholas Georgescu-Roegen, working out of the University of Tennessee, came forth with the principle of increasing material 'entropy', an idea borrowed from the Second Law of Thermodynamics, to suggest that the excessive production and consumption of contemporary society was exhausting the finite material and energy resources of the earth to the point where soon the human species itself would be threatened. And the next year Jay Forrester, Donella and Dennis Meadows, and others at MIT hit the headlines with their pioneering computer models that

49

became the basis of the Club of Rome's initial admonitory book, *The Limits to Growth*.

After that, a continual drumroll. Herman Daly's explorations, *Essays Toward a Steady-State Economy*, became available in 1972, Fritz Schumacher's path-breaking and deservedly popular *Small is Beautiful* appeared early in 1973 as did Leopold Kohr's *Development Without Aid*, and the influential scholarly quarterly *Daedalus* came out with a full issue on 'The No-Growth Society' in the Autumn of 1973. Other important writers joined in: Hazel Henderson, Rufus Miles, Rene Dubos, Barbara Ward, Robert Theobold, Howard Odum, William Ophuls. And by the end of the 1970s the notion of the no-growth, stationary, steady-state economy was sufficiently well established to be creating ripples throughout the circles of both academic and governmental economists.

Though the visions of the steady-state economy inevitably vary according to the particular proponent, they all agree on certain basic points. A growth economy uses up scarce resources, emphasizes consumption over conservation, creates pollution and waste in the process of production, and engenders inter- and intra-national competition for dwindling supplies. A steady-state economy minimizes resource use, sets production on small and self-controlled scales, emphasizes conservation and recycling, limits pollution and waste, and accepts the finite limits of a single world and of a single ultimate source of energy.

Herman Daly puts it this way: 'If the world is a finite complex system that has evolved with reference to a fixed rate of flow of solar energy, then any economy that seeks indefinite expansion of its stocks and the associated material and energy-maintenance flows will sooner or later hit limits'. Nothing can expand forever. By contrast, the steady-state economy is one 'with constant stocks of people and artifacts, maintained at some desired, sufficient levels by low rates of maintenance 'throughput', that is, by the lowest feasible flows of matter and energy'. Or as Boulding says it more elegantly: 'The essential measure of the success of the (steady-state) economy is not production and consumption at all, but the nature, extent, quality and complexity of the total capital stock, including in this the state of human bodies and minds included in the system'.

It is the art of living, not the art of getting on.

Perhaps the best way to envision the steady-state economy is in terms of ecological harmony. For in a real sense a steady-state economy takes the stable ecosystem as a model, since a properly balanced environment is in fact essentially stationary. Growth occurs in fluctuations from season to season and species to species, but overall any given system may be basically unchanged for eons; and however fierce may be the competition between some species, however preoccupied each may be with its own survival, none of the elements within it (except, in self delusion, the human) has found any particular advantage in the growth of the system as a whole.

Kirkpatrick Sale, *Human Scale*, 1982.

50

The Low-entropy Economy

Entropy is a measure of the amount of energy no longer capable of conversion into work. The term was first coined by a German physicist, Rudolf Clausius, in 1868. But the principle involved was first recognized forty-one years earlier by a young French army officer, Sadi Carnot, who was trying to better understand why a steam engine works. He discovered that the engine did work because part of the system was very cold and the other part very hot. In other words, in order for energy to be turned into work, there must be a difference in energy concentration (i.e. difference in temperature) in different parts of a system. Work occurs when energy moves from a higher level of concentration to a lower level (or higher temperature to lower temperature). More important still, every time energy goes from one level to another, it means that less energy is available to perform work the next time around. For example, water going over a dam falls into a lake. As it falls, it can be used to generate electricity or turn a water wheel or perform some other useful function. Once it reaches the bottom, however, the water is no longer in a state to perform work. Water on a flat plane can't be used to turn even the smallest water wheel. These two states are referred to as available or free energy states versus unavailable or bound energy states.

An entropy increase, then, means a decrease in 'available' energy. Every time something occurs in the natural world, some amount of energy ends up being unavailable for future work. That unavailable energy is what pollution is all about. Many people think that pollution is a by-product of production. In fact, pollution is the sum total of all the available energy in the world that has been transformed into unavailable energy. Waste, then, is dissipated energy. Since according to the first law energy can neither be created nor destroyed but only transformed, and since according to the second law it can only be transformed one way — toward a dissipated state — pollution is just another name for entropy; that is, it represents a measure of the unavailable energy present in a system.

In keeping with the dictum that the low-entropy economy is one of necessities, not luxuries or trivialities, production will centre on goods required to maintain life. To recognize the extent to which production will be diminished, we have only to take a tour through a suburban mall and ask ourselves 'How many of these products are even marginally useful in sustaining life?'. Any honest appraisal is sure to conclude that most of what is manufactured in our economy is simply superfluous.

The production that does continue should take place within certain guidelines in keeping with the low-entropy paradigm. First, production should be decentralized and localized. Second, firms should be democratically organized as worker-managed companies. Third, production should minimize the use of non-

51

renewable resources. All of these points are consistent with both the energy and ethical requirements of the entropic world view. Of course, adhering to these guidelines will necessarily mean that certain items will become impossible to produce. A Boeing 747, for instance, simply cannot be manufactured by a small company employing several hundred individuals. Thus, a new ethic will have to be adopted as the litmus test of what should be produced in the low-entropy society: if it cannot be made locally by the community, using readily available resources and technology, then it is most likely unnecessary that it be produced at all.

Many industries will not be able to withstand the new transition to a low energy flow. Unable to adapt to the new economic environment, the automotive, aerospace, petrochemical, and other industries will slide into extinction. Many of the workers will need to retrain for new labour-intensive trades vital to the survival of local communities. But again, we should not be beguiled into thinking that the transfer of workers from one mode of industrial production to another can be made easily. Like it or not, the shift in the mode of economic organization will mean hardship and sacrifice.

The move toward a low-entropy economy will spell the end of the reign of the multinational corporation. There are many reasons that these corporate behemoths will not be able to withstand the changing energy environment. They are too complex, and they are completely reliant for their maintenance on the extracting of nonrenewable resources from all over the world. The multinational corporation is the dinosaur of our energy environment. Too big, too energy consumptive, and too specialized, they will run into their own evolutionary dead end as production moves back to a localized, small-scale base.

The uses of technology will also change drastically in the future. Once technology is recognized as being essentially a transformer of energy from a usable to an unusable state, we will come to understand that the less we use complex energy-consuming technologies, the better off we are.

In a low-entropy society, big, centralized, energy- and capital-intensive techniques will be discarded in favour of what is called appropriate or intermediate technology. Futurist author Sam Love defines appropriate technology as 'locally produced, labour-intensive to operate, decentralizing, repairable, fueled by renewable energy, ecologically sound, and community-building'. E.F. Schumacher, credited as the father of the intermediate-technology movement, says that this low-entropy form of technique is 'vastly superior to the primitive technology of bygone ages but at the same time much simpler, cheaper, and freer than the super-technology of the rich'.

Jeremy Rifkin, *Entropy: A New World View*, 1981.

The Affluent Society

As a society becomes increasingly affluent, wants are increasingly created by the process by which they are satisfied. This may operate passively. Increases in consumption, the counterpart of increases in production, act by suggestion or emulation to create wants. Expectation rises with attainment. Or producers may proceed actively to create wants through advertising and salesmanship. Wants thus come to depend on output. In technical terms, it can no longer be assumed that welfare is greater at an all-round higher level of production than at a lower one. It may be the same. The higher level of production has merely a higher level of want creation necessitating a higher level of want satisfaction. There will be frequent occasion to refer to the way wants depend on the process by which they are satisfied. It will be convenient to call it the Dependence Effect.

Plainly, the theory of consumer demand is a peculiarly treacherous friend of the present goals of economics. At first glance, it seems to defend the continuing urgency of production and our preoccupation with it as a goal. The economist does not enter into the dubious moral arguments about the importance or virtue of the wants to be satisfied. He doesn't pretend to compare mental states of the same or different people at different times and to suggest that one is less urgent than another. The desire is there. That for him is sufficient. He sets about in a workmanlike way to satisfy desire, and accordingly, he sets the proper store by the production that does. Like woman's, his work is never done.

But this rationalization, handsomely though it seems to serve, turns destructively on those who advance it once it is conceded that wants are themselves both passively and deliberately the fruits of the process by which they are satisfied. Then the production of goods satisfies the wants that the consumption of these goods creates or that the producers of goods synthesize. Production induces more wants and the need for more production. So far, in a major *tour de force*, the implications have been ignored. But this obviously is a perilous solution. It cannot long survive discussion.

Among the many models of the good society, no one has urged the squirrel wheel. Moreover, as we shall see presently, the wheel is not one that revolves with perfect smoothness. Aside from its dubious cultural charm, there are serious structural weaknesses which may one day embarrass us. for the moment, however, it is sufficient to reflect on the difficult terrain which we are traversing. We have seen how deeply we were committed to production for reasons of economic security. Not the goods but the employment provided by their production was the thing by which we set ultimate store. Now we find our concern for goods further undermined. It does not arise in spontaneous consumer need. Rather, the dependence effect means that it grows out of the process of production itself. If production is to increase, the wants must be effectively contrived. In the absence of the contrivance, the increase would not occur. This is not true of all goods, but that it is true of a substantial part is sufficient. It means that since the demand for this part would not exist, were it not contrived, its utility or urgency, *ex* contrivance, is zero. If we regard this production as marginal, we may say that the marginal utility of present aggregate output, *ex* advertising and salesmanship, is zero. Clearly the attitudes and values which make production the central achievement of our society have some exceptionally twisted roots.

Perhaps the thing most evident of all is how new and varied become the problems we must ponder when we break the nexis with the work of Ricardo and face the economics of affluence of the world in which we live. It is easy to see why the conventional wisdom resists so stoutly such change. It is far, far better and much safer to have a firm anchor in nonsense than to put out on the troubled seas of thought.

John Kenneth Galbraith, *The Affluent Society*, 1979.

 # *The Technological Society*

The human race is beginning confusedly to understand at last that it is living in a new and unfamiliar universe. The new order was meant to be a buffer between man and nature. Unfortunately, it has evolved autonomously in such a way that man has lost all contact with his natural framework and has to do only with the organized technical intermediary which sustains relations both with the world of life and with the world of brute matter. Enclosed within his artificial creation, man finds that there is 'no exit', that he cannot pierce the shell of technology to find again the ancient milieu to which he was adapted for hundreds of thousands of years.

The new milieu has its own specific laws which are not the laws of organic or inorganic matter. Man is still ignorant of these laws. It nevertheless begins to appear with crushing finality that a new necessity is taking over from the old. It is easy to boast of victory over ancient oppression, but what if victory has been gained at the price of an even greater subjection to the forces of the artificial necessity of the technical society which has come to dominate our lives?

In our cities there is no more day or night or heat or cold. But there is overpopulation, thraldom to press and television, total absence of purpose. All men are constrained by means external them to ends equally external. The further the technical mechanism develops which allows us to escape natural necessity, the more we are subjected to artificial technical necessities.

The aims of technology, which were clear enough a century and a half ago, have gradually disappeared from view. Humanity seems to have forgotten the wherefore of all its travail, as though its goals had been translated into an abstraction or had become implicit; or as though its ends rested in an unfore-seeable future of undertermined date, as in the case of Communist society. Everything today seems to happen as though ends disappear, as a result of the magnitude of the very means at our disposal.

Comprehending that the proliferation of means brings about the disappear-ance of the ends, we have become preoccupied with rediscovering a purpose or a goal. Some optimists of good will assert that they have rediscovered a Humanism to which the technical movement is subordinated. The orientation of this Humanism may be Communist or non-Communist, but it hardly makes any difference. In both cases it is merely a pious hope with no chance whatso-ever of influencing technical evolution. The further we advance, the more the purpose of our techniques fades out of sight. Even things which not long ago seemed to be immediate objectives — rising living standards, hygiene, com-fort — no longer seem to have that character, possibly because man finds the endless adaptation to new circumstances disagreeable. In many cases, indeed, a higher technique obliges him to sacrifice comfort and hygienic amenities to the evolving technology which possesses a monopoly of the instruments necessary to satisfy them. Extreme examples are furnished by the scientists isolated at Los Alamos in the middle of the desert because of the danger of their experiments; or by the would-be astronauts who are forced to live in the discomfort of experimental camps.

But the optimistic technician is not a man to lose heart. If ends and goals are required, he will find them in a finality which can be imposed on technical evolution precisely because this finality can be technically established and cal-culated. It seems clear that there must be some common measure between the means and the ends subordinated to it. The required solution, then, must be a technical inquiry into ends, and this alone can bring about a systematization of ends and means. The problem becomes that of analyzing individual and social requirements technically, of establishing, numerically and mechanisti-cally, the constancy of human needs. It follows that a complete knowledge of ends is requisite for mastery of means. But, as Jacques Aventur has demon-strated, such knowledge can only be technical knowledge. Alas, the panacea of merely theoretical humanism is as vain as any other.

'Man, in his biological reality, must remain the sole possible reference point for classifying needs', writes Aventur. Aventur's dictum must be extended to include man's psychology and sociology, since these have also been reduced to mathematical calculation. Technology cannot put up with intuitions and 'literature'. It must necessarily don mathematical vestments. Everything in human life that does not lend itself to mathematical treatment must be excluded — because it is not a possible end for technique — and left to the sphere of dreams.

Who is too blind to see that a profound mutation is being advocated here? A new dismembering and a complete reconstitution of the human being so that he can at last become the objective (and also the total object) of tech-niques. Excluding all but the mathematical element, he is indeed a fit end for the means he has constructed. He is also completely despoiled of everything

55

that traditionally constituted his essence. Man becomes a pure appearance, a kaleidoscope of external shapes, an abstraction in a milieu that is frighteningly concrete — an abstraction armed with all the sovereign signs of Jupiter the Thunderer.

In 1960 the weekly *l'Express* of Paris published a series of extracts from texts by American and Russian scientists concerning society in the year 2000. As long as such visions were purely a literary concern of science-fiction writers and sensational journalists, it was possible to smile at them. Now we have like works from Nobel Prize winners, members of the Academy of Sciences of Moscow, and other scientific notables whose qualifications are beyond dispute. The visions of these gentlemen put science fiction in the shade. By the year 2000, voyages to the moon will be commonplace; so will inhabited artificial satellites. All food will be completely synthetic. The world's population will have increased fourfold but will have been stabilized. Sea water and ordinary rocks will yield all the necessary metals. Disease, as well as famine, will have been eliminated; and there will be universal hygienic inspection and control. The problems of energy production will have been completely resolved. Serious scientists, it must be repeated, are the source of these predictions, which hitherto were found only in philosophic utopias.

The most remarkable predictions concern the transformation of educational methods and the problem of human reproduction. Knowledge will be accumulated in 'electronic banks' and transmitted directly to the human nervous system by means of coded electronic messages. There will no longer be any need of reading or learning mountains of useless information; everything

High tech, high rise, high human cost. (*WHO*)

56

will be received and registered according to the needs of the moment. There will be no need of attention or effort. What is needed will pass directly from the machine to the brain without going through consciousness. . . . Perhaps, instead of marvelling or being shocked, we ought to reflect a little. A question no one ever asks when confronted with the scientific wonders of the future concerns the interim period. Consider, for example, the problems of automation, which will become acute in a very short time. How, socially, politically, morally and humanly, shall we contrive to get there? How are the prodigious economic problems, for example, of unemployment, to be solved? And how shall we force humanity to refrain from begetting children naturally? How shall we force them to submit to constant and rigorous hygienic controls? How shall man be persuaded to accept a radical transformation of his traditional modes of nutrition? How and where shall we relocate a billion and a half persons who today make their livings from agriculture and who, in the promised ultrarapid conversion of the next forty years, will become completely useless as cultivators of the soil? How shall we distribute such numbers of people equably over the surface of the earth, particularly if the promised fourfold increase in population materializes? How will we handle the control and occupation of outer space in order to provide a stable *modus vivendi?* How shall national boundaries be made to disappear? (One of the last two would be a necessity.) There are many other 'hows', but they are conveniently left unformulated. When we reflect on the serious although relatively minor problems that were provoked by the industrial exploitation of coal and electricity, when we reflect that after 150 years these problems are still not satisfactorily resolved, we are entitled to ask whether there are any solutions to the infinitely more complex 'hows' of the next forty years. In fact, there is one and only one means to their solution, a world-wide totalitarian dictatorship which will allow technique its full scope and at the same time resolve the concomitant difficulties. It is not difficult to understand why the scientists and worshippers of technology prefer not to dwell on this solution, but rather to leap nimbly across the dull and uninteresting intermediary period and land squarely in the golden age. We might indeed ask ourselves if we will succeed in getting through the transition period at all, or if the blood and suffering required are not perhaps too high a price to pay for this golden age.

If we take a hard, unromantic look at the golden age itself, we are struck with the incredible naivete of these scientists. They say, for example, that they will be able to shape and reshape at will human emotions, desires and thoughts and arrive scientifically at certain efficient, pre-established collective decisions. They claim they will be in a position to develop certain collective desires, to constitute certain homogeneous social units out of aggregates of individuals, to forbid men to raise their children, and even to persuade them to renounce having any. At the same time, they speak of assuring the triumph of freedom and of the necessity of avoiding dictatorship at any price. They seem incapable of grasping the contradiction involved, or of understanding that what they are proposing, even after the intermediary period, is in fact the

57

harshest of dictatorships. In comparison, Hitler's was a trifling affair. That it is to be a dictatorship of test tubes rather than of hobnailed boots will not make it any less a dictatorship.

When our savants characterize their golden age in any but scientific terms, they emit a quantity of down-at-the-heel platitudes that would gladden the heart of the pettiest politician. Let's take a few samples. 'To render human nature nobler, more beautiful, and more harmonious.' What on earth can this mean? What criteria, what content, do they propose? Not many, I fear, would be able to reply. 'To assure the triumph of peace, liberty, and reason.' Fine words with no substance behind them. 'To eliminate cultural lag.' What culture? And would the culture they have in mind be able to subsist in this harsh social organization? 'To conquer outer space.' For what purpose? The conquest of space seems to be an end in itself, which dispenses with any need for reflection.

We are forced to conclude that our scientists are incapable of any but the emptiest platitudes when they stray from their specialities. It makes one think back on the collection of mediocrities accumulated by Einstein when he spoke of God, the state, peace, and the meaning of life. It is clear that Einstein, extraordinary mathematical genius that he was, was no Pascal; he knew nothing of political or human reality, or, in fact, anything at all outside his mathematical reach. The banality of Einstein's remarks in matters outside his speciality is as astonishing as his genius within it. It seems as though the specialized application of all one's faculties in a particular area inhibits the consideration of things in general. Even J. Robert Oppenheimer, who seems receptive to a general culture, is not outside this judgement. His political and social declarations, for example, scarcely go beyond the level of those of the man in the street. And the opinions of the scientists quoted by *l'Express* are not even on the level of Einstein or Oppenheimer. Their pomposities, in fact, do not rise to the level of the average. They are vague generalities inherited from the nineteenth century, and the fact that they represent the furthest limits of thought of our scientific worthies must be symptomatic of arrested development or of a mental block. Particularly disquieting is the gap between the enormous power they wield and their critical ability, which must be esti-mated as null. To wield power well entails a certain faculty of criticism, dis-crimination, judgement and option. It is impossible to have confidence in men who apparently lack these faculties. Yet it is apparently our fate to be facing a 'golden age' in the power of sorcerers who are totally blind to the meaning of human adventure. When they speak of preserving the seed of outstanding men, whom, pray do they mean to be the judges. It is clear, alas, that they propose to sit in judgement themselves. It is hardly likely that they will deem a Rimbaud or a Nietzsche worthy of posterity. When they announce that they will conserve the genetic mutations which appear to them most favourable, and that they propose to modify the very germ cells in order to produce such and such traits; and when we consider the mediocrity of the scientists them-selves outside the confines of their specialities, we can only shudder at the thought of what they will esteem most 'favourable'.

None of our wise men ever pose the question of the end of all their marvels. The 'wherefore' is resolutely passed by. The response which would occur to our contemporaries is: for the sake of happiness. Unfortunately, there is no longer any question of that. One of our best-known specialists in diseases of the nervous system writes: 'We will be able to modify man's emotions, desires and thoughts, as we have already done in a rudimentary way with tranquilizers'. It will be possible, says our specialist, to produce a conviction or an impression of happiness without any real basis for it. Our man of the golden age, therefore, will be capable of 'happiness' amid the worst privations. Why, then, promise us extraordinary comforts, hygiene, knowledge and nourishment if, by simply manipulating our nervous systems, we can be happy without them? The last meagre motive we could possibly ascribe to the technical adventure thus vanishes into thin air through the very existence of technique itself.

But what good is it to pose questions of motive? of Why? All that must be the work of some miserable intellectual who balks at technical progress. The attitude of the scientists, at any rate is clear. Technique exists because it is technique. The golden age will be because it will be. Any other answer is superfluous.

Jacques Ellul, *The Technological Society* 1964.

Prospect for the Year 2000

If present trends continue, the world in 2000 will be more crowded, more polluted, less stable ecologically, and more vulnerable to disruption than the world we live in now. Serious stresses involving population, resources, and environment are clearly visible ahead. Despite greater material output, the world's people will be poorer in many ways than they are today.

For hundreds of millions of the desperately poor, the outlook for food and other necessities of life will be no better. For many it will be worse. Barring revolutionary advances in technology, life for most people on earth will be more precarious in 2000 than it is now — unless the nations of the world act decisively to alter current trends.

This, in essence, is the picture emerging from the US Government's projections of probable changes in world population, resources, and environment by the end of the century, as presented in the Global 2000 Study. They do not predict what will occur. Rather, they depict conditions that are likely to develop if there are no changes in public policies, institutions, or rates of technological advance, and if there are no wars or other major disruptions. A keener awareness of the nature of the current trends, however, may induce changes that will alter these trends and the projected outcome.

Rapid growth in world population will hardly have altered by 2000. The

59

world's population will grow from four billion in 1975 to 6.35 billion in 2000, an increase of more than 50 per cent. The rate of growth will slow only marginally, from 1.8 per cent a year to 1.7 per cent. In terms of sheer numbers, population will be growing faster in 2000 than it is today, with 100 million people added each year compared with 75 million in 1975. 90 per cent of this growth will occur in the poorest countries.

While the economies of the less developed countries (LDCs) are expected to grow at faster rates than those of the industrialized nations, the gross national product per capita in most LDCs remaining low. The average gross national product per capita is projected to rise substantially in some LDCs (especially in Latin America), but in the great populous nations of South East Asia it remains below $200 a year (in 1975 dollars). The large existing gap between the rich and poor nations widens.

World food production is projected to increase 90 per cent over the thirty years from 1970 to 2000. This translates into a global per capita increase of less than 15 per cent over the same period. The bulk of that increase goes to countries that already have relatively high per capita food consumption. Meanwhile per capita consumption in South Asia, the Middle East, and the LDCs of Africa will scarcely improve or will actually decline below present inadequate levels. At the same time, real prices for food are expected to double.

Arable land will increase only 4 per cent by 2000, so that most of the increased output of food will have to come from higher yields. Most of the elements that now contribute to higher yields — fertilizer, pesticides, power for irrigation, and fuel for machinery — depend heavily on oil and gas.

During the 1990s world oil production will approach geological estimates of maximum production capacity, even with rapidly increasing petroleum prices. The Study projects that the richer industrialized nations will be able to command enough oil and other commercial energy supplies to meet rising demands through 1990. With the expected price increases, many less developed countries will have increasing difficulties meeting energy needs. For the one-quarter of humankind that depends primarily on wood for fuel, the outlook is bleak. Needs for fuelwood will exceed available supplies by about 25 per cent before the turn of the century.

While the world's finite fuel resources — coal, oil, gas, oil shale, tar sands, and uranium — are theoretically sufficient for centuries, they are not evenly distributed; they pose difficult economic and environmental problems; and they vary greatly in their amenability to exploitation and use.

Nonfuel mineral resources generally appear sufficient to meet projected demands though 2000, but further discoveries and investments will be needed to maintain reserves. In addition, production costs will increase with energy prices and may make some nonfuel mineral resources uneconomic. The quarter of the world's population that inhabits industrial countries will continue to absorb three-fourths of the world's mineral production.

Regional water shortages will become more severe. In the 1970–2000 period population growth alone will cause requirements for water to double in nearly half the world. Still greater increases would be needed to improve standards of living. In many LDCs, water supplies will become increasingly erratic by 2000 as a result of extensive deforestation. Development of new water supplies will become more costly virtually everywhere.

Significant losses of world forests will continue over the next twenty years as demand for forest products and fuelwood increases. Growing stocks of commercial-size timber are projected to decline 50 per cent per capita. The world's forests are now disappearing at the rate of eighteen to twenty million hectares a year (an area half the size of California), with most of the loss occurring in the humid tropical forests of Africa, Asia and South American. The projections indicate that by 2000 some 40 per cent of the remaining forest cover in LDCs will be gone.

Serious deterioration of agricultural soils will occur worldwide, due to erosion, loss of organic matter, desertification, salinization, alkalinization, and waterlogging. Already, an area of cropland and grassland approximately the size of Maine is becoming barren wasteland each year, and the spread of desert-like conditions is likely to accelerate.

Atmospheric concentrations of carbon dioxide and ozone-depleting chemicals are expected to increase at rates that could alter the world's climate and upper atmosphere significantly by 2050. Acid rain from increased combustion of fossil fuels (especially coal) threatens damage to lakes, soils, and crops. Radioactive and other hazardous materials present health and safety problems in increasing numbers of countries.

Extinctions of plant and animal species will increase dramatically. Hundreds of thousands of species — perhaps as many as 20 per cent of all species on earth — will be irretrievably lost as their habitats vanish, especially in tropical forests.

The future depicted by the US Government projections, briefly outlined above, may actually understate the impending problems. The methods available for carrying out the Study led to certain gaps and inconsistencies that tend to impart an optimistic bias. For example, most of the individual projections for the various sectors studied — food, minerals, energy, and so on — assume that sufficient capital, energy, water and land will be available in these sectors to meet their needs, regardless of the competing needs of the other sectors. More consistent, better-integrated projections would produce a still more emphatic picture of intensifying stresses, as the world enters the twenty-first century.

At present and projected growth rates, the world's population would reach ten billion by 2030 and would approach thirty billion by the end of the twenty-first century. These levels correspond closely to estimates by the US National Academy of Sciences of the maximum carrying capacity of the entire earth. Already the populations in sub-Saharan Africa and in the Himalayan hills of Asia have exceeded the carrying capacity of the immediate area, triggering an erosion of the land's capacity to support life. The resulting poverty and ill health have further complicated efforts to reduce fertility. Unless this circle of interlinked problems is broken soon, population growth in such areas will unfortunately be slowed for reasons other than declining birth rates. Hunger and disease will claim more babies and young children, and more of those surviving will be mentally and physically handicapped by childhood malnutrition.

Indeed, the problems of preserving the carrying capacity of the earth and sustaining the possibility of a decent life for the human beings that inhabit it are enormous and close upon us. Yet there is reason for hope. It must be emphasized that the Global 2000 Study's projections are based on the assumption that national policies regarding population stabilization, resource conservation, and environmental protection will remain essentially unchanged through the end of the century. But in fact, policies are beginning to change. In some areas, forests are being replanted after cutting. Some nations are taking steps to reduce soil losses and desertification. Interest in energy conservation is growing, and large sums are being invested in exploring alternatives to petroleum dependence. The need for family planning is slowly becoming better understood. Water supplies are being improved and waste treatment systems built. High-yield seeds are widely available and seed banks are being expanded. Some wildlands with their genetic resources are being protected. Natural predators and selective pesticides are being substituted for persistent and destructive pesticides.

Encouraging as these developments are, they are far from adequate to meet the global challenges projected in this Study. Vigorous, determined new initiatives are needed if worsening poverty and human suffering, environmental degradation, and international tension and conflicts are to be prevented. There are no quick fixes. The only solutions to the problems of population, resources, and environment are complex and long-term. These problems are inextricably linked to some of the most perplexing and persistent problems in the world — poverty, injustice, and social conflict. New and imaginative ideas — and a willingness to act on them — are essential.

With its limitations and rough approximations, the Global 2000 Study may be seen as no more than a reconnaissance of the future; nonetheless its conclusions are reinforced by similar findings of other recent global studies that were examined in the course of the Global 2000 Study.

It is probable that all the world's governments will be more or less completely totalitarian even before the harnessing of atomic energy: that they will be totalitarian during and after the harnessing seems almost certain. Only a large-scale popular movement toward decentralization and self-help can arrest the present tendency toward statism. At present there is no sign that such a movement will take place. — Aldous Huxley, Foreword to *Brave New World*, 1932.

All these studies are in general agreement on the nature of the problems and on the threats they pose to the future welfare of humankind. The available evidence leaves no doubt that the world — including this Nation — faces enormous, urgent, and complex problems in the decades immediately ahead. Prompt and vigorous changes in public policy around the world are needed to avoid or minimize these problems before they become unmanageable. Long lead times are required for effective action. If decisions are delayed until the problems become worse, options for effective action will be severely reduced.

Council on Environmental Quality and the US Department of State, *The Global 2000 Report to the President*, 1982.

Giantism, Complexity and Violence

The metaphysical errors of the modern age have, as it were, incarcerated themselves in our way of life, including our technology, the latter having some very clear and very destructive features. Everybody can take his own choice in describing them. My choice is this: — it has become too big; it has become too complex; and it has become too violent. These three factors taken together make it incompatible with human nature; with the rest of living nature around us; and with the resource endownment of the world. So, if we want to help ourselves we must work to use the fullness of our modern knowledge, consciously and with the utmost determination, to create, or perhaps to recreate, a technology which has the opposite features, — which is small, that is, adapted to the human scale; which is simple so that we do not have to become too specialized to be wise; and which is non-violent, in the sense of working with nature instead of bludgeoning her all the time. As far as the first point is concerned, the question of scale, all material things including organizations have to have their proper human scale or they become anti-human. This has been known since Genesis. Aristotle knew it and so did Dr. Karl Marx. I think the best bit in the *Limits to Growth* report is a quotation from Aristotle, where he says 'To the size of states, there is a limit, as there is to other things, plants, animals, implements; for none of these retain their natural power when they are too large or too small. But they either wholly lose their nature or are spoiled'. Small units can find their resources within a short radius and also their markets. There is no need for monster transport. Small is beautiful. It is also simple and non-violent. Intelligent small-scale technology can use, for instance, income energy instead of fossil fuels; for income energy is produced all the time, being sent from the sun to the earth. It is humanly right, because the human mind can encompass it; is socially right, because being small it is accessible to small people, not only to those who

63

are already rich and powerful. There can be no genuine development without small-scale technology. Nor can there be a correction of over-development.

But size is not all. So, secondly, keep things simple or, rather, use the fullness of modern knowledge, no matter how sophisticated, to recapture a degree of simplicity. Man's needs are essentially simple. Only his extravagances and frivolities are highly complex. Complexity — any fool can make things more complicated; it takes a touch of genius to make them simple again. Complexity, by itself, is highly capital intensive and expensive. Complexity is produced by giantism but also leads to it. Complexity excludes the majority of the people who are not highly trained. Complexity demands people too specialized to be wise, and they are dangerous people. Assuredly, we are now far too clever to be able to survive without wisdom. Wisdom is subtle, but at the same time it is as simple as a little child. We now possess such superlative scientific knowledge and technical ability that we can make things simple again.

Thirdly, non-violence. Let us use our great knowledge to find non-violent solutions to our many problems. Biological processes are non-violent compared with mechanical or chemical processes. For instance, the energy required to create artificial polymers in factories is immense, whereas nature creates polymers just with sunlight. Income energy is non-violent, but to use it presupposes smallness. Organic gardening and farming is non-violent. Prevention is non-violent compared with cure. Pollution is the result of violence — almost a measure of it. The Club of Rome accepts nuclear energy as a breakthrough extending the limits to growth. To quote: 'The technology of controlled nuclear fission has already lifted the impending limits on fossil fuel reserves'. The authors give statistics about the expected accumulation of radioactive wastes, which are horrifying. But they are evidently so wedded to pure quantity that they cannot appreciate the awesome qualitative impact of what they are talking about. The quantitative approach, which almost totally neglects the essence of things and knows nothing but units, whether people, or capital, or acres, cannot help us to find solutions because our problems have arisen precisely from this approch. What is required is detailed, honest, painstaking work, to create a technology with a human face.

E.F. Schumacher in M.M. Hoda (ed), *Future is Manageable*, 1978.

Schumacher's influence has been definitely harmful as it has tended to reinforce 'downright common sense', and, as all good economists know to their cost, there is nothing so consistently wrong as 'common sense', especially when it is 'downright'. — Lauchlin Currie

Drift, Distention or Decentralism

We are at — in the middle of, if you will — a turning point in American and probably world, history. I know no better than you what is to come. But the choices are clear: drift, distention, or decentralism.

We can go on as we are, tryng to muddle through (rather more muddle than through), patching up disintegrating and propping up decaying states, squabbling and warring incessantly over depleting resources and the last few tolerable environments, and coping and groping with increasingly anxious and uncertain lives. Or we can hope for rescue in ever-larger and ever-more-complex systems — 757s and 797s after 747s, Models 1199 and 2199 after Model 499 — and ever-stronger and more grandiose governments, giving up our liberty for an anticipated security, our initiative for an anticipated welfare system, and all the while moving closer to nuclear and environmental disaster. Or we can work to achieve systems and organizations of a size where we may regulate them, to reshape our landscapes to permit ecologically sound and locally rooted settlements, to create for ourselves a world in which our societies, our economies, our politics are in fact in the hands of those free individuals, those diverse communities and cities, that will be affected by them — a world, of course, at the human scale.

Pentagon...Pyramid...or Parthenon.

Kirkpatrick Sale, *Human Scale*, 1980.

What is Simplicity?

Bicycles are superbly efficient and successful machines. Most people would agree that they are indeed simpler than cars and airplanes. But bicycle technology is mature. Bicycles evolved during the last half of the nineteenth century to reach their present state of development by about 1905. To be successful, they needed the invention, development and production of ball bearings, sprockets, roller chains, the freewheel and gear-changing mechanisms. The pneumatic tyre required advances in the processing of rubber. Lightweight frames needed thin-walled drawn steel tubing. (The most expensive bicycles today use tubing with carefully graded wall thickness to give extra strength near the ends.) If you think that bicycles are simple, try building one with the tools and materials in a blacksmith's forge. These would be an accurate example of the resources available to the bicycle pioneers. The plain fact is that bicycles were complicated solutions to the problem of providing cheap

Easy to use, but not easy to design

personal transport to people who would otherwise have walked or ridden horses. They would have appeared totally unbelievable and impossibly complicated to the leading engineers of the preceding age.

Stephen Salter, *New Scientist*, 1982.

Ingenious Simplicity

I believe it was Charles Kettering of General Motors — or was it Henry Ford — who said 'Build simplicity into it'. Whichever said it, they both practised it with great ingenuity. However, as one looks around at modern engineering products the wisdom of ingenious simplicity seem to have been forgotten and in its place complexity rules.

I felt sorry for the man who was unable to obtain spare batteries for his pocket calculator; he would have avoided that problem if he used a slide rule. I was not surprised that the owner of an £11,000 Jaguar was almost apoplectic when he was told that his car would be unserviceable for a week because his electric windows would not work and spare parts were not available. I am sure he would have agreed with Dr. Ernst Fuhrmann, President of Porsche, who wondered 'if we have been right in making cars more complicated...all these things do is become a possible source of future defects'. He went on to say: 'If I were a big company I would bring out a very primitive car and I think people would buy it. I would tell my engineers to

design a car that could be repaired with five or ten tools. It would not be a bad car — but it would be a reliable one'. Thank goodness that the wisdom of Kettering and Ford is not entirely forgotten.

A really outstanding example of ingenious simplicity is the Humphrey pump in which the water to be pumped also acts as the piston of the power source. Several very large examples were installed early in this century. Lately, interest in various sizes has begun to develop and the Intermediate Technology Development Group has supplied designs for several overseas organizations. It is a particlarly appropriate device for a Third World country. It is extremely simple to make, there is no piston or cylinder to machine, it requires no lubrication and suffers no wear. Consequently it can easily be made, operated and repaired locally. These same attractive properties are equally valuable here in Britain.

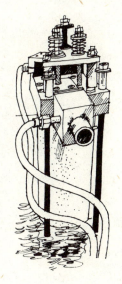

For more than a century very little attention has been paid to the very considerable number of small-scale water-power sources that exist in almost every country. During the Victorian era small water turbines were to a considerable extent scaled-down versions of big hydrosystem turbines, using the same principles of mechanical control. A different design approach has been adopted in the development of modern small-scale (1 to 10kW) water-turbine units. All attempts to control water flow by mechanical means have been abandoned in favour of an electronic black-box controller. As a result there has emerged a very much simpler machine to build, maintain and operate, which of course is very much cheaper per kilowatt capacity than its Victorian predecessors. The use of modern technological developments — self-lubricating bearings, electronic controllers — when properly applied can contribute to the recapture of simplicity.

Those engineers who are engaged in finding simple solutions to human problems very soon discover that the work is much more fascinating and challenging than the conventional approach, which seems remorselessly to lead to increasing complexity for even the simplest of tasks. It really is not very clever to use a powered auger costing several thousand pounds to dig a hole in which to set a telegraph post. We marvel at the wonders of a pocket eletronic calculator but is it, in its simplest form and for most applications, an advance on the slide rule or mental arithmetic?

John Davis,
Engineering, 1979.

A Fable for Tomorrow

There was once a town in the heart of America where all life seemed to live in harmony with its surroundings. The town lay in the midst of a checkerboard of prosperous farms, with fields of grain and hillsides of orchards where, in spring, white clouds of bloom drifted above the green fields. In autumn, oak and maple and birch set up a blaze of colour that flamed and flickered across a backdrop of pines. Then foxes barked in the hills and deer silently crossed the fields, half hidden in the mists of the fall mornings.

Along the roads, laurel, viburnum and alder, great ferns and wildflowers delighted the traveller's eye through much of the year. Even in winter the roadsides were places of beauty, where countless birds came to feed on the berries and on the seed heads of the dried weeds rising above the snow. The countryside was, in fact, famous for the abundance and variety of its bird life, and when the flood of migrants was pouring through in spring and fall people travelled from great distances to observe them. Others came to fish the streams, which flowed clear and cold out of the hills and contained shady pools where trout lay. So it had been from the days many years ago when the first settlers raised their houses, sank their wells and built their barns.

Then a strange blight crept over the area and everything began to change. Some evil spell had settled on the community: mysterious maladies swept the flocks of chickens; the cattle and sheep sickened and died. Everywhere was a shadow of death. The farmers spoke of much illness among their families. In the town the doctors had become more and more puzzled by new kinds of sickness appearing among their patients. There had been several sudden and unexplained deaths, not only among adults but even among children, who would be stricken suddenly while at play and die within a few hours.

There was a strange stillness. The birds, for example — where had they gone? Many people spoke of them, puzzled and disturbed. The feeding stations in the back-yards were deserted. The few birds seen anywhere were moribund; they trembled violently and could not fly. It was a spring without voices. The mornings that had once throbbed with the dawn chorus of robins, catbirds, doves, jays, wrens, and scores of other bird voices there was now no sound; only silence lay over the fields and woods and marsh.

On the farms the hens brooded, but no chicks hatched. The farmers complained that they were unable to raise any pigs — the litters were small and the young survived only a few days. The apple trees were coming into bloom but no bees droned among the blossoms, so there was no pollination and there would be no fruit.

The roadsides, once so attractive, were now lined with browned and withered vegetation as though swept by fire. These, too, were silent, deserted by all living things. Even the streams were now lifeless. Anglers no longer visited them, for all the fish had died.

In the gutters under the eaves and between the shingles of the roofs, a white granular powder still showed a few patches; some weeks before it had fallen like snow upon the roofs and the lawns, the fields and streams.

No witchcraft, no enemy action had silenced the rebirth of new life in this stricken world. The people had done it themselves.

This town does not actually exist, but it might easily have a thousand counterparts in America or elsewhere in the world. I know of no community that has experienced all the misfortunes I describe. Yet every one of these disasters has actually happened somewhere, and many real communities have already suffered a substantial number of them. A grim spectre has crept upon us almost unnoticed, and this imagined tragedy may easily become a stark reality we shall all know.

Rachel Carson, *Silent Spring*, 1962.

The Acid Rain Impact

The environmental and social costs of acid rain are skyrocketing. Yet smokestacks continue to belch millions of tons of sulphur and nitrogen oxides. North America receives over thirty-three million tons of sulphur oxides and twenty-four million tons of nitrogen oxides per year. About 85 per cent of that originates in the US. The major sources, coal and oil-fired generating stations, have nearly quadrupled their output of the pollutants over the past twenty-five years.

In Europe similar amounts of sulphur and nitrogen are released: more than thirty-three million tons of sulphur in 1978 alone. As in North America, most of these oxides come from burning sulphur-laden coal. These pollutants

can travel thousands of miles to other countries where they eventually fall as acid rain. Canada, for example, receives most of its acid rain from south of the border. Norway, one of the most hard-hit areas receives over 90 per cent of its acid rain from Britain and Germany.

Technology does exist to stop these pollutants. According to the US EPA, a 'desulphurization technology can now screen out up to 90 per cent of sulphur dioxide emissions'. Nevertheless companies have been reluctant to adopt preventative technology, citing high costs as the main objection.

Technology to curtail pollution from metal smelters also exists. In 1975, the International Nickel Company (INCO) in Sudbury, Ontario (the world's largest single source of sulphur emissions) developed a plan to reduce their daily 2,500 tons sulphur output by 200 tons. The proposal was later rejected as 'uneconomical'. At the same time a government report estimated that sulphur pollution from the INCO stack had caused $465 million worth of damage in the Sudbury region — damage that INCO would not have to pay for. Still the government continues to treat INCO with kid gloves, periodically delaying deadlines for meeting emission standards.

Instead of adopting pollution-control measures, companies choose to invest where they can make a profit. The enormous social and environmental costs are left to the public purse. 'From the companies' point of view' says an Ontario environmental official 'there are always better

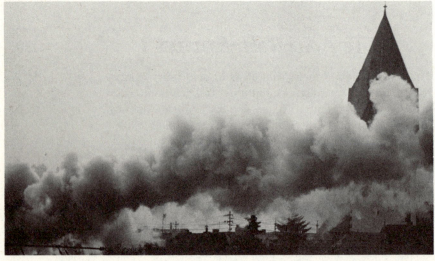

Can this be the height of civilization? (*UNITED NATIONS*)

and more productive uses for available cash than pollution abatement'.

INCO is a case in point. At the same time it was pleading poverty on pollution controls, the company took over the largest battery manufacturer in the US — E.S.B. Ray-O-Vac. For INCO, the $238 million deal was a sound business investment. For the dying lakes in northern Ontario and other downwind areas, the deal made no sense at all. Quipped INCO Vice President, pollution control would have only been contemplated as a social investment and who knows how to evaluate a social investment.

Both government and industry continue to avoid the enormous social and environmental costs their pollution creates. According to an Ontario Environment Ministry report, over 20,000 jobs in the tourist and resort industry will be lost if acid rain continues at the present pace. In Norway a major portion of the fishery industry has already been eliminated.

In Britain, Germany and the Netherlands governments have been equally lax in forcing industry to control polluting emissions. In Britain, where so-called 'super smoke-stacks' have been a major part of pollution 'control' programmes, increased use of coal will drastically accelerate acid rain in Scandinavia. The giant smoke-stacks don't solve the problem: they only spread it further afield.

Fortunately, alternatives do exist. Pollution control equipment is available. And much research and ingenuity has been expended to find ways of generating energy that do not create pollution which leads to acid rain. The wind, the sun, the use of conservation to reduce the need for energy are all non-polluting alternatives to fossil-fuelled generating stations.

But acid rain is also a political problem — it requires public pressure against polluting companies and against hesitant governments which are unlikely to move without being pushed. In Canada, the Canadian Coalition on Acid Rain, a collection of public interest groups, was formed to pressure both Canadian and US politicians. US environmental and labour groups under the banner of the Clean Air Coalition have also been actively challenging business efforts to relax pollution standards. And in May, 1981 a number of environmental groups gathered in Sweden for a European Conference on Acid Rain.

The problem has been diagnosed and the message is clear. The main task now is to make sure it's loud enough for government and industry to hear.

Phil Weller, *New Internationalist*, 1982.

71

Acceptance and Integration

In the ten years since the publication of *Small is Beautiful*, Schumacher's ideas have gained an extraordinarily wide acceptance both in the 'counter-culture' and the 'establishment', and I would submit that the revolutionary nature of Schumacher's message is not so much its content, as the fact that it has been understood by the establishment and gradually integrated into its systems of values. When heads of state, prime ministers, chairmen of atomic energy commissions, bank presidents and chief executive officers of the world's largest industrial corporations quote Schumacher or acknowledge that they have read his work, this is not simply a heartening success, but an important political phenomenon: revolutions do not occur when the existing political and social system is challenged by outsiders — be they poor farmers of the urban proletariat — but when it is put into question by large segments of the ruling elites.

Nicolas Jéquier,
*Appropriate
Techology*, 1983.

Appropriate Technology — Ten Years On

The critics who derided intermediate technology fifteen years ago or dismissed Schumacher's book in 1973 as an idealistic pamphlet may feel vindicated when they look at today's wonders of high technology. In the same way, AT proponents may justifiably feel a certain disappointment at how little the world of technology has apparently changed under the impact of Schumacher's ideas. Both groups, in fact, vastly underestimated the amount of time it takes before major reorientations in the technological system become visible, and failed to appreciate that innovation patterns are governed by long-term trends extending over several generations. It was, for example, only around 1900, more than 150 years after the beginning of the industrial revolution in Great Britain, that coal overtook wood as the most important source of primary energy in the world economy; as for oil, which started to be exploited commercially on a large scale in the 1880s, it is still today a less important source of energy in the world than coal.

If AT is so inconspicuous, it is not because of any failure on the part of the AT movement to push its ideas, develop new products or devise new forms of organization, but simply because the time-scale of technological innovation and institutional change is extremely long. Such misjudgements about the time element are not accidental. They are a normal, and indeed necessary component of the innovation process: if innovators knew from the beginning how long it would effectively take for their new institutions to leave a mark on society, they would never have the confidence and psychological stamina which are crucial to the success of innovation.

In the same way, the inconspicuousness of AT in the economies of industrialized and developing societies should be viewed as a perfectly normal phenomenon: it will take time — not just a few years, but several decades — before anyone could expect to see (even if they wanted to) a 'General Electric of AT' selling millions of village stoves, a 'General Motors of AT' manufacturing hundreds of thousands of low-cost vehicles for developing countries, or even an ITDG skyscraper in New York!

It could well be that such large AT firms will never develop, not because of any inherent inability of entrepreneurs in the AT field to make it big, but because the social patterns of innovation in AT may turn out to be rather different from what we are used to. One of the implicit assumptions in the philosophy of the AT movement was that in order to prove the validity of AT as a concept, it was first necessary to develop new products and new ways of doing things. This pragmatic approach was remarkably successful in generating a momentum of technological innovation and showing that there were indeed many promising alternatives to the dominant technologies which Schumacher so rightly put into question in his book.

This blossoming of innovation has tended however, to overshadow Schumacher's deeper and ultimately more important contribution to society: what *Small is Beautiful* did was to contribute in no small way to changing our dominant values and culture and creating a social demand for new types of technology. The patterns of technological innovation in any society are determined not so much by what is technically feasible, as by what is socially or culturally desirable. If technology can be considered as the product of a culture, it means that changes in the culture, or in the social demand for innovation are the precondition for fundamental changes in the technological system.

The cultural revolution ignited by Schumacher's word is still far from complete, and it may be interesting to reflect on the next ten years of the AT movement, and try to identify some of the big challenges which are likely to face AT groups throughout the world. The first of these, paradoxically enough, is the technological challenge. The majority of AT groups are actively involved in the development of new technologies, the improvement of traditional technologies or the scaling-down of modern technologies, and the public visibility of the AT movement rests essentially on its innovations in hardware, from biogas plants and small-scale agricultural machinery to solar heaters and low-cost water purification systems, to name but a few examples. These innovations all have a very important symbolic function to play as the expression of alternatives to the dominant technological system. Most of them however are still far from having reached the degree of reliability, cost-effectiveness and simplicity of use which characterizes the more sophisticated technologies which they are seeking to replace. Many AT groups, throughout the world, seem to have vastly underestimated this problem, and generally

73

attribute their difficulties in the diffusion of new products or new technologies to bureaucratic obstacles or lack of funds, and not to poor reliability or ineffective maintenance.

This indirectly suggests that AT groups, despite their substantial investments in research and development, their ability to identify a need for innovation and the originality of the technical solutions they propose, are perhaps not the most adequate institutions to carry out the complex technical work of testing, de-bugging and improving, without which a good prototype cannot be produced successfully on a large scale. This is the sort of work which is carried out extremely effectively by industrial firms, with their research departments, their production engineering teams and their after-sales services.

In this perspective, the growing interest of industrial firms in AT is a phenomenon of major significance: it could, in the long run, help to establish an effective linkage between the industrial production system and the innovative efforts of the AT groups. These AT groups would continue to play their vital role as 'opportunity identifiers', experimenters, networkers and generators of ideas, without falling prey to the temptation of doing what industry does so much better with its production experience, its access to large financial resources and its marketing skills.

A second important challenge facing the AT movement is of a political nature. If the ideas of the AT movement are now so widely known, and almost as widely accepted, this is not only the result of the publication of *Small is Beautiful* but also of Schumacher's own missionary work. Most AT groups still spend a considerable amount of effort on missionary work of this nature. As a result, consciously or not, they have come to acquire a political influence which is quite out of proportion to their small size, their limited financial resources or their effective strength as technological innovators.

This ability to propagate new ideas and influence policy-makers in government or decision-makers in industry is perhaps the most important intangible asset of the AT movement, and the one that deserves to be exploited in the most systematic way. This however calls for a significant revision in the political outlook of the AT movement, in the sense that what may be required of them is to work not against the system, but within the system. In their early days, many AT groups relished their position as outsiders and as critics of the existing order of things, and this marginality was in fact the necessary condition for innovation and originality.

Now that Schumacher's ideas are so widely known, and indeed so actively embraced in many of the ruling elites throughout the world, this political marginality of the early days is no longer necessary, and could indeed be counterproductive. Several AT grups have already sensed this intuitively, and now work more closely with large industrial firms, international development banks, government ministries and national planning agencies. What makes this co-operation so much easier today than it would have been ten years ago, is the new, large network of AT sympathizers within the establishment. These sympathizers are for the most part people who have read *Small is Beautiful* and understood its message, and who, like the termites in a building are inconspicuously eating away the wooden certitudes of high technology and the belief that 'Big is Better'.

My impression, no doubt somewhat subjective, is that the majority of AT groups have yet to appreciate the importance of the 'AT termites' and their crucial role in transforming Schumacher's ideas into practical realities. In the same way, many groups may have some difficulty in accepting the fact that in the long run, the effectiveness of their work could depend upon their ability to operate within the system rather than against the system.

Working within the system need not be a betrayal of the ideals of youth, but the means of their achievement. I believe that only in this way will it be possible for the AT movement to make its transition to what I like to call its second generation, or its age of maturity. And only in this way will it be possible for AT to reach the hundreds of millions of underprivileged people to whom Schumacher's important message was addressed.

Nicolas Jéquier, *Appropriate Technology*, 1983.

Economics As If People Mattered

Both in the rich and the poor countries, the idea of intermediate or appropriate technology is now entering the consciousness of economists, administrators and politicians. There would seem to be four stages in the process. It starts with the widespread rejection of the concept, because it means a radical break with conventional behaviour. The second stage is general acceptance of the idea, but little support from government, or international institutions. The third stage would be active involvement on a considerable scale to mobilize knowledge of technological choices and to test them under operating conditions; and the fourth would be the application of this knowledge on a scale that makes it not exceptional or 'alternative' but a normal part of administrative, business and community activity.

As far as developing countries are concerned the idea is now in its second stage and hovering on the edge of the third. Most poor countries are now aware of the extent to which they have become economically dependent upon the industrialized countries, and most see the need, if not to shake themselves completely free, at least to become much more self-reliant and regain their integrity. But as the recent United Nationals Conference on Science and Technology has revealed, most governments of developing countries are still far too preoccupied with questions of international trade, and with securing the technology of the multinational companies as easily and as fast as possible. This is a sort of tunnel vision that leads to aid and development policies which, in the event, bypass the rural areas of developing countries and thus bypass the very source and centre of their poverty.

Developing countries could help

themselves to change the 'rules of the game' by setting up their own technology assessment units. Their task would be to advise their governments about the choices of technology available, and about the implications of different choices. Thus a developing country intending to introduce cement manufacture, say, or sugar refining, would be advised on the choices open to it — in these instances, one huge factory or forty or fifty small units. What would these options imply in terms of availability, foreign exchange costs, running costs, social impact, employment, local development and income distribution, and so on? Ultimately it is only when developing countries start to demand that real choices of technology should be offered to them, and to assert this as their right, that they can begin to break away from their dependence on industrialized countries: they would then be free to choose technologies that maximized their self-reliance.

If the developing countries are well into the second stage of our four stages of appropriate technology, the industrialized countries are still at stage one. With a few exceptions in Britain, and a few more in the US, the official attitude towards appropriate technology and the alternatives groups that are applying it is one of rejection or antagonism.

It is true, of course, that for highly industrialized countries a changeover to technologies that are smaller, more decentralized, more humane and less demanding on non-renewable resources represents a major departure from orthodox thinking and conventional practice. The developing countries, after all, still have many of their options open. The rich countries are well along the road towards an economic and industrial system that looks less and less sustainable.

One of the principal characteristics of industrialization, as most people have experienced it, is its overriding tendency to create a more and more dependent population. This dependence, this external direction of people's lives, is most evident in the case of their employment, over which they have virtually no control, either as regards its availability or its quality. It also applies to other aspects of daily life. There are always two systems by which we support ourselves — the 'self-care' system and the market system; the latter requires us to earn money, in order to buy goods and services produced by others. The self-care system has declined to near-vanishing point, and the result is a great deal of waste and expense, and a loss of independence for the family and the community. The same applies to education, health care and recreation.

All this is very discouraging. But its positive aspect is that it has caused a large and growing number of people in Britain and in North America, as we have seen, to start reversing this trend by launching experiments of all kinds in the direction of greater individual and community self-reliance. This has produced what can only be described as a flowering of creative activity on the part of tens of thousands of people.

The origins of these groups working on alternative technologies and life-support systems are varied, and their attitudes to conventional insti-

76

tutions and ways of doing things are by no means invariably sympathetic. Some have come together out of a growing concern for the environment; others have their primary focus on energy conservation, housing, health, agriculture, local manufacture. There are now few branches of human activity where the conventional mode of doing things does not have its counterpart within the alternatives movement. The variety of their origins and methods of working should never blind us to the fact of what they have in common, namely, a recognition that conventional industrial society is on a collision course with human nature, with the living environment, and with the world's stock of non-renewable resources.

So long as there is no general acknowledgement that we are on this collision course, this upsurge of groups working on alternatives of all kinds will tend to be written off, by government and other power groups, as at best a way of life for eccentrics, at worst a serious threat to economic order and discipline.

It would be idle to pretend that the transition from the 'limitless growth' economy of the recent past, or the 'stagflation' economy of the present, to the conserver society of the future will be plain sailing. But it would be equally misleading to argue that policies favouring the widespread adoption of appropriate technologies would imply some drastic and unacceptable collapse of living standards and life as we know it.

Consider, for example, a few of the steps that Britain, or any other highly industrialized country for that matter, could well take now that

would in practice set us on the way to a more sustainable future.

One would be to start a major programme of energy conservation, with initial emphasis on reducing the 40 per cent or so of primary energy production that now goes into space heating: and at the same time to further the development of renewable energy resources on a really significant scale. (In Britain, this would enable us to avoid incurring the enormous costs and unimaginable dangers of nuclear power.)

A second part of such a policy would be to start a broadly based programme of research and experiment aimed at liberating agriculture and food production from its present very substantial dependence on fossil energy, and making Britain self-sufficient in food to the maximum possible extent.

A third component would be to promote and facilitate the localization and decentralization of manufacturing and service activity, both as a means of creating local employment, and of cutting down the escalating costs of long transport hauls and over-centralized services.

Could really effective action along these lines do anything but enhance the quality of life?

It will also be evident that if such deliberately decentralist and conserver policies were to be followed, then appropriate technologies, and most of the current activities of the alternative groups, would fall into place as perfectly obvious ways of attaining these objectives.

In fact this relatively unknown part of the economy represents an important part of present-day reality. It is opening up the way to what James

Robertson calls the sane, human and ecological society, which recognizes and encourages the variety of human resources and abilities that exists in all communities. In contrast, the mechanistic 'single-solution' approach of conventional economics — modelled on the physical sciences — typified by the centralized manipulation of aggregate demand, and by the concept of gross national product, is by its very nature incapable of recognizing this variety or of freeing the creative energies of large numbers of people. No amount of juggling with the monetary, fiscal or price system can provide useful and satisfying work for the people who are losing their jobs through the increasing capital-intensity of large-scale industry, or through growing shortages of resources; nor can it provide any answer to the alienation of the workforce, to inflation, or to the degradation of the environment.

Can we not recognize that there is really no other choice than to create a new technology and economic system designed to serve not a continuously escalating spiral of production and consumption, but to serve people by enabling them to become more productive? This is precisely what is being attempted by such groups as the Local Enterprise Trusts in Britain, for example, and by the Lucas Combine, by small co-operatives and common ownership firms on both sides of the Atlantic, by Sudbury 2001, by, in short, the alternatives movement. The rich countries need more of this kind of work at least as urgently as the developing countries.

George McRobie, *Small is Possible*, 1981.

'Appropriate technology' is a concept which will play an increasingly important role as a catchword in international conferences when discussing technology and development. . . .

'There can be no doubt that the question of appropriate technology for developing countries has in the last few years gained international acceptance. Consequently, the United Nations Conference on Science and Technology for Development (UNCSTD) is bound to consider it.' — The Advisory Committee on the Application of Science and Technology to Development, 1977.

The era of only the best for the few and nothing for the many is drawing to a close. — Halfdan Mahler, Director-General, WHO

II

Technology Choice: Theory and Practice

IN CHAPTER I, it was implied that a range of techniques exist and that it is important for society that the techniques actually used are those which are most appropriate in terms of enabling a sustainable pattern of development. But how are choices of technology made, and who does the choosing?

The concept of the appropriateness of a technology implies that a set of criteria exists by which a choice can be made. In practice, of course, different technologies will be chosen according to who sets the criteria and who does the choosing. Transnational corporations may have very different motives from small rural enterprises, and both may differ in their motivation from government planners. Schumacher himself acknowledged that 'appropriate' technology begged a question — appropriate for whom, appropriate for where and appropriate for when? This is why he chose the term 'intermediate' technology which he felt answered the question in terms of being the technology appropriate to the masses of poor people in the rural areas of developing countries at that point in time.

From the economist's point of view, the best known way of looking at the issue of technology choice is within the context of the neo-classical model. At its simplest, this picks out two characteristics of techniques — the labour and the investment requirements — and regards the question as one of choosing between techniques of differing labour and investment intensity. The relative price of labour and investment is regarded as the determinant of this choice, with that technique being selected that maximizes profits, given the relative price, and the substitutability between labour and capital. This lies at the basis of policy recommendations by economists who seek to mould the pattern of development by altering factor prices. For instance, those who feel that the economy is unduly reliant on capital-intensive techniques would argue for raising the price of capital relative to that of labour, thus encouraging investors to choose more labour-intensive techniques.

In developing countries, capital tends to be scarce (and therefore expensive), while labour is abundant (and reasonably cheap). This combination suggests that investors should favour labour-intensive techniques. In practice, however, this is not necessarily what happens: as was seen in the previous chapter, many of the problems in the developing countries are related to the use of inappropriate capital-intensive techniques. There are two reasons for this. First, factor prices are often distorted in such a way as to reduce artificially the price of capital. This is a situation that can be easily dealt with within

79

the neo-classical model by recommending the removal of the distortion, which may, for instance, be the result of highly subsidized credit.

Second, there are many considerations which cannot be taken into account in the neo-classical model which nonetheless have a very significant impact on the choice of technology. Not all possible techniques available may be known to the decision-maker so that s(he) may be unable to choose the one which maximizes profits. Further, considerations other than maximization of profits may be important to the decision-maker. These can include ease of management, prestige and a desire for modernity.

A further factor influencing the choice of capital-intensive techniques in developing countries is the choice of products and the choice of projects. For example, if refined white sugar is demanded then very capital-intensive techniques are needed to produce it. Less capital-intensive techniques are required if consumers are prepared to accept brown or grey sugar. If top quality tar-macadamed roads are required then more capital-intensive techniques must be used than if lesser quality roads (often more than adequate for the low-traffic volume in rural areas) are chosen. Similarly, if a government decides to invest in a large dam or a new, modern hospital, more capital-intensive techniques are required than if a small-scale irrigation or rural health programme is launched.

This chapter seeks to give a better understanding of these and other issues involved in the theory and practice of technology choice. Section 1, which concentrates on theory, starts with an extract by Frances Stewart which, although a little difficult for those who are not economists, is included because of its useful description of the neo-classical approach and its limitations. Non-economists may find the account by Thomas and Lockett on choosing appropriate technology, easier to understand, but this covers fewer of the issues than the Stewart extract.

These are followed by extracts which expand on the themes of choice of product and choice of project. The short extract by Frances Stewart looks at some of the complexities of the product choice issue. That by Chris Baron and W. Van Ginneken is more empirically-oriented. It describes the actual effect of choice of technology and employment creation in Bangladesh of a redistribution in income leading to a rise in demand (from the poor) for basic goods (e.g. unwrapped washing soap) rather than the more sophisticated goods (e.g. wrapped, scented, coloured, toilet soap). The choice of project issue is illustrated by Luis de Sebastián with an example from El Salvador. Here he argues that once the government had decided to invest US$150 million in an international airport, the problem of technology choice became marginal. The real issue was in choosing an airport rather than the building of schools or hospitals for the rural people. He goes on to examine the role of foreign financing in influencing this decision.

The section ends with a look at the important issue of the methodologies used in choosing technologies or projects. Here too, a choice is involved — between conventional economic techniques developed during the era of output maximization development strategies, and alternative techniques devised by economists who are concerned about basic needs in rural areas as well as efficient use of resources. The extract by George Baldwin gives an excellent account of the limitations of the use of Present-Value calculations in choosing between alternative technologies for rural water supply projects where recurrent (maintenance) costs are high and may be difficult to meet. He argues that concern about meeting these future costs should be allowed to influence decision makers and should not be discounted heavily in a Present-Value calculation. Similarly, Robert Chambers argues about the wisdom of relying on social cost-benefit procedures when choosing between projects. These procedures are, he states, too complicated and too much subject to personal values and political pressures. As an alternative he suggests some simpler procedures more appropriate for use in poverty-focussed rural development.

A major factor suggesting that the labour-intensive techniques advocated by Schumacher and others are appropriate in developing countries is the abundance of labour and shortage of capital. There are, however, several other factors which, in a perfect world, would tip the balance in favour of small-scale, simple, cheap, labour-intensive techniques. These include small markets, scattered resources and population, large poor rural populations with need for low-cost basic needs goods, and poorly developed infrastructure. These important considerations are looked at in Section 2. The extract by Keith Marsden illustrates why large-scale factories don't work well in developing countries, while that by Malcolm Harper and T. Thiam Soon shows why intermediate technologies and small-scale industries do. Frances Stewart explodes the myth of economies of scale and points out that really large-scale production can only take place in an urban setting, whereas the majority of people in the Third World live in rural areas.

Finally, Section 3 of the chapter presents a selection of case studies on technology choice. These include an extract by M.K. Garg giving comparative economic data on alternative sugar processing technologies in India, and an account from Northern Nigeria of the diastrous socio-economic impact of the decision to invest US$550 million on a large dam. This is followed by an extract on the effect of technology choice in the fishing industry in India where 'improved' fishing boats have led to short-term gains for small fishermen followed by the threat of economic crisis due to over-fishing.

Also included is the account by Thomas of the choice of technology for irrigation tube-wells in East Pakistan (Bangladesh) which gives a frank analysis of how and why decisions were made. It shows that although economic analysis indicated that smaller low-cost technologies would have been more in accord with the stated objectives of the government, both the government and the donor chose a less optimal, medium-cost technology for a variety of organizational and perceptual reasons.

The section ends with further extracts which examine why inappropriate choices are made. The case studies in this chapter are heavily biased towards examples of inappropriate choices and inappropriate technologies; this does not imply that appropriate technologies are rarely found in use in the developing countries. On the contrary, with the increased awareness of the issues

81

involved, decision makers are beginning to take a more appropriate approach towards technology choice. Some of the many examples of application of appropriate technology in the developing (and developed) world are given in the following five chapters.

1 CHOICE OF TECHNOLOGY: THEORETICAL ASPECTS

The Technological Choice

Technology consists of a series of techniques. The technology available to a particular country is all those techniques it knows about (or may with not too much difficulty obtain knowledge about) and could acquire, while the technology in use is that subset of techniques it has acquired. It must be noted that the technology available to a country cannot be identified with all known techniques: on the one hand weak communication may mean that a particular country only knows about part of the total methods known to the world as a whole. This can be an important limitation on technological choice. On the other hand, methods may be known but they may not be available because no one is producing the machinery or other inputs required. This too limits technological choice.

Each technique is associated with a set of characteristics. These characteristics include the nature of the product, the resource use — of machinery, skilled and unskilled manpower, management, materials and energy inputs — the scale of production, the complementary products and services involved etc. Any or all of these characteristics may be important in determining whether it is possible and/or desirable to adopt a particular technique in a particular country and the implications of so doing.

$$wT = \{Ta, Tb, Tc, Td, ... Tn\}$$

(where 'known' means known to the world) as constituting world technology. For a particular country, the technology available for adoption is that subset of world technology known to the country in question and available. Say, $cT = \{\bar{T}a...\bar{T}n\}$, where c denotes the country and the bar indicates that only techniques known to the country and available are included. Thus $cT \subset wT$.

Each of the techniques Ta, Tb...etc. is a vector consisting of a set of characteristics, ai, aii, $aiii$, bi, bii, $biii$... Thus technology can be described in matrix form, with each column representing the characteristics of each technique, as follows:

82

Matrix of World Technology = wT

Characteristics	Ta	Tb	Tc	Td	Te
Product type					
Product nature					
Scale of production					
Material inputs					
Labour input:					
skilled					
unskilled					
Managerial input					
Investment requirements					

The technology in use in a particular country is that subset of the technology available to it that has been selected and introduced, or $uT = \{\bar{T}a...\bar{T}n\}$ where $uT \subset cT \subset wT$.

The processes by which world technology is narrowed down to an actual set of techniques in use may be crudely described as follows:

of characteristics — consisting of the inputs required, quantitatively and qualitatively, the nature of the product, the scale of production, productivity of the various factors, the organization to which it is best suited and so on. These characteristics tend to reflect — or at least be in tune with — the ecomomic/historic circum-

The actual technology in use is thus circumscribed first by the nature of world technology, then by the availability to the country of known techniques, and finally by the choice made among those available. If the technology in use is thought to be inappropriate, it may be inappropriate because world technology is inappropriate, or because an inappropriate subset is available to the country, or because an inappropriate selection is made, or for some combination of the three reasons. Confusion is caused by failing to distinguish between the three.

It was argued earlier that each technique is associated with a vector

stances of the economy where the technique was first introduced. Those who introduce techniques into underdeveloped countries thus make a choice among the techniques available; the choice actually made depends on the nature of the decision makers and their objectives, the economic circumstances in the economy concerned, and the characteristics associated with different techniques, bearing in mind that their choice is confined to the techniques they know about, and that their knowledge may often be incomplete or inaccurate.

Decision makers differ as to motive, knowledge and circumstances, so who takes the decision may determine what decision is made. For example, a subsidiary of a multinational firm may have, as prime motive, maximization of profits after tax, on a worldwide basis. Locally and privately owned firms may aim to maximize local profit after tax. This difference can make a considerable difference to choice of technique in terms of nature of output, scale, specialization, type of inputs used, price of such inputs, etc. A government-owned corporation may aim to maximize local profits before tax; or it may also include other aims that are given little weight by the private sector e.g. employment expansion, or the spread of opportunities to the rural areas. The aims of those taking the decisions may differ from those of the corporations for whom they decide — individual income and/or prestige maximization may alter decision making, sometimes allowing corruption to be decisive in choice of technique . The aims of family enterprises are likely to be in terms of total income of the enterprise, rather than profits.

The circumstances in which firms operate also differ as between the type of operator. For example, access to funds for investment, in quantity and quality, differs between firms. Multinational firms may obtain more or less unlimited funds at the cost of funds for the group as a whole. Local large-scale firms may borrow from the banks, often at interest rates which are held low to encourage investment. Smaller-scale enterprises, including family enterprises, may find it difficult to raise funds in any quantity and may have

to pay high prices. Different types of firms tend to serve different markets: for many subsidiaries of multinational firms the world is their market. Locally owned firms tend to be more confined with the larger-scale firms serving the upper income groups, and doing some exporting. Family enterprises, particularly in the informal sector, tend to produce for the consumption of those in the immediate vicinity, particularly among the lower income groups. Scale of operations is a function of organization, availability of funds and the nature of the market. The scale of operations is often the decisive characteristic in determining selection of technique, with only one technique that is efficient at each scale. Another characteristic that is often decisive, as suggested by empirical studies, is product specification. Product specification depends on the nature and income levels of consumers, and the structure of the economy as a whole. Different types of firms tend to have different product requirements — mainly because their markets differ, with the larger-scale serving the higher income groups and competing on the world market, while the small-scale and rural cater for local low-income consumers. Keeping up standards, the prestige of the firm generally and maintaining the value of the brand name also help determine product standards.

The price and availability of other inputs also differ between types of firm within any economy: it has been shown that raw materials may be obtained at a lower price by the large-scale than the small. Firms with foreign technology contracts, and multinational firms, have access to inputs at different prices from those without. A major difference between types of enterprise is their access to different types of labour, and the price they pay for it. Factors holding up wages — such as government regulations, and trade union activity — are confined to the large-scale; family enterprises may generally obtain labour at a much lower price.

While there are major differences in objectives and circumstances between different types of firm within an economy, within each category selection of technique also depends on the way in which the economy as a whole operates. This is partly a question of price and availability of different inputs, including labour and investment goods; partly of income distribution determining the nature of markets; partly of the openness/closedness of the economy determining the extent to which products have to compete internationally. As argued above, the package aspect of technology means that any one technique which, looked at by itself, may appear efficient and appropriate, may be inefficient in the context of the technology in use. For example, a decision on the technique to be adopted in tyre manufacture will depend on the nature of the economy and income distribution within it — whether cars are being consumed locally, whether they are produced locally, or whether it is a bicycle economy, the standard of roads, the extent to which the tyres have to compete with other tyres manufactured locally or imported, the standard and prices of the competitive goods, the availability and price of inputs required, including energy,

labour of different skills, materials and so on. The decision has to be made in the light of, and may be uniquely determined by, the nature of the economic structure as a whole. A system in which private firms compete, each with free access to foreign technology, may lead to oligopolistic competition via product differentia- tion — as it has in capitalist advanced countries — rather than competition via price. Such a structure may force each firm to adopt the most recent techniques in order to secure its market by providing the most recent product; in such a situation the technique is determined by the market in the context of the technology in use, although taken together the decisions also determine that technology. This is why it is difficult to induce marginal changes in technology.

To formalize, we may say that the process of selection of techniques consists of selecting from the matrix of known technically efficient and available techniques, which we described as cT, the nature of which was determined by the history of technological development, discussed above. Each technique within this matrix is represented by a vector describing its characteristics: $cT = (Ta, Tb, Tc, ... Tn)$ where $Ta, Tb, ...$ are the different techniques, and each technique consists of a vector of characteristics $ai, aii, aiii...$associated with $Ta; bi, bii, biii, ...$associated with Tb, and so on.

Decision makers may be categorized into groups, each of which has an objective function representing its aims. Suppose the decision makers are categorized into M (multinationals), L (large-scale local

85

firms), G (government-owned enterprises), F (family enterprises) and so on. Corresponding to each group is an objective function, which we may represent as m,l,g,f. In trying to maximize their objective function, the decision makers are subject to a series of constraints, some of which are common to all of them, and some of which vary according to the category of decision maker. One such constraint is the nature of technology available — this may vary somewhat between decision makers, since knowledge about and access to different types of technology varies between groups — for example, a subsidiary of a multinational will have access to a different selection of techniques from a small family enterprise. In addition, as argued, markets, scale, factor availability and price may vary between the groups of decision makers. The underlying conditions in the economy, in contrast, tend to be common to all decision makers. We may describe the constraints of each group as Cm, Cl, and those common to the economy as a whole as C^*. Selection of techniques then consists in the attempt by each group to maximize its objective function subject to the constraints. The overall balance of techniques within the economy depends on the size of the different groups. So $uT = f(m,l,g,f..., Cm, Cl, Cf...C^*)$, where uT is that subset of total available techniques that are selected. The neo-classical model of choice of technique picks out two characteristics of techniques — labour and investment requirements — and regards the question of choice of technique as consisting of choosing between techniques of dif-

fering labour and investment intensity. The relative price of labour and investment is regarded as the determinant of this choice, with that technique being selected that maximizes profits, given the relative price, and the substitutability between labour and capital. The approach, at its textbook simplest, may be shown with a smooth, convex isoquant representing different methods (in terms of I/L ratios) which may be adopted to produce the 'same' output. In the figure below R represents the profit-maximizing equilibrium point, so that technique would be selected.

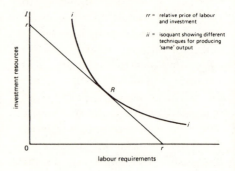

Technical progress may be introduced without altering the basic model. Technical progress is assumed to be 'neutral', affecting all techniques equally. Thus the entire isoquant is shifted inwards over time, so that any one time choice of technique is between techniques as represented by an isoquant even though the isoquant is shifting over time. The figures below illustrate how choice of technique is affected by technical progress.

Developments of the neo-classical approach in application to underdeveloped countries have concentrated on two aspects: first, that savings generated per unit of investment

86

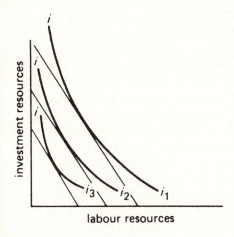

labour resources

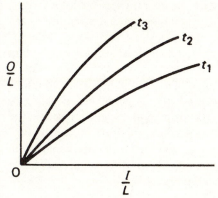

investment being the determinant of choice.

In one way the neo-classical model may be regarded as a special (and, in practice somewhat insignificant) case of the general model described above. The neo-classical model picks out just two characteristics of the manifold characteristics associated with each technique — investment and labour requirements — and completely ignores the others, such as scale of output, nature of product, skilled labour requirements, material inputs, infrastructural requirements etc.,etc. In terms of selection mechanisms it concentrates on just one — relative prices of labour and investment — corresponding to one type of decision maker, the profit-maximizing entrepreneur with unlimited access to finance at constant rates of interest. In order to achieve this simplification the model makes a *ceteris paribus* assumption about the many other influences over decision making, and about characteristics of techniques that determine technical choice.

should be a (sometimes the) criterion for choice of technique; secondly, that the ruling relative price of labour and investment may be 'distorted', with the consequence that socially inappropriate techniques are selected (generally speaking excessively investment-intensive) with resulting un- and underemployment. Both are premised on the same basic approach to choice of technique as the neo-classical approach, at its simplest, as described above: that is to say they are premised on the existence of a wide range of techniques of varying labour and investment requirements as shown in the isoquant above, with the relative price of labour and

It might be argued, none the less, despite the overwhelming empirical evidence that labour and investment intensity are not the most significant variables nor their relative price the sole critical determinant of choice, that the approach highlights an important subsection of technical choice — that concerned with the investment/labour intensity aspect of producing given output. Even this much cannot be granted because of the nature of technological development over time, which makes nonsense both of the assumption of a range of techniques of varying labour and investment intensity, and most

87

significantly, of the *ceteris paribus* assumption about other characteristics of techniques.

As argued above, techniques have to be developed historically and the techniques available for selection at any one time are those that have been developed at some time in the past. Generally speaking the techniques developed at one time reflect the resources of the economy when developed. In terms of investment and labour intensity this means that more labour-intensive techniques are likely to have been developed at earlier periods, when savings were lower in relation to labour supply, while more investment-intensive techniques were developed later when more savings were available in relation to labour supply. In so far then as a neo-classical type range of techniques is available (i.e. techniques of varying I/L ratios), it consists of techniques developed over a historical period with the more labour-intensive techniques dating back to an earlier period than the investment-intensive. But because the earlier techniques originate at an earlier time, they have less scientific and technical knowledge to back them up, and therefore tend to be of lower productivity. Many of them have become technically inefficient, as argued above, using more investment as well as labour in relation to output produced. Thus, far from there being a complete isoquant corresponding to each moment of time, for each scientific and technical age, there is a series of techniques developed at different times, with a tendency for the earlier ones to become technically inefficient.

In some ways this view of technical development is similar to that of Salter and of Atkinson and Stiglitz. Salter believed, with the neo-classicists, that potentially there was a whole range of techniques of varying investment and labour intensity, corresponding to each level of scientific and technical development, or 'state of technique' as he puts it. But of this large potential range, only the immediately profitable were actually developed; he contrasted the 'relatively narrow range of developed techniques which could be designed with the current state of knowledge'. The former — actual machines — are confined to a narrow range of investment and labour intensity reflecting the resources when they were developed. They might therefore be represented by a point on a diagram. Atkinson and Stiglitz discuss the localized nature of part technical progress, consisting of improvements in techniques already in use, as shown in the Figure below.

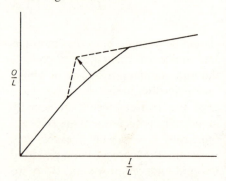

Atkinson/Stigilitz localized technical progress

Both Salter and Atkinson Stiglitz stick to the neo-classical framework in two respects; first, showing a production function whether potentially (Salter) or as the starting point (Atkinson and Stiglitz), despite the fact that the conclusion of a historical approach to technical development must be that there is little, if any, meaning to the assumption of an almost continuous range of techniques reflecting possibilities at a single 'state of knowledge'; all there is are developed techniques which have been developed at different times. Whether or not it makes sense to talk of a potential production function, along with Salter, is really a matter of metaphysics. More relevant for our discussion is their concentration, as in the basic neo-classical approach, on the two neo-classical characteristics of techniques of labour and capital intensity, ignoring the other characteristics, and selection mechanisms. This is significant because these other characteristics are systematically related to each other and to the investment intensity of techniques, and tend therefore to invalidate any conclusions which ignore them. One aspect discussed at length earlier is product development which occurs systematically over time, and which means that the later techniques are associated with different — more efficient and higher-income — products than earlier techniques. Hence, to assume that techniques of varying labour and capital intensity exist which produce the same product does not make sense, because the technical developments that have increased investment have also been associated with changed products. This also means that some early techniques which might appear to offer a labour-intensive alternative, are ruled out because associated with obsolete products. In a way this conclusion is ironic because many

neo-classicists have used product choice to rescue the idea of technical choice in terms of investment intensity — arguing that while empirical evidence supports the idea of relative coefficient fixity for each product, product choice allows coefficient variability. Other systematic changes in characteristics of techniques over time are scale changes, and input (and particularly skill) requirements. Again this means that later techniques are designed for larger scale than earlier so it rarely makes sense to think of them as producing on the same isoquant. Changes in the skill availability and requirements, and other changes in the technology in use, also mean that later techniques are associated systematically with different characteristics than earlier ones. Since the only genuine choice of technique along neo-classical lines is the result of the survival of earlier labour-intensive techniques along with later more investment-intensive techniques, any choice between them also involves a choice of the other characteristics that have changed systematically with time. So the early techniques are designed, on the whole, for production at smaller scale, use fewer skills, require less technologically advanced inputs, and so on and so forth, as compared with the later techniques. In view of this it is scarcely surprising that investment intensity often becomes of subsidiary relevance to the choice, and the relative price of labour and capital also only of minor importance, compared with the other characteristics that are ignored in the neo-classical approach.

Frances Stewart, *Technology and Underdevelopment*, 1977.

Choosing Appropriate Technology

In practice, the use of the concept of appropriateness of technology means a deliberate use of criteria for choosing which technology is appropriate in a given set of circumstances. We can expect the criteria to be related as closely as possible to the meeting of human and social needs. The majority of technological decisions, in the Third World as in the United Kingdom, are not made as self-consciously as this. However, let us look at a real example to see how choices can be made deliberately using criteria of appropriateness. Details of the example are outlined in the panel. Three points can be made immediately from this example.

First, if a simple criterion is chosen, such as 'lowest total cost per can', then there is no question which technology should be adopted. In this case the automatic processes are the ones that would be appropriate on this criterion. However, constructing tables of 'key figures' or indicators like ths can only give an unambiguous result if a single, measurable quantity (such as 'total cost per item produced') is chosen as a criterion, in place of the sort of multitude of possible, vaguely desirable attributes of technology indicated by lists like those given in Chapter I.

Second, the same simple criterion can lead to different results in different circumstances. Thus, in a country where monthly wages for unskilled labour were considerably less than the equivalent wages in Kenya, and if other factors remained equal, the semi-automatic sealing process could become the 'appropriate' technology on the same criterion of least cost per unit output. Alternatively, one might think of using Schumacher's criterion of capital cost per workplace. This quantity, according to Schumacher should be kept down to roughly one year's average wages. The third point, then, is that different criteria lead to different choices.

To get beyond these fairly obvious points we need to consider reasons for using any particular criterion in the first place, and the likely effects of making a choice in this way.

Different criteria reflect different social goals, though this may be a question of implicit values rather than explicit choice. So the idea of minimizing cost, reflected in the criterion of least total cost per can, may derive consciously from a goal of profitability for the individual canning enterprise, or it may simply be the unthinking result of cultural bias towards capital-intensive techniques on the part of design engineers or other technical personnel. Similarly, the idea of keeping the capital cost per workplace low could be derived logically from a policy aim of self-reliant development (keeping industrial innovations within the financial reach of the average worker means that there is some chance at least of such innovations spreading spontaneously) or it could just come from uncritical acceptance of Schumacher's doctrine of intermediate technology.

These social goals or policy aims can be expected to reflect efforts to

90

meet basic human needs. However, the link between social goals and basic needs almost certainly depends on a particular ideological view, and entirely different policy aims may be justified by reference to the same basic need. So, for example, we might take the creation or maintenance of permanent jobs as a basic human need in the above example. The traditional development economist's view may be that the automatic processes, being more profitable, would create more wealth generally,

Canning in Kenya

Sealing of cans is the process of closing one end of the can with a lid. This process can be done with two kinds of production technology.

One way is with an automatic machine that seals four cans at a time. The manual work is minimal, except for control of the machine. The workers have to be specialized to operate the machine. The quality of the sealing process is very high and there is thus no need for quality control of the sealed cans.

The other way of doing the job uses a machine that is semi-automatic and which makes one can at a time.

There is not much skill required to operate the machine, but the quality is poor, so that 5 per cent of the cans do not reach the desired quality. Therefore it is necessary to have a supervisor for every third or fourth machine.

Packing the finished cans in boxes can likewise be done two ways: either the cans are automatically placed accurately on a conveyor belt and from there fall into the boxes; or the packing is done by hand, using a simple wooden tool to lift the cans from the conveyor belt where they are in no particular order.

Key figures for the choice of technology	Sealing		Packing	
	Automatic	Semi-automatic	Automatic	Semi-automatic
costs of machine/£	10,000	500	6,000	600
number of workers per machine	2	1.3	3	5
monthly pay per machine/shilling	1,050	500	1,050	1,750
production/(cans per min)	270	38	460	280
lifetime of machine years	15	10	15	6
capital: output ratio/ (£ per can per min)	37	13	13	2.1
output: labour ratio/ (cans per min per worker)	135	29	153	56
capital: labour ratio/ (£ per worker)	5,000	385	2,000	120
capital costs/(sh per 10,000 cans)	3.71	1.65	1.30	0.34
wages (sh per 10,000 cans)	2.70	9.14	1.58	4.34
total costs/(sh per 10,000 cans)	6.41	10.79	2.88	4.68

which would filter down to all areas and classes, creating jobs indirectly through development and economic growth. On the other hand, the intermediate technologist would argue that unless capital cost per workplace is low enough, technology cannot spread to more than a few centres, so that it is the adoption of the semi-automatic process that will bring jobs to all areas, including the poorest rural parts.

It is worth pointing out here that it is not necessarily possible to find a way of implementing a choice made like this. It is all very well deciding in the abstract that a technology must meet certain criteria, but perhaps it just cannot be done. For example, none of the options in the example has a labour productivity of 200 cans per worker per minute, so, if that were a criterion, it could not be met. This would be an example of a technical constraint; a subsequent innovation or technical development might make this criterion realistic. More intractable, perhaps, are economic and political constraints. An example of the former would be a shortage of capital. So, for example, although it would be cheaper to produce cans in the long run by borrowing £16,000 at 20 per cent interest for one automatic sealing machine and one automatic packer, this money may not actually be available. On the other hand, once an automatic plant is in existence, its products will have a cost advantage which will mean that any subsequent semi-automatic plant could hardly compete, unless protected.

It is less easy to find a straightfor-ward example of a political constraint, but one can easily imagine various political reasons why certain options may in fact not be available. For example, certain developments may depend on the goodwill, or the investment or supplies, of particular foreign companies — and they may be prepared to make available only certain technological options. In the canning example, one can certainly speculate that a large overseas company would prefer to install automatic machinery, and be able to exert powerful control, through access to exclusive knowledge on maintenance etc., rather than make available semi-automatic machinery that could easily be maintained and copied by independent local operators.

Finally, adopting a criterion does not guarantee implementation of a preferred course of action. For example, the semi-automatic canning process might be chosen for a certain place, but perhaps no local people in fact take up the idea, and the goal of local self-reliance would hardly be met by forcing it on them. Indeed, such entrepreneurs as there are may prefer more profitable options. And there may be completely unanticipated consequences or unquantifiable effects that could not possibly appear in a table of 'key figures' drawn up in advance. The main areas to look for these unexpected developments are usually 'human' areas, such as the local cultural context or the area of management and promotion of schemes, but unanticipated ecological consequences may also be important in some cases.

A. Thomas and M. Lockett, *Choosing Appropriate Technology*, 1978.

Choice of Products

An important dimension of choice of techniques is choice of product. Many different products fulfil the same need. Such products may differ in most respects — for example, sausages and milk both fulfil a need for nourishment — or in a few respects, where the differences are usually called 'quality differences': finely spun and coarsely spun cotton both fulfil the need for thread. The extent to which different products or products of different quality do fulfil the same need depends on how the need is specified. The more generally needs are specified the greater the number of products of varying quality that may fulfil them. The more narrowly specified the fewer products; if very narrowly specified only one product may fulfil the need. For example, only Renault 16 car doors will do if one needs to replace a door on a Renault 16 car. Techniques of production very often differ in the quality of products produced, and they may also differ in the type of product produced. This raises two difficult questions. First, any study of techniques has to start by deciding which techniques are to be included. To do this it is necessary to decide how the needs are to be specified and hence the variety of products that would fulfil the needs. Secondly, since the different techniques are associated with different products, the study has to value these differences in coming to results.

A needs-based approach to choice of technique threatens to encompass a wide variety of different products if needs are broadly defined — thus raising the major difficulty of how to value such differences. On the other hand, excluding product variations, by making special efforts to find alternative techniques for producing a homogeneous product, as many studies have done to avoid valuation difficulties, restricts the comparison so that little choice of technique is likely to remain, while it begs the main question — that of the variety of possible ways of meeting given needs.

Economists have suggested two ways of valuing different products; a simple way round many of these problems is to take the consumers' sovereignty way out, valuing products according to market prices. But the conditions required to make this valid — desired income distribution, perfect competition, no externalities, no advertising — are so far from being met that consumers' sovereignty is a snare not a solution. The second method suggested for open economies is to take export price as a guide to the value of output, since this represents foreign exchange that might be acquired. This is legitimate where the items in question actually are exported. Where they are not, but satisfy previously unsatisfied domestic demands, act as import substitutes, or substitutes for other domestic output, the foreign exchange price is not a guide. To take the foreign exchange price to compare the value of e.g. mud bricks with concrete walls implies accepting the 'world' valuation of mud bricks and walls — a valuation which reflects world income distribution and taste patterns. Where home production acts as a direct and exact substitute for imports then import prices may be a correct guide to foreign exchange saved. But in choice of technique studies the central question is the different ways in which needs may be fulfilled. Exact import reproduction is only one way, and hence import prices often provide no guide to relative valuation of different qualities of goods.

One way of approaching the problem is to try to specify the need that is to be met as precisely as possible. Any output that meets this need may then be classified as of the same value, for the purpose of the exercise. If it overfulfils the need — producing e.g. stronger bricks than specified — this should not add to the value, at least in the initial assessment. If it can be shown that this overfulfilment — or meeting of non-specified needs — is of value, then this must be weighed in the final con-clusion. In a study of can sealing a major reason why the automatic method was preferred was that less supervision was required to get the quality of cans necessary for export markets. In so far as the cans are in fact going to be exported, then this extra quality presumably is necessary. But it may not be necessary for domestically consumed cans, or for exports within East Africa. There is an interaction between the results of the studies and needs: most needs are not autonomously defined but are related to the costs of meeting them. For example, if it is shown that two-storey building is going enormously to reduce employment and increase costs compared with one-storey, this might lead to a rethinking of the building programme in terms of one-storey accommodation. Hence prior definition of needs, to rule out single-storey accommodation and therefore study of hand-block makers, would misleadingly limit the scope of the study.

Frances Stewart, *Technology and Underdevelopment*, 1977.

Appropriate Products and Technology

Are the products which are most appropriate those which are produced by 'appropriate technologies'? It is impossible to answer this question unequivocally. The economists' definition of appropriate technology is 'the set of techniques which makes optimum use of available resources in a given environment. For each process or project, it is the technology which maximizes social welfare if factors and products are shadow-priced'. The reference in this definition of 'available resources' implies that in developing countries where labour is in relatively abundant supply an appropriate technique will tend to be labour-intensive, assuming that such a technique does exist and is available to those willing to apply it. However, the definition of appropriate technology quoted here, like most others, takes the product as already specified. Yet there is no obvious a priori reason to suppose that a product which is appropriate in the basic-needs sense may be manufactured or assembled by several technologies among which the most appropriate is also relatively labour intensive. On the other hand, some impressionistic evidence is suggestive. Wicker furniture, raffia mats and earthenware cooking utensils are household durables made by labour-intensive methods which are generally cheap and can be regarded as efficient in the basic-needs sense. One can generalize from these examples and suppose that traditional handicraft shops are not only employment-generating but also produce goods which are cheap and consumption-efficient, the more so because the craft worker is familiar with the needs of the people among whom he lives and works.

But the argument can scarcely be regarded as robust and examples in the contrary sense are not difficult to find. A moulded plastic chair may be aesthetically displeasing but would otherwise seem to embody the essential characteristics of a chair, although not much more, and may certainly be preferable to sitting on the floor. However, such chairs are manufactured by applying a very capital-intensive technology although unit costs may be low in long production runs. Whether similarly Spartan wooden chairs can be made as cheaply would depend on wage rates and the cost of timber in the country concerned.

Soap is regarded as a basic need in most human societies, for the purpose of personal hygiene, and also for washing clothes and other household purposes. In Bangladesh, for example, the consumer may choose between attractively wrapped, scented, toilet soaps of various sizes and colours; and usually scented, but unwrapped, washing soap sold in the forms of bars or balls. The latter may be used for toilet purposes as well as for the washing of clothes. Washing soap is usually produced in small-scale or cottage industries whereas toilet soap is manufactured only in a dozen or so large-scale modern factories, which also produce a limited quantity of washing soap.

Technically there are considerable differences between the larger factories applying modern technology and the cottage-industry type of soap producer. The modern factories are dependent on imported technology, and they also import many of their raw materials. Heating oil is a significant material input requirement; in contrast, the small-scale factories tend to produce soap in small lots, the handling of materials between processes is manual and firewood is generally the energy source.

The large factories are more capital-intensive than the small ones. Abstracting from the data collected in the production survey in Bangladesh, it may be concluded that, for an equal volume of soap manufactured, the capital-intensive approach implies very roughly three times more investment than small-scale production, whereas the latter employs about forty times more people.

The number of job opportunities created by soap production therefore depends on the choice which consumers make (in aggregate) between

95

relatively sophisticated toilet soaps and washing soap. A survey of consumers was carried out in order to identify the determinants of demand for the two types of soap. This survey revealed that the total volume of purchases of soap trends to increase with income, demand for toilet soap tending to be more predominant among high-income groups. Households in urban areas tended to consume more toilet soap than those living on the same incomes in rural areas. (A key reason for this was the greater availability of the sophisticated soaps in urban areas.) There were some differences between occupations: interestingly, for example, manual work was found to be associated with a higher level of consumption of toilet soap at the same level of income.

Consumers were asked questions about the attributes they seek in soap. The key attributes were identified as washing ability, durability, resistance to breakage, scent, colour and wrapping. Most consumers regarded the first three as the most important, although scent, colour and wrapping were accorded greater importance by households in higher income groups. This is what one might expect but it is still useful confirmation of the rationality of consumers. However, the replies given by consumers to the questions about attributes were not entirely consistent with their purchases.

There were several reasons for this. The most important was that the small-scale sector of the soap industry cannot make sufficient supplies of the simpler soap product available. Its marketing capability is weak.

Because of import restrictions it lacks raw materials, and production is often intermittent. The large-scale sector tends to receive more favourable treatment as regards the importation of raw materials, and it also has an advertising capability which may have a persuasive effect on consumers that overrides their objective perceptions of product attributes.

Would consumer demand for different types of soap, and hence employment generation in the industry, be affected by a redistribution of income? Estimates of the effects of a (static) redistribution of income are necessarily open to question, since they depend on many assumptions. Very roughly, an overnight shift of Tk 100 million of income (or approximately 2 per cent of GNP) from households with incomes of Tk 400 per month and over to households below this level would have the effect of shifting Tk 1-3 million worth of consumer demand for soap from the capital-intensive to the labour-intensive sector. Between thirty and one hundred new jobs could thus be created. This rather modest change is due to the small size of the industry in Bangladesh and the relatively low demand for soap per capita in this particularly poor country.

Other computations concerning the long-term future of the demand for various types of soap in Bangladesh suggest that the impact of increasing demand due to rising incomes and a rapidly growing population is likely to be much more significant than that of a static (overnight) redistribution of income.

Chris Baron and W. Van Ginneken, *International Labour Review*, 1982.

Choice of Projects

The following diagram shows the linkages between four types of choices, each of them conditioning the ensuing choice.

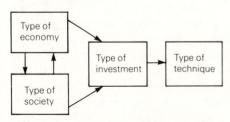

Most discussions of appropriate technology focus, rather narrowly, on the last of these categories, as if an autonomous choice were really possible. Fortunately, however, there is a growing body of knowledge which analyzes the technological problem as only a limited aspect of the whole social system.

The real and the false aspects of the problem are illustrated in El Salvador, where a new international airport is being constructed with the latest technology. Many have raised angry voices against the use of capital-intensive technology in a country with so much unemployed labour. But the technology used in the construction of the Cuscatlan Airport is not the major problem. It is, at most, only a marginal one. The real issue is that $150 million has been expended on a new airport instead of building more hospitals and schools for the rural populace. The problem lies, then, in the choice of the investment project and not in the type of technology used in the construction of a modern airport, an activity which does not allow for much freedom of technological choice anyway.

Ostensibly, a modern airport was chosen over equipment and services urgently needed in the countryside because the exporting segment of the economy required an airport to increase non-traditional exports and the dependent segment of society needed to promote tourism. Moreover, convenient financing for the project was offered to the Salvadorian government by Japan, apparently in view of the logistical advantages that the new airport will provide to Japanese trade along the Pacific coast. It is inconceivable that a foreign country would have granted financing on the same terms for a rural hospital. Thus, the discussion of capital-intensive construction techniques in this case fails to recognize the true nature of the problem.

Luis de Sebastiàn in J. Ramesh and C. Weiss, *Mobilizing Technology for World Development*, 1979.

The division of labour. . . . it is not, truly speaking, the labour that is divided; but the men:— Divided into mere segments of men — broken into small fragments and crumbs of life . . . You are put to stern choice in this matter. You must either make a tool of the creature, or a man of him. You cannot make both. — John Ruskin

A Choice of Methodologies

In any specific field there are always different technologies that can be used to produce a desired output, and these technologies often differ greatly in their capital costs, in their subsequent requirements for operating and maintenance costs, and in their lengths of life. Engineers and economists, and the people who employ them, naturally want to choose that technology which will 'cost the least', consistent with achieving the desired output and agreed standards of service and safety. But when competing technologies have quite different proportions of capital and recurrent costs, and use equipment with very different lives, the task of finding out which technology has 'the lowest cost' can be difficult. To cope with this problem, the technique of converting all values into 'Present-Value' terms has been widely employed. Present-Value (PV) analysis is simply the application of compound interest arithmetic to values that occur in the future to derive their PV equivalent. When future values, spread over a number of future years, have each been discounted back to their PV equivalent, they can then be added up to give a single value that tells us how much that whole stream of future values is worth in Present-Value terms. If we did the adding before discounting each value back to the present, we would be adding 'apples and oranges', since $1 that occurs seven years in the future has a lower PV than $1 that will occur three years in the future. Discounting gives all values the same weight and thus gets rid of the 'apples and oranges' problem.

The standard way of identifying the lowest-cost alternatives between two or more technologies is to construct a table in which all cash expenditures (capital and recurrent) are listed according to the years in which they will occur. Each annual figure is then converted into a Present-Value by discounting. There are many arguments as to what discount rates should be used, and these arguments can become important since different rates can change the outcome of the calculation. The purpose of the calculation, of course, is to see which technology has the lowest present value. This then becomes the technology of choice. In the Rural Water Supply (RWS) field PV calculations today play a minor role in choice-of-technology decisions. They should not be expected to play more than a minor and occasional role, and economists have far more important contributions to make to RWS projects and programmes than testing technologies with Present-Value calculations.

Table 1 works out the Present Value of a RWS project for 300 deepwell hand-pumps estimated to have a capital cost of US$600 per well (cost of well plus hand-pump, drop-pipe, pumping rods and cylinder). These are assumed to be built in equal numbers over a three-year period. Operating costs are zero but maintenance costs — which are not well known in this still-to-be-built programme — are assumed to vary between $75 and $150 per well (these costs will start during the second year for the first 100 wells built during the preceding year and build up to a constant annual cost of between $22,500 and

$45,000). The stream of total costs is then shown at three discount rates: zero, 5 and 10 per cent. The three resulting Present Values are shown, each with a range which reflects the uncertainty over maintenance costs. At the bottom of the table are figures showing the proportion of Present Value accounted for by maintenance costs. The latter are seen to vary from a high of 50-66.7 per cent at zero discount to a low of 40-57 per cent at 10 per cent discount. This modest conclusion is already of some interest; maintenance costs are likely to account for at least half of total present values in many hand-pump pro-

Hand-pumps: the low-cost choice. (*UNICEF*)

grammes. The proportion will naturally be lower at higher discount rates. But if it appears that resources to cover recurrent costs will be difficult to find, then common sense tells us that these future costs should not be discounted heavily — they deserve to retain a high weight in the calculation; this concern about the future should be allowed to influence the decision on what technology to use. Consultants, donors, or sector officials may not be free to choose a discount rate that reflects their estimate of the amount of weight they want to give future recurrent costs, since there may be standing instructions from a

Table 1.
Illustrative capital and recurrent cost of three RWS technologies

A. Present value of deep-well hand-pump scheme (in thousands) (using 5 and 10 per cent discount rates).

Assumptions: Three hundred wells are to be built and fitted with hand-pumps, the total cost of well plus pumps being $600 per well. The work is to be spread evenly over 3 years and the pumps are estimated to last 10 years. Average maintenance cost is estimated at anywhere between $75 and $150 per well per year; minimum and maximum values of $22,500-$45,000 per year have therefore been used.

| | | | | PV of total costs | |
Year	Capital	Recurrent	Total $	at 5%	at 10%
1	60	–	60	–	–
2	60	7.5-15	67.5-75	61.4-68.3	56.0-62.3
3	60	15.0-30	75.0-90	64.5-77.4	56.3-62.3
4	–	22.5-45	22.5-45	18.5-36.9	15.3-30.6
5	–	22.5-45	22.5-45	17.6-35.1	14.0-28.0
6	–	22.5-45	22.5-45	16.9-33.8	12.6-25.2
7	–	22.5-45	22.5-45	16.0-32.0	11.5-23.0
8	–	22.5-45	22.5-45	15.3-30.6	10.6-21.2
9	–	22.5-45	22.5-45	14.4-28.8	9.5-19.0
10	–	22.5-45	22.5-45	13.7-27.4	8.8-17.6
	180	180-360	360-540	295.3-427.3	249.3-427.3

Recurrent costs as % of total costs:

				50-70%	40-57%

B. Cost of diesel and electric alternatives to above hand-pumps scheme (diesel replaces five hand-pumps).

Diesel: Sixty diesel sets to pump 60 bore holes, and housed in simple structures. Total capital cost of $330,000. Diesel pumping will also require a storage tank for each set plus a pipe network to neighbouring standpipes.

Operating cost: Each set will cost over $1 per hour for fuel and attendant. With six hours pumping per day, annual operating cost will come to around $2,000. Maintenance will cost $30,000-$50,000 additional per year for the system as a whole. Thus total O&M costs of $140,000-$160,000 are estimated.

Electric: Sixty bore holes costing around $1,000 each plus an electric submersible pump costing around $400, for a total at-well investment cost of $84,000. Each well will need a storage tank ($30,000 for 60 wells) plus a reticulation system to serve neighbourhood standpipes. Investment cost is therefore substantially lower than diesel.

Operating cost: Annual operating cost per well is estimated at about $465, or $27,900 for 60 wells. The maintenance cost should be lower than diesel, say $20,000-$40,000 for the system. Total recurring cost would therefore run roughly $60,000 per annum.

central authority on the rate to be used. If there is a significant difference between the prescribed rate and a rate which sector officials think makes more sense, they should use both rates and present their results as a sensitivity analysis, with supporting arguments.

Table 1 also presents some illustrative figures for the capital and recurrent costs of two additional competing technologies, diesel-pumped boreholes and boreholes pumped with electric submersible pumps. The diesel estimate shows capital and recurrent costs that are both higher than those for the hand-pumps. On a straight cost-effectiveness calculation, therefore, the diesel alternative would be a 'non-starter'.

The electric-pump borehole alternative in Table 1 makes a more interesting and more difficult comparison with the hand-pumps. The capital costs are estimated to be one-third cheaper than hand-pumps but the operation and maintenance (O&M) costs substantially higher. The preference is not obvious, so the Present Value of the electric-pump option was calculated. This shows that at all discount rates from zero to 10 per cent the hand-pumps have a lower Present Value. It is also obvious, even without discounting, that electric pumping would be much lower-cost than diesel pumping. Indeed, all over the world electric pumping is almost invariably preferred to diesel pumping, on both cost and other grounds — provided electricity is available so that electric pumping does not have to include the investment cost of providing the electricity supply.

The examples summarized in Table 1 introduce us to typical numbers involved in technology comparison but do not say much about when it is helpful to use discounting and when it is unhelpful. One balanced and moderate answer to this question is provided in a Swedish consultant's 1977 report on SIDA-assisted RWS schemes in Kenya. The consultant was bothered by the perverse results which the use of high (10-15 per cent) discount rates was producing in the choice of technology in Kenya's RWS sector. High discount rates were doing what they always do, i.e., favouring projects with relatively low capital costs and relatively high running costs over those with a reverse cost structure. But most of the engineers working in the sector had strong reasons for preferring the technologies with the higher capital costs but lower O & M costs. Experience had taught them that recurrent resources are more difficult to obtain than capital resources and that pumped schemes are very much more troublesome than gravity schemes and that thermal power is vulnerable to upward oil price shifts.

The outstanding discussion of appropriate technology in RWS in *Water for the Thousand Millions* addresses itself (near the end) to the role of cost-benefit analysis in the sector. The principal message of this sensitive and comprehensive pamphlet is that while a cost-benefit calculation (the authors really mean a cost-effectiveness calculation) can 'in principal' capture all the considerations that need to be taken into account, a large part of the relevant factors are incapable of quantification. The authors therefore conclude that 'the basic technology choice is largely dictated by the other criteria', i.e., criteria that cannot be captured in a cost-benefit calculation. But such calculations are not put aside completely: 'rational choices between technological alternatives do sometimes depend in an important way on the question of how costs and benefits occurring at different points in time should be compared, i.e., how they should be discounted'. They go on to note, without comment, that high discount rates tend to discourage capital-intensive technology while low discounts tend to favour them. Capital intensity does not refer to whether the capital costs of one technology are larger than another; it refers to the proportions of capital and recurrent costs in two or more technologycal alternatives. Two ways of measuring capital-intensity are shown in Table 2. Method 1 simply adds up all capital and recurrent costs over the estimated life of the project and sees what proportion of these total costs are accounted for by the capital expenditures. The second method involves converting the capital costs into an 'annualized value' and then adding this annual capital value to annual recurring cost to get a total annual cost; the annual capital component of this cost is then expressed as a percentage. Although the two methods will give

Table 2
Measuring capital intensity

Year	Diesel pumped scheme Capital	Recurrent	Piped gravity-flow scheme Capital	Recurrent
1	6,000	—	12,000	—
2		2,500		250
3		2,500		250
4		2,500		250
5		2,500		250
6		2,500		250
7		2,500		250
8		2,500		250
9		2,500		250
10		2,500		250
		22,500		2,250
Total cost: 28,500			Total cost: 14,250	

Capital intensity:
Method 1: (capital cost as a percentage of capital plus all recurrent costs over the project's life)

Diesel: $\frac{6,000}{28,500} = 21\%$ Gravity flow: $\frac{12,000}{14,250} = 84\%$

Method 2: (straight-line depreciation over annual recurrent cost)

$\frac{600}{2,500} = 24\%$ $\frac{1,200}{250} = 484\%$

Recurrent cost intensity: (annual recurrent cost over annual straight-line depreciation)

$\frac{2,500}{600} = 4.2$ $\frac{250}{1,200} = 0.2$

different proportions, they will both correctly rank the capital intensity of alternative technologies.

However, paying too much attention to capital-intensity can be downright misleading, since the usual assumption is that LDCs should avoid capital-intensive technologies because they do not have the savings needed to pay for capital. In fact, however, it is often easier for poor countries to acquire the initial capital (through aid programmes or liberal financing terms) than to find the recurrent resources needed to keep a scheme in operation. A simple way of measuring recurrent-cost intensity is to calculate the ratio of recurrent costs to one year's depreciation, using the straight-line method. When this is done for the two technologies of Table 2, the diesel pumps show a recurrent-cost intensity of 4.2, the gravity-flow scheme an intensity of only 0.2. When you divide the second figure by the first you conclude that the gravity-flow scheme has a recurrent-cost intensity less than one-twentieth that of the diesel pumps. An even simpler direct comparison of the recurrent cost streams (250/2500) shows an advantage for gravity-flow that is only half that found when each scheme's recurrent costs are first related to their respective annual capital costs. The second method is intellectually more appealing — but seems less relevant than the first, which focusses all attention on the critical budget problem.

The upshot of this discussion is that the gravity-flow scheme would be rejected if the aim were to avoid the more capital-intensive scheme but would be preferred if the aim were to use the scheme with the lowest recurrent costs. Since low recurrent costs is the right criterion, the gravity scheme would be a much better choice. The mechanics of discounting are such that the process removes from consideration a higher and higher proportion of values that fall in the future. For example, if one is thinking of installing the diesel pumping scheme represented by the first set of figures in Table 2, discount with a 5 per cent discount rate will extinguish (and therefore leave out of the PV result) 5 per cent of the investment cost but 25 per cent of the next 10 years' costs (the calculations are not shown). At 15 per cent, almost half the recurrent costs are omitted from the PV figure. In 'collapsing' all costs into a single figure (the Present Value), the distinction between capital and recurrent costs is removed and all costs over the life of the project are treated as if they were capital costs. This does not seem a particularly useful thing to do if a primary basis for choice is the desire to avoid high recurrent costs. If that is a major criterion for choice — as I believe it should be — then the sensible way of comparing different technologies is simply to make a straightforwad comparison of annual recurrent costs without extinguishing any of these costs by discounting.

Choice of technology through the PV technique asks the question: 'Which method involves the least use of resources — undifferentiated resources — over its lifetime?' That is not what we usually want to know when comparing the costs of competing RWS technologies. What we usually want to know is: which technology will minimize our future recurrent costs? PV calculations can tell us nothing about the relative attractiveness of different technologies with respect to their demands on recurrent budgets. That is the method's fatal flaw.

In projects whose recurrent costs can easily be covered by sales of project outputs (every factory, bank, or store is this kind of a project), ability to meet recurrent costs will be a much less important problem than minimizing life-cycle costs. But in non-revenue-earning projects, which require heavy budgetary subsidies, ability to meet recurrent costs is usually a far more important consideration than minimizing life-cycle costs as measured in a single PV figure. What this amounts to saying is that the widely-used present value technique is simply inappropriate for the analysis of a large class of non-revenue-earning projects. Indeed, for such projects the technique can be downright misleading.

If investment decisions in RWS should be governed primarily by undis-counted calculations of financial O & M costs, does this mean that capital costs should play no role in choice of technology? That would be saying too much; but their proper role should be determined qualitatively, not by making use of traditional PV calculations, which remove the distinction between capital and O & M costs.

Some people may believe that there is a systematic inverse relationship between capital and recurrent costs, suggesting that one can save on recurrent costs if one is willing to spend more on capital costs. While this is a familiar phenomenon in many fields, I doubt that any such general law holds for RWS technology. My own view is that the best way to take capital costs into account is to do this qualitatively, by seeing how much coverage one will be able to achieve with a given investment budget. The designers of any scheme, if they are fully aware of the range of technology that should be considered, should narrow the choice down to a very small number of alternatives on the basis of technical considerations and the respective recurrent costs. One then examines the relative capital costs to see if they are significantly different. If there is a trade-off to be made, then one simply has to decide, qualitatively, what weights to assign to capital and to recurrent costs. The argument of this paper is that differences in recurrent costs ought to be given a much higher weight than differences in capital costs. One might argue that PV calculations could be made using, e.g., a 10 per cent discount for capital costs and a 3 per cent (or zero) discount for recurrent costs. My strong preference is not to go this route but to carry out separate, undiscounted comparisons of recurrent and capital costs on the technological 'short list' and then to base decisions on qualitative discussions of whatever trade-offs may exist. But, as noted, there may often be no trade-off at all: the technology with the lowest recurrent costs may also have the lowest (or at least a very low) capital cost. Cost tables constructed to justify choice of technology should give far more emphasis than is normally done to estimates of O & M costs — and who will pay for them.

George B. Baldwin, *World Development*, 1983.

Simple is Optimal

Whatever has happened to the economies of the poor countries, the literature of project appraisal has an impressive record of growth. The observer may be forgiven for wondering where it will all end, as some try to develop appraisal methods which will keep pace with changing criteria of appraisal (new criteria being added rather than old ones beng subtracted) and practitioners struggle to follow their advice. One question here is whether the addition of employment and poverty criteria to social cost-benefit analysis will lead to a net improvement in resource allocation. To answer this question would require a major study. A positive case can be argued at both theoretical and practical levels. Certainly, in practice, the questions asked of a project during appraisal can influence the 'yes-no' decision, and also design. The negative case, however, often goes by default because it does not fit into the cost-benefit paradigm. In presenting parts of the negative case, the purpose here is to raise issues of concern rather than to pretend to have definitive answers.

Any evaluation of a method of project appraisal should be based not on its appearance, nor on the theory of how it should be applied, but on what happens in practice. It is not the study of manuals and procedures that is relevant but the study of behaviour. Analyses from the standpoints of public administration and political science are valuable not least because they admit forms of evidence about behaviour which some mathematicians and some economists might be inclined to disregard or discount. It is common to find that practitioners of social cost-benefit analysis admit in private that what appears as a clinical and objective procedure is in practice a compound of judgement about future events which are very difficult to predict, and judgement about discount rates and shadow prices within limits which allow for wide variation. The uncertainties and difficulties are especially acute with agricultural projects. In one case the same agricultural project appraised by three different teams was accorded rates of return, 19 per cent, 13 per cent and minus 2 per cent, respectively, much of the variation being explicable in terms of differing estimates of rates of implementation and/or the adoption of innovations, both of which are inherently difficult to anticipate.

It may be asked to what extent the combination of uncertain judgement and methodological complexity exposes social cost-benefit analysis to political pressures. Ironically, appraisal techniques developed to make decision-making more rational may be used to legitimize decisions arrived at in other ways. Partly this is possible because of the obscurity of the calculations when final data are presented to a decision-maker. Partly it may occur because decision-makers know that the results are easily manipulable. Far from defending appraisers from political pressures, the procedures may then expose them all the more. In practice, rates of return are sometimes determined first and the calculations done later to produce them; and there are more subtle personal and political interactions between calculations and desired results. The danger is that the addition of employment and poverty criteria to social cost-benefit analysis will have little effect because the procedure itself is so sensitive to judgement and so vulnerable to personal factors and to political pressure.

Complex procedures may also contribute to and sustain dependence and delay. The combination of pressure to find projects, shortage of good projects, and the demand of donors for complex apparaisals, creates congestion. The response of many international agencies is to intervene in project preparation. But as has been argued in an examination of the World Bank, USAID and UNDP,

The direct intervention of international agencies in project preparation is in part a response to the severe deficiencies in planning and project analysis skills in developing nations, but the 'deficiencies' are, in a sense, artificially created by the complexity of international procedures. Project preparation guidelines are designed to ensure that proposals are compatible with lending institution policies, procedures and requirements: and as such have become instruments of control rather than of aid. And as those procedures become more numerous and complex, further demands are placed on the limited planning and administrative capacity of developing nations, making them more dependent on foreign expertise...the imposition of international requirement ...may in fact, have aggravated the problem of preparing relevant and appropriate investment proposals.

The argument is not that there are no benefits from such procedures. The question is to what extent the costs of following the procedures are justified by the benefits. For the costs can be high, especially in the poorest countries which are precisely those in which the procedures are most difficult to carry out. Donors are liable to respond to these difficulties in ways which either sustain dependence (by posting in their own staff to do the job) or which reduce benefits to the poorer countries and to the poorer people within countries, by concentrating on other countries and on groups other than the poorest.

There may thus be a syndrome in which what passes for sophistication in project selection actually hinders aid to the poorest. Donors bring to bear 'an imperious rationality' on recipients. The laborious procedures required delay projects. Delays to projects increase pressures for donors to spend. Pressures to spend exert biases towards the less poor developing countries, towards larger projects, towards urban areas, towards the more accessible rural areas, and, within rural areas, towards those who are better off. In short, complex procedures can divert development efforts away from the poorer rural people.

The essential part of any poverty-focussed rural development is the devizing and use of simple procedures. There is an almost universal tendency for procedural overkill.

Procedures are almost always additive: new ones are introduced, but old ones are not abolished. Procedures drawn up by committees, or through consultation with various people or departments, tend to be longer and more complicated than those drawn up by one person — and participative management may reinforce this tendencey. It is often safer to add a requirement for an additional item of information than to leave it out. Promotions go to bright people who can devize and answer questions, and not to those who tell their superiors that they did not consider the benefits of being able to answer their questions justified the costs of collecting the information necessary.

A first step is then to have the insight to see what it is not worth knowing, and the courage not to find it out. Courage is needed because optimal simplicity looks naive.

Simple procedures are also necessary if decisions are to be kept in the open, making it clear to the decision-maker what criteria are being used, and how the method works. A practice has not much to recommend it if the working of the method and the decision criteria are not evident to the decision-makers. The obscurity of some social cost-benefit analysis exposes it to abuse. It is easy, and known to be easy, to adjust assumptions (discount rates, shadow prices, rates of implementation or adoption, etc.) to produce a wide range of results. Rationality may be defended through selection procedures in which the assumptions are always clear and which so far as possible can be understood by a non-economist layman decision-maker.

Five simple approaches are suggested. Probably none is new. Most or all of them may be used in governments and aid agencies already, especially for small projects. But curiously, while social cost-benefit manuals are published and widely distributed, these simpler aids to selection are rarely written about.

Decision matrices: Decision matrices can be used to present alternatives clearly, keeping factors separate instead of conflating them into a single numeraires. They enable the decision-maker to assign his own implicit weights and to understand more clearly the implications of his decisions. They can be used to present the implications for the poorer people of alternative projects or alternative approaches to the same project.

Poverty group rankings: Poverty group rankings are a device for concentrating thought and attention on which groups in the society will benefit from a project. They require those preparing a project to ask the crucial 'who benefits?' question, and to rank groups according to their degree of benefit. The question should make low administrative demands on those who have to answer them. The result should be to force officials, whether in ministries or in decentralized administrations, to think at an early stage about beneficiaries; and the procedure can be designed so that those originating a proposal for a project have to defend the rankings which they have given it. Such a system should benefit the poorer rural people by affecting the thinking, behaviour and choices of those who identify, design and select projects.

Checklists: Checklists of factors to consider are widely used but little written about. They may be used specifically to alert appraisers to considerations such as poverty, employment and administrative capacity. Some officials have their own checklists. Checklists do, however, run the risk of becoming too long. As with other procedures it is optimal to stay simple.

Listing costs and benefits: Where some sort of cost-benefit appraisal is needed for a small project, a simple approach is to list anticipated costs and benefits putting figures on them as appropriate. . .

Unit costs and cost-effectiveness: Unit cost and cost-effectiveness criteria are widely applicable and useful. They are especially useful with projects for health, education, water supply and the provision of other services.

These five procedures are open to criticism by perfectionists. The traditions and methods of mathematics value precision. But in practical decision-making there are optimal levels of imprecision and ignorance. The key to optimizing procedures is to realize that the cost-effectiveness of the procedures themselves relates to low costs in staff time and in demands for information as against high benefits in improving the quality of the decisions. The danger is that 'intelligent' criticism of simple procedures will consider only the benefit side and neglect the costs, leading to improvements which make the procedures more laborious, less practical

more costly to carry out, and counter-productive. Complexity and sophistication are not synonymous; on the contrary, complexity can be crude and naive. The true sophistication is to see how far it is optimal to be simple.

A danger remains that demands for information by bilateral and multilateral donors will develop a galloping elephantiasis which will paralyze administrations, reduce aid to the poorest, and perpetuate and increase dependence on foreign expertise. The danger is that more and more highly trained and experienced people will be sucked or enticed into the prestigious, well-paid, urban-biased business of project identification, appraisal, monitoring and evaluation. Thus at a time when rural development has become a priority, especially the much more difficult objective of rural development which benefits the poorer rural people, there may perversely be less and less contact between those responsible for rural projects and policies on the one hand and poor rural people on the other. These trends can be moderated by the decentralization and the simple procedures advocated above. But there is one more measure to be taken: a conscious and determined drive to counteract the effects of the urban and elite life styles, experiences and perceptions of many of those concerned with rural policies and programmes. The seriousness of the need varies by country and region. But the reform proposed is a requirement by every donor agency, and selectively by governments, that their officials should be systematically exposed to and encouraged to learn about rural life and especially rural poverty. This could mean, for donors, that each official would be required to spend two weeks of every year actually living in a village, not making the easier, more congenial visits of a rural development tourist, thereby learning how rural people, and especially the poor rural people, live, and so trying better to understand their needs.

The benefits would be many. Some officials would resign. Others would work harder and better the asymmetry of the aid relationship would be mitigated, since 'donors' would have to go cap-in-hand to 'recipients' and ask them to allow their 'donor' staff to be recipients of experience in villages.

The main benefit would be improved judgement. However carefully procedures are devised, training undertaken, and feasibility appraised, the element of judgement always has a major part to play in project selection. With poverty-focused rural development, judgement must be based upon an understanding of rural realities. Direct exposure to village life, if sensitively managed, should enable officials better to assess rural needs, better to appreciate the capabilities of rural people and their potential for participation, and better to understand and counteract the tendency for projects to be captured by rural elites. Officials should become better judges of the implementability and of rates of change. They might repeatedly learn and relearn the lesson that simple is optimal. The outcome should, indeed, be that more projects would be selected and implemented which would truly benefit the poorer rural people in ways which they would welcome.

Robert Chambers, *World Development*, 1978.

2 CHOICE OF TECHNOLOGY: SOME CONSIDERATIONS

Why Big May Be Bad

Much modern technology has been designed for use in large-scale plants. Certain prerequisites of large-scale operations have therefore to be fulfilled before it can be employed economically. These are not always present. Large-scale capital-intensive production is not efficient if markets are small, scattered, highly seasonal or fragmented; if distribution channels are not well organized; if workers are not used to factory discipline; if management does not know the necessary managerial techniques or, though aware of them, cannot implement them because they conflict too strongly with accepted customs, beliefs, systems of authority, etc., of the employees; or if there are no service engineers who can get the complicated machinery going again when it breaks down. Scarce capital is wasted if it is invested under such circumstances. And these are factors that cannot be changed overnight by a simple planning decison.

There are many examples of this:—

— The large public-sector shoe factory which operated at 20 per cent of capacity because it had no means of reaching the small private shoe retailers who handled 90 per cent of the shoe trade.

— The battery plant which could satisfy a month's demand in five days.

— The woollen-textile factory which had a 10 per cent material wastage figure (costing precious foreign exchange) because its management did not know how to set and control material usage standards.

— The $2 million date-processing plant which had been out of action for two years, ever since a blow-out in the cleaning and destoning unit, because there were no service engineers who knew how to repair it.

— The confectionery plant which was inactive for most of the year because 80 per cent of sales were made during the month of a religious festival.

— The ceramic factory whose quality was poor as it had to use up two years' stocks of the wrong glaze, imported in error, because the general manager had too much work to attend to all details satisfactorily. This was attributable to a lack of middle management and supervisory personnel willing to accept responsibility, combined with a reluctance to delegate authority on his part—both reflections of prevailing social attitudes.

— The radio assembly factory whose production line broke down repeatedly because of the high rate of absenteeism among key workers.

These economic and social realities explain why large-scale factories in some developing coutnries achieve a much lower level of labour productivity and capital productivity (which is more important) than do identical plants in the long-industrialized countries. The relevance or importance of these factors obviously varies from country to country. An assessment needs to be made by those who frame policy. Economies of scale are not just the product of technical coefficients. They cannot be realized unless the co-operating economic and social factors are favourable. This does not imply any inherent inferiority or inability, but merely the lack of opportunity in many countries in the past to acquire the particular skills and habits required in all strata of society or to build up the infrastructure which is essential for smooth business operation.

Keith Marsden, *International Labour Review*, 1970.

With conditions as they are in our country, co-operation must precede the use of big machinery. — Mao Tse-Tung

The Advantages of Intermediate Technology

The advantages of intermediate technology are:

— Intermediate technology often utilizes local and therefore accessible raw materials thereby avoiding additional imports which consume scarce foreign exchange. Advanced technology involves massive importation of machinery and sometimes also raw materials resulting in a large drainage of foreign exchange reserves. One school chalk factory in Nigeria not only uses foreign machinery but also imports gypsum, the material from which chalk is made, although the raw material and locally manufactured equipment is available.

— Intermediate technology employs more labour per unit of scarce investment capital, whereas advanced technology is capital-intensive and does relatively little to solve the unemployment problem.

— Intermediate technology employs existing local skills and introduces new skills which can be easily acquired in a short period of time. It therefore builds a technology base which is essential for eventual technological takeoff. Advanced technology, on the other hand, employs foreign experts with few local operatives who may only push buttons and never really acquire significant skills, and hence there is very little effective technology transfer.

— Machines made with intermediate techniques can be serviced locally and the spare parts made in local workshops, thus providing valuable training for local artisans while at the same time ensuring continuous production. Advanced machines may not have local servicing facilities and breakdowns lead to long work stoppages as spare parts or experts have to be flown in from foreign countries, usually at a very high cost. In one developing country with a sophisticated oil refinery, business activities are frequently affected because of lack of petrol. The refinery often has to wait for spare parts to be flown in from Germany. . . .

— Intermediate technology usually fits into the existing social structure while advanced technology can sometimes be socially disruptive. The installation of large sophisticated sugar mills in Kenya has been accompanied by the massive relocation of villagers in order to make room for cultivation of sugar cane around the mill. This scale of cultivation is absolutely necessary if the mill is to be profitable. On the other hand, small-scale mills with simple, locally-made equipment are fed with cane from existing mixed farms. The mills employ more people and the social structure is not disrupted.

— Sophisticated factories in developing countries rarely operate at more than 30 per cent of their capacity because of the poor supporting infrastructure, so that products become very expensive. For example, when a glass factory was established in Nigeria two large standby generators had to be installed for use during power failures. Small factories using intermediate technology techniques may utilize local power sources and interruptions are less serious since the capital commitment is lower.

Malcolm Harper and T. Thiam Soon, *Small Enterprises in Developing Countries*, 1979.

Economies of Scale

Some believe economies of scale to be invincible technological truths — facts of life, technologically speaking. And within existing technology they are: to produce on a smaller scale may involve losses in output which are often substantial. But the economies of scale are themselves formed by the history of technological development, in which particular organizational forms have favoured the development of large-scale techniques. In the competitive struggle, firms have to grow in order to survive. Market imperfections have added a premium on to size, additional to and independent of production technology and production economies of scale: thus advertising has introduced a new source of overhead and scale economy, while market power tends to increase with size, adding to the advantages of size. The growth in the size of the firm can be explained in terms of the competitive struggle between firms combined with market imperfections. It also undoubtedly has a technological counterpart, in economies of scale and specialization. But these are themselves in part the product of the economic system, in which large firms dominate, which gave birth to them. In an alternative system in which small units dominated the process of technological development, techniques suited to such units would be developed. Today, it is impossible to decide how much of the undoubted economies of scale are due to the organization of the economy and therefore of technological development, and how much to genuine physical facts which must be present in any organizational system. What is known is that very little of modern scientific and technical effort has been devoted to producing techniques suitable for small-scale units, and that where minification has been attempted its success has surprised engineers and economists, who have assumed that what is, must be.

Frances Stewart, *Technology and Underdevelopment*, 1977.

112

The Rural/Urban Split

The rural/urban question is, to some extent, tied up with that of scale, as really large-scale production can only take place in an urban setting. Those who have emphasised the need for small-scale techniques have also tended to emphasize the need for additional techniques in the rural areas.

The majority of people in underdeveloped countries live in the rural areas, as shown below — so, potentially, techniques designed for rural use might extend to more people than urban techniques. Moreover, despite considerable urban squalor and poverty, poverty is generally concentrated in the rural areas in greater proportion than population. The manifold problems arising from rapid urbanization also suggest that more attention needs to be paid to creating rural opportunities. All the theories of rural/urban migration agree that rural/urban opportunities (or lack of them) play a significant role in influencing the movement to the towns. None the less, the rural aspects of appropriate technology may have been overemphasized. Urban poverty and lack of productive opportunities is also widespread.

Urban/Rural Population, 1970 (millions)

	Urban	Rural
Africa	75	277
Latin America	161	123
China	167	605
South Asia	231	880

While the balance of numbers suggests that comprehensive improvements in rural techniques would affect more people, the problems of developing such techniques and communicating them throughout the rural areas might well be greater. In so far as modification of F-sector technology is concerned, this largely has to be urban because that is where the existing techniques are. External economies are often greater in an urban setting.

The choice between urban and rural techniques, in so far as there is a choice, is not therefore simply a matter of absolute numbers, or absolute numbers in poverty in the two sectors, but also how many the new techniques are likely to reach, and their income-creating implications.

Frances Stewart, *Technology and Underdevelopment*, 1977.

A reasonable estimate of economic organization must allow for the fact that, unless industry is to be paralysed by recurrent revolts on the part of outraged human nature, it must satisfy criteria which are not purely economic. — R. H. Tawney

Mini Sugar Technology in India

The Planning Commission of India studied mini sugar technology in 1977 and included the following policy decision in the Indian draft five-year plan for 1978-1983:

> The alternative technologies available for the production of sugar consistent with desirable capital employment parameters show that future demand for sweetening agents, after allowing for fuller utilization of the existing and licensed sugar-mills, can be met by necessary expansion through Open Pan khandsari plants. It is proposed to work out the policy framework for the further expansion of the sugar industry in the light of these studies. For the time being, therefore, no new sugar-mills will be licensed, although expansions of existing units may not be ruled out where this is necessary for maintaining their viability.

Table 1
Comparative Data on Large-scale and Mini Sugar Technology

Item	Large-scale	Mini
Total capital available for investment (Rs million)	60	60
Capital required for installation of one unit (Rs million)	60	1.3
Number of units which can be set up with available investment capital	1	46
Working days	120	100
Total sugar output (t/a)	14,550	34,500
Persons employed	900	9,292

Table 2
Comparative Analysis of the Productivity of the Two Technologies

Item	Large-scale	Mini	
Working days	160	120	100
Cost of 100 quintals of cane (Rs)[a]	1,250	1,250	1,250
Cost of processing 100 quintals of cane (Rs)	1,032	585	632
	2,282	1,835	1,882
Total sugar produced (quintals)	9.6	7.5	7.5
Cost of sugar per quintal (Rs)	238	243	251

[a]1 quintal = 100 kg.

Some comparative economic data on large-scale and mini sugar technology under Indian conditions 1977/78 are shown in Table 1. The data show that mini technology would produce 2.37 times more sugar and create 10.3 times more jobs for the same capital cost than would large-scale technology. However, experience indicates that the efficiency of large-scale sugar technology improves if the unit working periods are extended.

A comparative analysis of the productivity of the two technologies is presented in Table 2.

To produce one unit of sugar, mini technology is superior to large-scale technology in all respects but cane consumption. Mini technology requires 13.33 units of cane to produce one unit of sugar as against 10.4 units for large-scale technology but this difference will be minimized when recent technological innovations are put into commercial use. The development of an expeller for increasing crushing efficiency will increase yield by 0.7 per cent to 8.2 per cent and manufacture of liquid sugar out of khandsari molasses will raise it to about 9.2 per cent. It should also be emphasized that the sugar produced by mini technology is of identical quality with the sugar produced by large-scale technology. Mini sugar units now crush 10 per cent of the total cane grown in India and produce about one million t/a.

The socio-economic benefits to India of mini technology include:

—Capital amounting to Rs.800 million has been invested by the unorganized sector, primarily in rural areas;

—Potential employment for 150,000 persons in the rural sector has been created in the agricultural slack season when these persons formerly migrated to cities in search of work;

—Tax revenues to federal and state governments have increased by Rs.40 million;

—A manufacturing industry producing mini technology components has developed whose annual turnover has reached almost Rs.100 million;

—Mini sugar equipment requires only about 60 per cent of the iron and steel needed to fabricate large-scale equipment producing the same quantity of crystal sugar. This saves 40 per cent on iron and steel;

—More than 60 per cent of the cane for large-scale units is transported by vehicles consuming fossil fuel. Practically no fossil fuel for transport is utilized by mini units. They are near the cane growers and cane is transported by bullock carts;

—Mini units provide repair services for other mechanical agricultural equipment;

—Centralized large-scale technology draws capital away from rural areas. This weakens the rural capital base and makes the introduction of improved agricultural technology difficult. Rural mini units help build up local capital resources;

—Mini sugar equipment can be manufactured and maintained by small workshops in rural and semi-urban areas;

—Mini units employ about 2.3 times as many labourers for the same capacity as those employed by large-scale units;

—The capital required to manufacture an equivalent quantity of crystal sugar by mini technology is only 42 per cent of the cost of doing so by large-scale Vacuum Pan technology;

—The price mini units pay for cane is at par with that paid by large-scale units. This price is at least 25 per cent higher than small-scale traditional farmers, who grow cane as a cash crop, can obtain from other cane buyers; it can be said that the establishment of dispersed mini units has added about Rs.400 million to agricultural income;

—Some mini technology innovations have filtered through to the gur and khandsari industries. The introduction of power crushers to gur manufacture, crystallization in motion, and improved boiling furnaces, for example, have improved the output and efficiency of these traditional industries.

The question is often asked why mini sugar technology should be introduced rather than improving the traditional technologies for gur or khandsari sugar. These traditional technologies are low-cost and the skills needed to operate them are already available in rural India. The answer is that mini sugar technology is more economical to operate and that its product is more acceptable to consumers.

Gur is a concentrated product of the whole cane juice. It contains about 80 per cent sucrose and other substances having nutritive value. It is used by the rural pupulation more as a food than as a sweetening agent. The gur yield is 10 per cent on cane processed, that is, to manufacture one quintal of gur, 10 quintals of cane are required. The standard price of 10 quintals of cane fixed by the government for large-scale mills is Rs.125-135. The market price of gur fluctuates sharply; during the last five years it has ranged from Rs.80-150 per quintal. Even at the peak price of Rs.150, a margin of only Rs.15-25 is left to cover processing. This means that gur manufacturers cannot pay the standard cane price and that returns to cultivators who must sell to them will be low. Any improvement in the technology will raise the manufacturing cost of the product and its marketability will be reduced. Improved technology has in fact been tested but was given up because the gur, although of higher quality, did not find buyers at higher prices. Attempts to increase the yield of gur from cane necessitated higher investment which again affected the cost of the product and defeated the purpose of trying to keep gur manufacture as a cottage industry.

Khandsari sugar is a creamy white powdery sugar containing 94 to 98 per cent sucrose. Nutritively it is said to be superior to white crystal sugar. The price range is Rs.200-300 per quintal. Its recovery is 5.5 per cent, that is, to produce one quintal of khandsari sugar, eighteen quintals of cane are required. The price of cane to large-scale units is Rs.225-245 per quintal; this leaves very little margin to pay for technological improvements.

One present strength of large-scale technology is that its standardized products are acceptable to Indian society. On the other hand, consumer acceptance of gur and khandsari produced by traditional low-cost technologies has

Construction work on a mini-sugar plant, in India.

declined. This fact is usually not given enough weight in schemes for reviving and improving traditional technologies by introducing new equipment and techniques; governments have concentrated on providing marketing and financial aid. These efforts, however, have not succeeded in putting low-cost technology on its feet.

For example, gur production and consumption are gradually declining. In 1930, 65-70 per cent of the total cane crop in India was made into gur; currently only 45-50 per cent of the cane crop is being processed in this way. The manufacture of traditional khandsari sugar has also been declining; at present the khandsari industry utilizes barely 2 per cent of cane grown in India.

The declining production of gur and khandsari sugar reflects the demand for white crystal sugar, a product of large-scale technology. At present the output of large-scale sugar technology in India is 5-6 million t/a. Mini sugar technology, which was introduced as a pilot project in 1956/57, is now producing 1 million t/a. The potential market for white crystal sugar can be judged from the fact that annual per capita consumption is only 6kg in India, while in developed countries the annual per capita consumption averages 50kg. Therefore, from the point of view of product selection, the only alternative was the development of a suitable small-scale technology for manufacturing white crystal sugar.

M.K. Garg in UNIDO, *Monographs on Appropriate Industrial Technology*, 1982.

117

The Costly Lessons of a $550m Dream

Bakolori is a technologists' dream. It is the name of a dam and agricultural project in the dry and impoverished north of Nigeria. If the application of modern technology to developing countries means anything, then the idea at Bakolori of damming a seasonal river to give farmers two certain crops a year, instead of one uncertain one, ought to be a triumphant success.

It was certainly hailed as such by President Shehu Shagari when he inaugurated Bakolori at the start of the long campaign which last month brought him re-election for another four years. For him it was 'a dream come true' in his home state of Sokoto. The $550m investment is a major element in the Nigerian 'Green revolution' which aims at reversing the collapse of the country's agriculture.

Yet though most of the 23,500 hectares of irrigated land at Bakolori are now ready for farming by the 100,000 Hausa peasants of the district, no one has forgotten that the scheme was only completed after the bloody suppression of a revolt against the whole idea by the very people it was designed to benefit. Even now wholehearted acceptance of the scheme by the farmers is uncertain and its future depends heavily on the Nigerian government pouring in a lot more of the one commodity it badly lacks at the moment — money.

The huge scheme is an impressive technical achievement for its Italian builders, Impresit, the construction arm of Fiat, Italy's largest private enterprise. But apart from demonstrating the enormous social upheaval caused by such schemes, it raises the question of whether such sophisticated technology is either economic or even appropriate for raising food production in primitive and environmentally delicate areas like the sub-Saharan savannah belt of Africa.

From an engineer's point of view, Bakolori is the classic example of a place where it would have been a crime not to build a dam. The far north of Nigeria is flat, densely populated with clusters of mud houses, but extremely poor. The land is fertile only after the mid-year rains, and they occasionally fail.

Sokoto state is crossed by the Sokoto and Rima rivers which rise in the wetter south. They swell again in the rainy season to flood their valleys before turning south again to join the mighty Niger river. From colonial times it has seemed an obvious idea to store the flood-water behind dams, and release it gradually, partly to supplement the rains in the wet season, but mainly to enable the farmers to grow a second crop in the dry season. A further incentive to the idea of concentrated irrigation has been the fear of desertificaton spreading south from the Sahara.

Studies by FAO in the 1960s favoured the concept of river basin schemes, starting with Bakolori. Impresit got the message, and began a feasibility study in 1972. In 1974 the company received a letter of intent for the contract to build both the dam and the irrigation works. The deal was signed the following year.

Drawing on considerable experience of dam-building all over the world (Impresit built the Kariba dam in Zambia), the Italian company finished the 3.5 mile dam in only thirty months, creating a lake with a capacity of about 450m cubic metres of water. The Sokoto-Rima River Basin Development Authority (SRBDA), set up to handle this and other schemes, was slower off the mark.

Initially the problem of what to do with the 14,000 people who lived in the area to be flooded by the lake was almost ignored, and little was done to explain to farmers downstream what the project would mean to them: that their tiny plots would be expropriated to be levelled for irrigation, then reallocated to them in the form of regular sized units, 20 per cent smaller than the total area they had farmed before, and demanding a whole new way of farming.

As the lake water rose, the then military government finally leapt into action and the displaced farmers were resettled; but the land given to them — never previously farmed — was poor, there was little financial compensation and the unhappy new settlement (which today is almost derelict) erupted into riots in August 1978.

As for the farmers on the irrigation scheme proper — the main intended beneficiaries — they became exasperated for a different reason: they had to stop planting their land in the wet season to allow the contractor to level it for irrigation. That work often took longer than expected

(Impresit had much less experience of irrigation than of dam building), there was no compensation for loss of crops, and in some cases the precious topsoil simply blew away.

The explosive result was that from early 1979 to April 1980 the farmers staged a revolt. Gangs of them sealed off construction sites so that work was forced to a standstill. 'The government wanted us to go on working, but the farmers were so well organized that they managed to shift their roadblocks very quicky to wherever we were' says Dr. Enrico Tasso, managing director of Impresit Bakolori, and the man regarded as father of the project. 'We couldn't do anything.'

He blames much of the trouble on political agitators in the tense period of Nigeria's elections for civilian rule. Though Bakolori became a national issue, many observers still think the protest was a largely spontaneous response to inefficient planning and heavy-handed action by the military government.

In the end, President Shagari's new civilian government agreed to pay generous compensation and, when a number of farmers still refused to co-operate, sent in a very large force of police, which put down the revolt at the cost of an official death toll of nineteen. Impresit won 23m naira in compensation for the delay, to be added to the contract price which, with inflation and extra work, had already soared from the original 110m to about 400m ($550m) today.

More recently, the contractors have had to face formidable delays in payments, as Nigeria has suffered the effects of the international oil glut, and government revenues have been drastically reduced.

Despite the payment delays, work has gone ahead more smoothly since April 1980. Rice has been grown with some success, though most farmers grow maize. Wheat, which had been envisaged for the scheme, has not proved very satisfactory.

In order to lessen the disruption caused to the farmers by land preparation, there has been a switch from area irrigation to sophisticated sprinkler systems, which require less levelling: instead of taking water from the canals with syphons, the farmers attach pipes to hydrants fed by electric pumps. The initial investment is less, but running costs are higher. They are considered better for the farmers, who often could not be bothered with syphons, and broke down the canal walls to get water.

A basic problem remains: persuad-ing farmers to use irrigated techniques to gain a second crop - the main objective of the scheme. Most of the farmers on the newly irrigated land are still producing only one crop a year, mainly by traditional methods. Indeed, the amount of land being farmed for a second crop actually fell from 4,000 hectares in 1981-82 to 2,600 hectares in the current dry season.

The chairman of the river basin authority, says this is because of a dispute over the ending of subsidies on some inputs, like fertilizer: the farmers held back from planting to call the authority's bluff until it was too late.

A more fundamental reason is that many farmers are not attracted to the type of agriculture the scheme offers, especially the unfamiliar irrigation which requires going out at night, when superstition makes them

Big dams can be big trouble. (*UNITED NATIONS*)

afraid. Moreover, most northern Nigerian farmers are used to having a less active dry season, using it to repair their homes, or go on pilgrimages to Mecca. The economic incentives of the second crop are not obvious enough to persuade them to change their habits.

The feasibility study reckoned that it would take five years from the completion of the project for it to come fully into operation, and the authority claims that the farmers are adapting fast. Yet even assuming full operation, the capital cost of the second crop, will be more than $20,000 per hectare — a formidable sum to recover by farming, particularly when the farmers are currently reckoned to be obtaining less than 30 per cent efficiency in water use (against the 70-75 per cent obtained in the developed countries).

In assessing the overall value of the project one must take into account the serious loss of yields caused to farmers further down the river who no longer enjoy the abundant and fertilizing — if destructive — floodwaters. No one mentioned this in the preliminary study.

Sophisticated projects like this can only be economic on a day-to-day basis if they are well-run and well-maintained. The water supply must be administered fairly and efficiently, the dam, canals and pipes kept in good order. If not, the project will gradually die, as has nearly happened to several irrigation schemes in Sudan, requiring very expensive rehabilitation operations. The river basin authority has a frightening responsibility, but like many concerns in Nigeria it is pathetically short both of skilled personnel and money as budget cuts rain down.

Bakolori may still be a success. But if not it will be another warning that what seems politically and technically attractive in Africa does not necessarily win the acceptance of the people it is supposed to benefit.

James Buxton, *Financial Times*, 1983.

Small-scale Fisheries in Kerala

The village of Sakthi was the first traditional village in Kerala to become exposed to a process of modernization of its fishing technology. The impulse to this process was actually given in the early 1950s by a comprehensive development project known as the Indo-Norwegian Project (INP). Conceived as a bilateral aid programme of technical assistance for the development of fisheries in Kerala, INP's purpose was 'to bring about an increase in the return of the fishermen's activity, to introduce an efficient distribution of fresh fish and improvement of fish products, to improve the health and sanitary conditions of the fishing population, and to raise the standard of living of the community in the Project Area in general.' Furthermore, 'the Project called for mechanization of the fishing boats, provision for repair facilities, introduction of new types of fishing gear, improvement of processing and curing methods, building one or more ice producing plants, supplying of insulated vans and motor-crafts for transport of fresh fish', etc. In short, the INP aimed at radically transforming the whole set-up of the Project Area from above and at revolutionizing the traditional techniques and methods of cashing, preserving and distributing fish. The objective of INP was to equip active fishermen with more efficient technology in order to raise their standards of living.

The INP management selected traditional fishermen from the area to be trained on the Project's motorized boats and to become boat owners upon completion of their training period. Mechanized boats and modern nets were issued to these fishermen on a heavily subsidized basis. Nevertheless, a number of these initial owners were not able to provide for the proper maintenance and repair of their boats so that ownership was transferred to local money-lenders and fish merchants.

The early 1960s marked the beginning of a profound change in the orientation of the Project. This coincided with the discovery of the immense profit opportunities offered by the international market for crustaceans (prawns, shrimps and lobsters), especially in the United States and Japan.

In the INP boatyard, the construction of 32-ft trawlers fitted with powerful 84-90hp engines was started while previously built 22/25-ft boats were now equipped with large nylon drift nets designed for catching quality fish varieties. The whole landscape of Sakthi underwent fast and almost incredible changes from the 1950s and throughout the 1960s, so much that the area became known as 'little America'. A number of large processing and export companies of marine products sprang up alongside ice plants and peeling sheds which employed several thousands of people on a permanent or casual basis. The total catch landed by mechanized boats in Sakthi increased from 0 in 1953 to an average of around 90,000 tonnes during the period 1974-1979.

One of INP's main purposes was to create a new and efficient system of fish distribution. To this end, a co-operative sales organization was established as an alternative to the old network of private fishmongers, auctioneers and middlemen. However, the co-operative never reached the expectations of the Project's authorities (catches continued to be sold mainly to private dealers) and in 1962, for reasons that are not clear, it went bankrupt soon after a sudden increase in its turnover. As a consequence, private fish merchants of the traditional type remained masters of the field. Parallel to the phenomenal increase in the fish landings of mechanized boats after the discovery of the export market for prawns and lobsters, the size of Sakthi's fish market increased tremendously. This enhanced market size resulted in both a larger scale of operation for most fish dealers and in a greater number of fish merchants participating in sales transactions.

With the gradual mechanization of fishing technology, Sakthi's economic system lost its basic homogeneity and became increasingly divided between two distinct sectors: the traditional and the modern (mechanized) sector. While in the former sector fishing assets are basically considered as instrumental in providing employment opportunities to the family unit, in the modern sector they are usually acquired with a view to yielding high private

122

returns in financial terms. Put in another way, with the mechanization of fishing technology, the craft has lost is character of a concrete means of production, employment and survival, to become an abstract economic factor whose handling is clearly dissociated from its ownership. The change in behavioural patterns wrought by the mechanization drive in Sakthi is tremendous: the purchase of a boat is no longer intended to provide its owner with a work-tool and to enable him to become his own master; instead, it is part of an economic strategy aimed at rapid enrichment of the investor. Among local people, the growing success and prosperity of most boat owners arouses new desires and creates strong aspirations to a higher material standard of living. Would-be owners want to buy boats because they wish to emulate successful neighbours under the irrestible influence of an all-pervading 'demonstration effect'.

The growth of the mechanized sector has been based both on a large inflow of migrant workers from bordering districts or from neighbouring states (especially Tamil Nadu), and on a massive transfer of workers from the traditional to the modern sector of fishing. If things continue to follow the course they have taken so far, the modern sector will soon engulf the entire fishing economy of Sakthi, leaving only one technology alive.

The impact of mechanization on the social structure of Sakthi beach village has been in the direction of increasingly sharp differentiation. At the top of this social structure, we find a privileged minority of fish dealers, businessmen, export agents and trawler owners, with the latter often being engaged in commercial or business occupations. Their beautiful terraced houses, many of which are concentrated near the jetty, form an undisputable mark of economic prosperity and contrast sharply with the thatched huts of most tra-

Better boats: short-term gains and long-term losses? (*Kireni Hemachandra*)

ditional fishermen who reside under the coconut trees on the edge of the beach. Crew labourers who have worked regularly on trawler boats and the owners of INP boats often occupy an intermediate position. True, there are fishermen who have entered the small elite and now possess a trawler boat acquired brand new and in full ownership; but their number is comparatively small. Moreover, one must keep in mind that most of the people who made the biggest profits from the new prawn fishing business are outside capitalists who do not belong to the Sakthi community: they had large amounts of capital to invest in the fishing trade (including the ownership of boats) and strong political pull to get institutional loans.

Downward mobility is not infrequent in the new social structure of Sakthi. Fish merchants and businessmen may suddenly go bankrupt due to miscalculation, imprudence, wrong anticipation or simply bad luck (as when a whole batch of frozen prawns is refused on the ground that it does not meet quality export standards). But the risk of bankruptcy is also high for new boat owners: boats may sink or may need heavy repairs before their owners succeed in repaying their loans, in which case they may lose the fixed assets they have mortgaged to the lender(s). In fact, a number of present crew labourers are fishermen who formerly owned a mechanized boat and were never able to regain their previous ownership position. . .

The market economy development of Sakthi exposes its inhabitants to the typical risks involved in any process of capitalist economic growth. This is true not only in the sense that individuals may go bankrupt due to specific circumstances, but also in the sense that market risks may affect adversely all economic agents at one time. There is, of course, the risk of over-production or of a slump arising from a slackening market demand and causing a sudden fall in prices. But in fisheries, given that fish is a replenishable resource subject to the possibility of (temporary) exhaustion, there is also the risk that over-investment — disproportionate growth of the fleet of mechanized boats and practices of indiscriminate fishing — will lead to over-exploitation of marine resources beyond the point of largest sustainable yield, thereby jeopardizing the conservation of the fish potential. Moreover, since fish is a common-property resource (the fish caught by a given fishing unit reduces the stock of fish available to other units), exploitation of the sea under conditions of individualistic competition may also entail immediate decreases in the catches per unit of fishing effort for all operating units.

Of late, one has actually noticed clear signs of prawn over-fishing in Sakthi, at least in the second sense given above. While prawn resources were at first largly unexploited and the yearly catch per boat was high, with the result that considerable income accrued to boat owners (as foreign demand was brisk, prices were also quite high), at a later stage the rate of profit began to decline due to the growing pressure of exploitation on prawn resources following a tremendous increase in investors attracted to the field. Today it seems that the private rate of return on a boat hardly exceeds the dominant rate of interest in the local unorganized credit market. It is revealing that big investors (mostly from outside the area) have already begun to withdraw their capital from Sakthi. The presumed losers are first the small local boat owners who did not correctly anticipate the down-swing of the cycle and based their profit expectations on the past performance of trawlers; and second, the boat crews whose employment opportunities are suddenly frustrated. Furthermore, if over-trawling in Sakthi's inshore waters has also encroached upon the available stock of fish varieties other than prawn, traditional fishermen are probably a third category on the losers' list. In many respects, the story of Sakthi is therefore the history of capitalist development in a nutshell.

Jean-Philippe Platteau, *Development and Change,* 1984.

Irrigation Tube-wells in Bangladesh

A World Bank appraisal team spent three weeks in East Pakistan (Bangladesh) in 1970 to examine the choice of technology for irrigation tube-wells. The following results were obtained:

	Low-cost*	Medium-cost*	High-cost*
Initial cost (Rs):			
Market prices	31,660	58,005	194,805
Shadow prices	35,660	94,457	334,727
Internal Rate of Return (%):			
Market prices	48	33	7
Shadow prices	54	75	4

*Low-cost = jet/percussion drilling, centrifugal pump, low speed diesel engine.
*Medium-cost = contractor/power drilling, turbine pump, high speed diesel engine.
*High-cost = contractor/power drilling, turbine pump, electric engine.

On balance, the arguments for the low-cost wells over medium- and particularly high-cost wells were impressive. The country's development objectives in terms of economic returns, employment creation, increased potential for the creation of domestic industry, and the distriubtion of benefits would have been better served by the low-cost wells. Ultimately, however, the low-cost wells were rejected in favour of medium-cost wells.

The choice of the medium-cost technology is puzzling on first examination. The World Bank, as well as the government were strongly committed to the development objectives of East Pakistan. Considerations of administrative feasibility did not offset these concerns. Yet a technology that was definitely not optimal was chosen. Since the economists, engineers and administrators who participated in the decision presumably shared East Pakistan's development objectives, it is necessary to look at other factors affecting the choice.

Perhaps the central characteristic of the procedure by which such a decision is arrived at is that, instead of being a process intrinsic to an institution or gov-

125

ernmental system, it is a bargaining process between the government and an external institution. When this process crosses national and cultural lines as well, the difficulties of arriving at an agreement on an optimal solution are greatly increased. The availability of external aid to finance new investment in itself affects the choice of technology, for the preference of the aid giver then becomes an important element in the decision-making process. The administrator in a developing nation may very rationally accept a technology that he considers second or third best if foreign financing is available only for that choice, since that may be the only way to receive this aid. Furthermore, the official may tailor his programme to the technology he considers most likely to attract foreign aid.

On the donor side, many considerations other than the needs of the recipient country may affect the type of aid or credits made available. In fact, the form of aid frequently depends more on the requirements of the donor country's economy than on those of the recipient's. National commercial interests affected the forms of assistance available under the GEC supplier's credit arrangement and the Yugoslav barter agreement. An international agency like the World Bank, however, is not directly tied to parochial interests or to profit-making objectives; nor are the developing countries seriously constrained from arguing their own interests. Yet both the Bank and East Pakistan concurred in a tube-well technology that was less than optimal for East Pakistan. The following discussion of economic factors, various perceptions of the issue, and organizational objectives provides some additional insights into the influences upon the tube-well decision.

Economic factors can play an important role in the choice of technology. Price distortions, caused by inconsistencies in the tax and duty structure, may influence choices as they did in East Pakistan where the market cost of imported high-speed diesels was less than that of the locally produced low-speed engines, even though the latter cost only 75 per cent as much to produce. Similarly, the contractors of government agencies implementing tube-well programmes will have a rational preference for scarce capital goods (machinery) over abundant human labour when selecting technology, if the local currency is overvalued, as was the Pakistan rupee. In that situation imported equipment, such as pumps or power drilling rigs, could be obtained for as little as half their true cost to the economy. The result is that the price of capital goods is subsidized, a benefit which any economizer, public or private, is happy to accept.

If institutionally established wage rates exceed the real cost of labour, as they did in East Pakistan in 1970, labour-intensive methods become less attractive. When the problems of management of labour crews are added, contractors will generally adopt capital-intensive methods despite the fact that this is inconsistent with the nation's need to utilize low-cost labour.

Although economic factors can bias the choice of technology, project analysis techniques such as accounting prices can offset these factors. The rate of return analysis demonstrated that despite subsidized capital, the low-cost technology yielded a better economic return.

126

Participants in a decision may have different perceptions of the same problem. Foreigners new to the country may perceive the choices very differently from economists, and they both may see things differently from the administrator or the farmer.

For the World Bank, emphasis in this initial tube-well programme in East Pakistan was on quality and reliability, concerns suggested by the past experience of the members of the appraisal mission and their institutional outlook. Their experience led most in the mission to favour reliable, imported technology. One of the Bank's engineers who was appraising tube-wells in Comilla remarked while watching the low-cost wells being installed in a sea of mud by a large group of villagers, 'You can't install reliable tube-wells this way'. Yet tests and specifications showed that the labour-drilled wells were of a quality equal to the power-drilled wells.

The reliability of the wells had to be judged both in terms of the tube-well itself and of the environment in which the wells had to operate. To most of the Bank's appraisal staff, the power-drilled, medium-cost wells appear to be more reliable. Yet, although it is evident that the preferred, high-spped, diesel engine may operate longer and more reliably in ideal conditions, in an East Pakistan village, where there are few trained operators or mechanics, it is less tolerant of misuse and much more difficult to repair than the low-speed engine. The well that may be inferior in technical and engineering terms may prove more reliable for the farmer, more suitable for the conditions in which he operates, and therefore more efficient in producing the ultimate objective of increasing agricultural output.

An evaluation of a well's efficiency has another dimension. Many technicians fail to see the step beyond efficient water production: the construction and economical maintenance of wells throughout the countryside. The critical issue concerns not only the type of well that produces water efficiently but also the type that has a mixture of cost and performance characteristics as well as maintenance potential that makes it suitable for widespread use in the area. Examples from both West Pakistan and Vietnam illustrate these perceptual problems. Prototypes for low-cost tube-wells in West Pakistan and for low-lift pumps in Vietnam were dismissed by engineers as having 'little technical validity' after they had been tested and proved inefficient (in an engineering sense). But in each case, the innovative prototypes proved to be of a cost and specification that made them extraordinarily profitable under the existing conditions and their use spread very rapidly.

Risk is another consideration that was perceived very differently. For the farmer, the risky technology was that which was installed by outsiders and which he could not operate or repair, and therefore might not use. The simple technology was more adaptable to changing conditions. In case of failure, much less was invested in an individual unit. The one successful pilot project — in Comilla — had been carried out with this technology. For the World Bank, the reliable installation agent was the foreign contractor, despite the fact that the pilot tube-well project at Thakurgaon carried out by foreign contractors was largely a failure. For the Bank, concentrated drilling locations

127

where they could observe operations implied less risk than a decentralized, locally supervized operation.

Finally, there was the view held by some Pakistani officials in which anything 'modern' was preferable. The appearance of modernity was important to this group; to them it was synonymous with 'the best' and the best was what in their view aid donors should provide.

These varying perceptions all contained important elements of reality. They led people to differ honestly as to the appropriate tube-well technology for East Pakistan. In ths case, the role of the aid donors was particularly influential and undoubtedly contributed to the choice of medium-cost technology.

The search for an explanation of why organizations behave in ways that appear contrary to their stated objectives would strike a student of organization theory or decision-making as very familiar. Indeed, much of the literature of administration, beginning with Chester Barnard, has touched on this issue, for people in organizations frequently act in ways that appear irrational.

In this case, the organizational requirements of the developing countries and the aid donor agencies appear to have superseded development policy objectives. Thus, organization theory provides more insights into the choice of technology than does economic theory and proves useful in explaining the choice of tube-well technology.

Work in this field has stressed the concept of 'bounded' or limited rationality in decision-making. According to the theory, the limits of man's capacity as solver of complex problems circumscribe the decision-making processes of individuals and organizations by forcing them to simplify problems in a variety of ways which result in replacement of the goal of finding the optimum solution with that of finding a course of action that appears satisfactory to them. Elaboration of the factors that limit the exercise of full rationality helps to explain the choice of tube-well technology.

1 Satisfactory, Rather than Optimal Solutions
   Since an optimal solution is rarely identifiable or ascertainable, a satisfactory one is accepted. In this case, established procedures, protection of bureaucratic domains and routines, and the need to avoid risk were all obstacles blocking the economically rational solution. In addition, limitations on information and time prohibit the decision-maker from considering all possible solutions to a problem and force him to choose among the first satisfactory alternatives. The World Bank Appraisal Mission was very short of both information and time. It was not able to make all the calculations or obtain all the information contained in this paper. As a result, the members chose the solution that seemed to fit their perception of what was needed, met their organizational requirements, and also had a satisfactory economic justification. The medium-cost technology was familiar and was considered reliable. Since the World Bank was to able to show an anticipated 32 per cent internal rate

of return for their medium-cost wells, a less than optimum solution was satisfactory.

As far as the government officials were concerned, aid was available only for medium-cost wells, and this technology conformed with their institutional requirements and perhaps their personal preferences as well. For both the Bank and the government, a tube-well project on which they could co-operate was a satisfactory basis for action.

2   Avoidance of Uncertainty and Reliance on Established Programmes of Activity

Minimization of risk characterizes organizations and individuals within them. This tendency is a deterrent to innovation and stresses repetition of procedures that are familiar and tested. For the Western engineer, machine power drilling, fibreglass screens, high-speed engines and turbine pumps represent a familiar and reliable technology unlike the low-cost technology. Foreign drillers who could be held contractually responsible for performance appeared safer (despite serious mistakes at Thakurgaon) than a decentralized programme involving a large number of low-cost rigs. Thus, familiarity and the avoidance of risk were preferable to optimum results.

For government officials, technology with the appearance of modernity is less subject to criticism than simple technologies. Furthermore, they could reduce personal responsibility by claiming they had accepted less than optimal techology under pressure of aid officials in order to obtain aid.

3   Operating Prodecures and Established Routines

Organizations are bound by their operating procedures and established routines. Because an agency like WAPDA is staffed with engineers trained in modern construction methods, supported by large groups of foreign consultants and a flow of foreign aid, has a large, well-capitalized construction division, and, finally, makes decisions on the basis of feasibility studies, it is almost incapable of employing low-cost technology. Its staff, equipment, and procedures are all oriented towards high cost, high-quality construction. Although the other East Pakistan sponsor of tube-well programmes, the Agricultural Development Corporation, was less tied to 'high' technology, the head of their water development division had just come from WAPDA and he and others had strong preferences for a 'modern' technology. For both the Bank and the government, existing organizational routines and physical procedures limit the options available to them in the short run.

4   Control

Control is an objective of most organizations. It is closely related to risk minimization and was an important factor both for the government agencies and the World Bank. Government agencies and their staffs derive power, prestige, and sometimes an opportunity for profit by attracting foreign aid and implementing large programmes. For the Bank, control was equally important to ensure proper implementation to protect against misuse of funds, and to ensure maximum return on the project. These objectives, were, in the Bank's eyes, best accomplished in a programme it could supervize closely and control. For government and Bank alike, a low-cost programme with up to 3,000 rigs operating at scattered locations throughout the country required a decentralized administrative system and a resultant loss of control. To avoid this, they were all willing to accept less than optimal economic returns and social benefits.

Unfortunately, most of these organizational factors were implicit, not explicit. Most of the participants in the decision-making process contended they were acting only on the basis of development goals and administative feasibility, although the evidence seems clear that they were not. The result was that there was no conscious effort to weigh all the real considerations and the trade-offs among them. Had such a weighing of trade-offs been possible, solutions to organizational goals might have been found that would have permitted the most desirable technology to be chosen. The low-cost wells might have been restricted to a tube-well field where closer supervision would be possible, thus satisfying the Bank's need for supervision and control. With more time a different foreign contractor might have been found who could deal with the low-cost technology, or at least part of the funds might have been allocated as risk capital for experimental wells that could test the merits of alternative technologies, and perhaps even develop new alternatives that might be preferable to those considered here. Much would have been gained if legitimate organizational objectives had been made explicit. Then consideration of the trade-offs between these and the more explicit development goals might have led to the selection of an optimal technology.

Ultimately, it was the organizational requirements of the implementing agencies, including the aid donors, that determined the choice of tube-well technology for East Pakistan. In the actual decision-making, such factors as risk avoidance, appearance of modernity, established procedures, familiar techniques, and by no means least, control, outweighed development policy objectives. It is in these factors that an understanding of decisions as to choice of technology must be sought.

John Woodward Thomas in C. Peter Timmer, *The Choice of Technology in Developing Countries*, 1975.

. . . unnatural effects, arising out of the resources of the country having been drawn unnaturally together into great heaps. — William Cobbett

---

On every hand, the living artisan is driven from his workshop, to make room for a speedier, inanimate one. The shuttle drops from the fingers of the weaver, and falls into iron fingers that ply it faster. — Thomas Carlyle

Why Inappropriate Choices are Made

Entrepreneurs in developing countries are not always free to choose the source of supply of their technology. In some cases they are forbidden to import technology from certain countries, or they are forced to use technologies imported from 'friendly' countries. Foreign subsidiaries and joint-venture firms are often obliged to use technologies specified in the technical assistance contract with their principal firms and licencers. This is also true of the integrated (i.e., totally self-sufficient) plants built under foreign assistance programmes. At the same time, foreign investors may find it easier to manage machinery than a large number of local workers, because they are unaccustomed to the local rules which govern labour relations.

The licensing system covering foreign investment and imports of machinery and technology can be an extremely effective instrument for preventing the adoption of technologies unsuitable to a country's development objectives and for promoting desirable ones. In developing countries, however, the criteria used for the screening of investment and import applications often conflict with development goals, political considerations dominating economic ones. Sometimes national prestige is thought to demand the use of the most modern technologies, and the authorities take offence when foreign investors try to import used machinery.

Even when the stated criteria are sound, the licensing procedure is subject to widespread 'irregularities' where decisions are influenced by the 'commissions' directly or indirectly paid to responsible officials. This subject is almost taboo as a research topic and is rarely mentioned in scholarly discussions, for obvious reasons, but there are grounds for believing it to be an important barrier to the choice of appropriate technologies in the Third World.

A rare empirical study in this domain is John Eno's work on the salt industry. According to his findings, foreign exchange allocations for the capital goods imported by this industry exceed the actual requirement by 50 per cent or more in Indonesia, the Philippines and Thailand. The excess is sold on the black market and provides a major source of profit. As the machinery becomes more sophisticated, it is more likely to be imported and expensive and the windfall profits tend to be larger. At the same time, the more influential the applicantis, the more generous the allocation becomes.

Susumu Watanabe, *International Labour Review*, 1980.

What Managers Say

The following quotes made in Indonesia are not untypical:

After noting that machines in the process of being scrapped were capable of producing a high-quality product, one foreign manager explained the replacement of the intermediate equipment with sophisticated equipment as follows: 'You have to modernize to stay ahead of the competition.' This statement was made in spite of the facts that there was a ban on the import of the product and that there were no other firms manufacturing the product locally. When asked what he was going to do with the old equipment, he explained that he was going to cut it up and scrap it. He would not sell it because 'some of these Indonesians can get any old equipment running'.

The manager of a hand-rolled-cigarette factory answered the question as to which technology was cheaper by assuring me that he did not introduce machines, he explained that 'the interest payments on the money we would have to borrow would exceed our wage bill'. When challenged with the possibility that this statement might be inconsistent with his claims on cost, he responded that the automated plant would be cheaper 'in the long run'.

Another manufacturer who was employing young girls to attach labels by hand to his products explained that he was ordering a machine to replace them. Asked whether it was cheaper to attach them by machine, he explained that he did not know, but the girls were a lot of trouble: 'They just cause management problems'.

The manager of a plant with both an automatic and a semi-automatic line explained that he was converting the semi-automatic to fully automatic as soon as possible. He wanted to produce a 'high-quality' product. The output of both lines was already meeting the standards of the foreign licenser and was considered among the best in Indonesia by the firm's competitors. It was not clear that further automation would improve the quality, at least in a way that consumers would notice.

Louis T. Wells in C. Peter Timmer et al, *The Choice of Technology in Developing Countries*, 1975.

If a man write a better book, preach a better sermon, or make a better mouse-trap than his neighbour, tho' he build his house in the woods, the world will make a beaten path to his door. — Ralph Waldo Emerson

---

The existence of a technology does not require its use any more than the existence of a gun requires us to shoot. — Tom Bender

III
Technology for Development
Agriculture, Food Processing, Livestock

In the preceding chapters, two messages come over loud and clear. First, if the basic needs of the masses of the rural and urban poor are to be satisfied, then technologies appropriate to their needs and means must be identified and applied. Second, despite the existence of a wide range of technologies for the production and provision of most goods and services, choices are constantly made which result in the application of technologies which are inappropriate to the masses of poor people in developing countries.

We shall return in detail, in Chapter VIII, to some of the issues involved in the generation and choice of technology. First, however, it is useful to build up an understanding of the range of appropriate technologies already in existence and to learn how they are actually being applied in developing (and some developed) countries to the benefit of rural communities and the urban poor. This chapter deals with agriculture, food processing and livestock. The following four chapters cover: health, water and sanitation; biomass and renewable energy; housing, construction, and transport; and manufacturing, small-scale mining and re-cycling. Given the growing amount of literature on the topic, an additional 'case study' chapter could easily have been compiled relating specifically to women and technology. It was decided that this might give the impression that technologies for women are a special category dealing only with those tasks such as child- and home-minding which are commonly thought of as being 'women's work'.

In practice, nearly all of the technologies in the 'case study' chapters are of direct relevance to women. In most countries women are almost totally responsible for crop and food processing and for the care of minor livestock, and, in some, they have major responsibility for food production too. Health, water and sanitation are nearly always areas of women's responsibility, as is transportation of domestic and farm inputs and products. In many parts of the world, women are involved in the construction and building materials industries and they are traditionally involved, to a greater or lesser extent, in many other rural industries such as textiles, soap making and recycling.

Thus, women are well represented in each of the following chapters, with the reader's attention being drawn, when appropriate, to their special role or contribution so that they do not become 'invisible' through being integrated in this way.

133

Although the examples in these chapters tend to be successful ones, there are also examples of failures, where 'appropriate' technologies have turned out not to be appropriate at all. Also, many of the successful examples are based on pilot projects: the numbers of people being helped by some of the proven technologies are, therefore, small relative to the scale of the need. The issue of replicating or expanding successful pilot projects is one which is examined in Chapter IX.

Measured globally there is more than enough food for every man, woman and child. Despite this, millions are severely undernourished or starving and the situation appears to be getting worse. After a respite from famines of over a quarter of a century, food deficits started widening in the early 1970s and, since then, famines have claimed hundreds of thousands of lives, providing a grim reminder of the fragility of food security even in an age of advanced technology. According to a recent World Watch Institute Report:

> There is no simple explanation of why efforts to eradicate hunger have lost momentum or why food supplies for some segments of humanity are less secure than they were, say fifteen years ago. Declines in food security involve the continuous interaction of environmental, economic, demographic, and political variables. Some analysts see the food problem almost exclusively as a population issue, noting that wherever population growth rates are low, food supplies are generally adequate. Others view it as a problem of resources — soil, water and energy. Many economists see it almost exclusively as a result of underinvestment, while agronomists see it more as a failure to bring forth new technologies on the needed scale. Still others see it as a distribution problem. To some degree it is all of these.[1]

Whatever the cause of hunger, there is obviously an urgent need to increase the amount of food supplies available per capita in the developing countries, and to ensure that these are evenly distributed throughout the entire population. But how can this best be done?

As will be gathered, the solution to world hunger is not as simple as growing more food through bringing more land under the plough, increasing yields per acre, or raising cropping intensities. If this were the case, then the new technologies of the Green Revolution would have solved the problem. The introduction and use of new HYV seeds, fertilizers, irrigation and improved cultivation equipment has in fact resulted in an enormous increase in food production. In 1950, the world's farmers produced 623 million tons of grain. In 1983, they produced nearly 1.5 billion tons. Despite these increases, however, there is more hunger than ever — a contradiction which needs some explanation.

The main constraint appears to be one of extreme inequalities in control over land, institutional credit and other resources. In their review of the pos-

---

[1] Lester R. Brown, *State of the World: A Worldwatch Institute Report on Progress Toward a Sustainable Society* (W.W. Norton & Co., 1984).

ition on world hunger, Frances Moore Lappe and Joseph Collins draw the following conclusion about the Green Revolution:

When a new agricultural technology — such as hybrid seeds that yield more in response to irrigation, fertilizers and pesticides — is introduced into a social system shot through with power inequalities, it inevitably benefits only those who already possess land, money, credit 'worthiness' or political influence or some combination of these. This is simply a social fact...

The potential productivity represented by the new technology attracts a new class of 'farmers' — money lenders, military officers, bureaucrats, city-based speculators and foreign corporations — who rush in and buy up land...As land values rise, so do rents, pushing tenants and sharecroppers into the ranks of the landless. Seeing new profit possibilities, landlords evict their tenants and cultivate the land themselves with the new agricultural machinery. The percentage of rural workforce that is landless has doubled in India (now over one-third) since the introduction of Green Revolution innovations. In northwest Mexico, the birthplace of the Green Revolution, the average farm size has jumped from 200 to 2,000 acres with over three-quarters of the rural labour force now deprived of any land at all.

And, while more landless are created by the expansion of the better-off growers, fewer jobs are available to them. The large commercial operators mechanize to maximize profits and avoid labour management problems...

In country after country where agricultural resources are still regarded only as a source of individual wealth, the narrow drive to increase production totals ends up excluduing the majority of rural people from control over the production process. And, we have found, to be cut out of production is to be cut out of consumption...

In-depth investigations by the United Nations Research Institute for Social Development (UNRISD) of the impact of Green Revolution techniques in twenty-four different underdeveloped countries have confirmed a consistent pattern — the decline in well-being for much of the rural majority even as agricultural production bounds ahead.[2]

Increased landlessness and unemployment have not been the only ill effects of the spurt in agricultural production using 'modern' agricultural techniques. In countries as diverse as Bangladesh and the United States, irrigated agriculture is threatened in some areas by falling water tables. Response of crops to the use of additional fertilizer is now diminishing, particularly in agriculturally advanced countries. Millions of acres of top-soil have been lost as land that was not suited to permanent cultivation of crops has been brought under the plough. These factors, combined with the increasing scarcity of new cropland and the end of the cheap energy on which modern agriculture was based, make an expansion in food production progressively more difficult — especially with continued reliance on existing 'energy' intensive techniques. The Worldwatch Institute Report points out that while world food output

[2] Francis Moore Lappe and Joseph Colins, *World Hunger: Ten Myths* (Institute for Food and Development Policy, 1979)

grew at an adequate 3 per cent per year between 1950 and 1973, it has achieved a growth rate of less than 2 per cent per year since then. 'Trends in Africa', it points out, 'are a harbinger of things to come elsewhere in the absence of some major changes in population policies and economic priorities. Between 1950 to 1970, per capita grain production in Africa has declined rather steadily. The forces that have led to this decline in Africa are also gaining strength in the Andean countries of Latin America, in Central America and in the Indian sub-continent. Whether the declining food production now so painfully evident in Africa can be avoided elsewhere will be determined in the next few years'.[3]

Further problems can be outlined to drive the point further home. One is the diversion of land from subsistence crops to cash crops (with assistance from foreign donors) with the resultant loss of the capacity of rural families to feed themselves. In Africa this has resulted in men earning cash from land once used by women to grow food for the family. With cash spent on consumer goods rather than food, nutrition levels decline. Another problem is the dependence of farmers (especially larger farmers and cash crop farmers) on imported inputs — chemical fertilizers, tractors, diesel oil, pumps, etc. Many countries are now cutting back on such imports due to foreign exchange problems, thus making it difficult for modern farmers to sustain production.

A change of strategy in the agricultural sector is needed. One which moves away from large-scale, mechanized, energy-intensive, environmentally harmful and non-sustainable systems to greater concentration on small-scale units using labour-intensive, less harmful and more sustainable techniques. The end of cheap oil, and bans on 'modern' imported inputs encourages this trend: land reform measures would encourage it even further. But can levels of food production be increased or even maintained under such cconditions?

Indications are that they could. Throughout the world, smaller landholders consistently produce more per acre than large producers. Natural alternatives to chemical fertilizers and pesticides are numerous and have been proved effective. Besides maintaining yields, they also increase the numbers who can be employed in agriculture (thus increasing the purchasing power of the poor for food) and reduce dependence on imported inputs.

It is with the small-farm sector and the food-related productive activities of the near-landless that this chapter mainly concerns itself. The first section looks at pre-harvest food production activities — the issues and problems; the choice between crops; the introduction of more ecologically sound crops and techniques; the introduction of irrigation technologies; and the choice and introduction of small-scale land preparation equipment.

Following some general extracts on the 'food production' problem, including one on recognizing the important role of the women food producers of Africa, a major theme in the section is the use of crops and techniques which are less demanding on the soil and better suited to very small land-holders. Several extracts look at the benefits of root/legume crops over cereals in terms of nutrients per acre and use of nitrogen reserves in the soil. McDowell's

[3] ibid.

extract gives statistics which show that, acre for acre, more calories and proteins can be provided from groundnuts, cassava, beans, sweet potatoes and plantains, than from maize or other cereals. They also require less expenditure of energy in harvesting, threshing, winnowing and grinding. He suggests that the rationale for promoting and choosing cereals rather than roots/legumes is the greater export potential of the former.

Despite this, however, projects to introduce more nutritious foods are becoming increasingly popular. One example has been chosen from the thousands of projects aimed at introducing vegetable growing to small landholders. In the Asian situation, the land involved amounts to no more than that surrounding the home — in other words — a kitchen garden. Bringing small, hitherto unused plots of land into use can help rural women (and men) provide useful supplements to the food supply of the near-landless. Other extracts refer to amaranth which has twice as much protein as wheat or maize, and to the winged bean which has higher nutritional value than soya bean, and thrives in very poor soil. References are also made to some of the ecologically sound farming techniques available to small-holders: these include no-till cropping systems, waste recycling and crop rotation using legumes to fertilize the soil by 'fixing' atmospheric nitrogen. As a contrast, and to emphasize the need for all of this, the extracts by Jeremy Rifkin and George McRobie point out the unsustainable nature and 'efficiency' of Western agricultural systems.

Irrigation is an important way of raising agricultural yields in many parts of the world. But the introduction of new technologies for irrigation — either to replace the traditional techniques or to provide water where it was previously unavailable — has been a mixed blessing. Diesel pumps and electric pumps have usually been acquired by the richer members of society, who have access to cash and credit. In the absence of checks on the quantities of water drawn by such cultivators, ground-water levels have dropped, thus reducing the ability of poor farmers to draw water using traditional irrigation methods. The Ceres article on pump projects in Mauritania illustrates what is a common phenomenon in many countries of the Third World. But pumps do not necessarily have to exclude poorer farmers or the landless; the extract by Clay shows how organizational and technical innovations can enable a greater number of people to benefit from new irrigation techniques. But all schemes to remove ground water are environmentally damaging in the long-run — whether the poor benefit in the short term or not. A more natural method of irrigation is water catchment — a system which has the added advantage of reducing silting, flooding and soil erosion. An example of this type of scheme in India is given in the extract entitled 'When the hills came tumbling down'.

Another type of technology which has tended to help large farmers at the expense of the small farmers and the landless is the tractor. In most countries, these machines, plus the spares and diesel require to run them, usually have to be imported, thus using up valuable foreign exchange. Increasing attention is now being given to the design and introduction of 'intermediate' level equipment for land preparation — particularly in Africa where choices between the hoe and the tractor have been virtually nil. Extracts examine experiments with the design and introduction of various devices — animal-drawn planters in Botswana, the 'Snail' tillage device in Malawi, the 'Tinkabi' tractor from Swaziland, and the one-ox plough in Ethiopia. Also included is an extract on experiments with the introduction of donkey-powered ploughs in West Africa — experiments which allow women farmers to cultivate larger areas.

Section 2 of the chapter deals with post-harvest systems. There are two major themes here. One is the extent to which food losses are experienced during the post-harvest phases. The other is the extent to which the mechanization of such activities is desirable in terms of output, employment and equity considerations.

A good introduction to the subject is given in the extract by Peter Muller. Besides describing the various types of post-harvest technologies in use in different parts of the world, it also introduces the point that food losses at the small farmer level are much lower than commonly believed. While it is not unusual to find quotes from sources such as the United Nations of post-harvest losses of up to 50 per cent, statistics based on detailed research find that actual losses experienced by small farmers are less than 10 per cent. Large losses only become common when huge surplusses (from larger farmers) come onto the market for centralized drying, storage and distribution. But even losses of 5 to 10 per cent are worth cutting, and research into improved post-harvest technologies for small farmers are still useful.

Most post-harvest activities such as threshing and winnowing are extremely labour-intensive and result in bottlenecks which can constrain levels of output. In addition, some of the new varieties of cereals and crops such as wheat, which have been recently introduced to Asian countries, are more difficult to thresh than indigenous crops. For both reasons, even relatively small farmers feel the need for some type of mechanical device to help with these operations. Much work has been done in southern Asia on the design of small-scale threshers, winnowers and dryers which can be locally manufactured. It should be remembered that even small machines displace some labour — unless groups of landless can be assisted to buy a thresher or winnower to perform custom work for farmers.[4] An additional problem with this type of equipment is the casualities it can cause. This is illustrated only too clearly in the extract by Radhakrishna Rao on 'The carnage of the Green Revolution'.

Another very labour-intensive food-chain activity is that of grinding cereals. Here again, the constraints imposed by traditional technologies have led to the rapid spread of rural mills (usually owned by entrepreneurs and large farmers) which are mechanically, though not nutritionally, superior and provide a cheaper service. Such mills have wiped out millions of jobs for landless

[4] It tends to be women who are displaced from threshing and winnowing activities when new machines are introduced, and men who get the jobs available as machine operators. In a few countries, such as Bangladesh, attempts have been made to assist groups of landless women by providing credit to purchase a machine.

women in countries such as Bangladesh and Indonesia while doing little to alleviate the burden of farm women who continue to pound their grain for their own needs by hand. Similarly, cereal mills in Africa charge high rates (increasing with the rising price of diesel), thus precluding their use by all but women from the better-off families. Many projects now aim to assist women to take advantage of the new technology, rather than be exploited by it. There are schemes to introduce cereal mills, oil presses and other crop or food processing technologies to women in many countries of the Third World. Not all of these have been totally successful. A small selection of the many case studies available are included here — a women's millet mill in Senegal, a groundnut oil press in Upper Volta and a women's banana-chip co-operative in Papua New Guinea.

The final section of the chapter covers livestock and animal health care. An important starting point here is that if food has to be grown specially to feed animals, on land that could otherwise be used to grow food for people, it is an inefficient use of resources. On the other hand, if animals are kept on a small scale, finding food by foraging from household scraps, they can provide an occasional, nutritious supplement to the diet. This point is well covered in the extract by McDowell. Keeping this in mind, extracts follow on the advantages of small animals over large animals for small farms, and the advantages of small-scale poultry units over large-scale ones.

The much publicized 'White Revolution' — the Indian dairy scheme which aimed to provide regular quantities of milk at reasonable prices to households in Indian cities — is explained and discussed, along with some of the criticisms which have been raised against this controversial programme. Alternatives to cattle are also looked at: goats which are frequently referred to as the 'poor man's cow', and camels which provide more milk and result in less grazing and trampling than cows.

Finally, there is an extract on the important, but often forgotten, area of animal health care. As with doctors, the costs of training veterinarians is enormous and there are not nearly enough vets to cater for the vast needs of livestock in the rural areas. 'Barefoot' vets could do much to relieve the suffering of animals and to help livestock owners to help themselves to protect their own assets. The topic is discussed again in Chapter X in an extract which points out the need to train female, as well as male, animal health workers.

The Overburdened Hectare

For those who enjoy contemplating global averages, there is a recent estimate that, as of 1975, each hectare of the globe's surface under cultivation had to provide for the nutritional needs of three people. By the end of this century, it is further estimated, the arable land-to-population ratio will be one hectare for six people. Is there a stronger case to be made for pampering the productive capacity of each available hectare? — Ceres, 1983.

1 FOOD PRODUCTION

 Food Situation Worsening in Africa

In Africa population increases far surpassed food production, which declined by 7 per cent in the 1960s and by 15 per cent in the 1970s. Food imports and food aid have met only a portion of the production shortfall, even with a doubling of grain imports in the 1970s to 11 million tons. Food consumption today is 10 per cent less than it was a decade ago. Dependence on imported food may triple by the mid-1980s if food production declines as projected. Considering the amount of attention Africa's hunger problems have received over the past decade, only a small proportion of agency assistance appears to have been specifically directed to the domestic production of food crops. There is an underlying assumption in some agencies that if agriculture is promoted generally, domestic food production and consumption will take care of themselves, but the experience of Africa does not bear out this assumption.

A United States Agency for International Development (USAID) analysis indicates that out of 570 projects in Africa, 218 were expected to have some impact on food production, but only 22 — representing 7 per cent of the total projects by value — were directly concerned with food crop production. Of 22 World Bank agriculture and rural development projects approved by sub-Saharan Africa in 1981, only two are specifically aimed at raising domestic food ouput while seven are directed to export crops. The African food situation has been allowed to worsen without sufficient policy attention for a decade.

The misdirected development policies that were documented again at the 1980 UN Conference on Women and Development, in Copenhagen, show that discrimination against women in development programmes, especially in rural areas and agricultural projects, is in large part to blame for the failure of food production to increase in Africa. Traditionally, food production — that is, subsistence farming as opposed to cash and export crops — has always been done by women and even today is the responsibility of village women in all rural areas.

The failure of development assistance to provide women with training, tools, fertilizers and the means to increase their production, and the failure of African men to engage in subsistence farming is directly to blame for the failure of Africa to feed iself. It is strange that African men refuse to give up their traditions — which decree that subsistence farming is women's work and beneath their dignity — they would rather let their women and children starve.

The failure of international agricultural programmes to address the real problem — the diminishing production of subsistence farming or local food in relation to soaring population growth — is no doubt in large part due to the male-headed

140

Another of woman's responsibilities.

African governments. Since African men have first call on food even if women and children starve, they are not interested in improving food production. After all, this is not their concern — subsistence farming is a women's concern.

This provides the male-dominated development agencies with the perfect excuse: they can only assist such programmes as they are asked to help with technical advice, financing and training. Predictably, African men do not ask for assistance with subsistence farming as it is a women's affair — and in any case, men do not starve, women and children do.

Diplomatic World Bulletin 1982 and *Win News* 1982.

Which Crops?

The overall tendency in agricultural 'development' over the past few decades has been towards the promotion of cereal production at the expense of systems based on roots or starchy fruits and legumes. In some cases the nutritionists have had a hand in this through their sometimes groundless condemnation of starchy crops. However, there are also grounds for suspicion that the desire to promote the export of food has

'It's a brand new seed. It kills bugs, waters and fertilizes itself and produces four corn crops a year. Unfortunately, it needs a monthly service by a man from New York.'

provided the main motivation since yams, sweet potatoes, plaintains and many local legumes are not readily exportable, whereas cereals are.

The use of combinations of legumes with roots or starchy fruits can be seen to possess many nutritional advantages over cereals, which require a heavy expenditure of energy in harvesting, threshing, winnowing and grinding as compared to other foods. Also, as far as childrens' diets are concerned, cereals often provide a bulky, low-nutrient porridge. Most important, in terms of yield of nutrients per acre, root/legume mixes possess a definite advantage and they are, of course, less demanding upon the nitrogen reserves of the soil than are cereals.

A simple calculation will demonstrate the advantage of nutrient yield. One acre of maize, grown in a traditional manner will yield about 1,000 lbs i.e., some 1,400,000 calories and 72 lbs of protein. This yield of calories and protein could be obtained from 0.65a. beans plus 0.22a. of plaintain, i.e. a total of 0.87a. Similarly, the same quantity of calories and protein could be obtained from 0.44a. of groundnut plus 0.1a. of cassava — a total of 0.55a., or from 0.4a. of groundnut plus 0.2a. of plaintain. For simplicity of calculation the above figures relate to crops grown in single stands, and assume that the soil and climate are suitable for all the crops. In mixed cultivation the advantage of root/legume mixes can be even greater.

In nutritional terms the root/legume mixes also show definite advantages over cereals. One hundred grams of maize, for example, when made into a fairly stiff porridge will have a total bulk of about 300g and will provide about 8g of protein and 360 calories. The same amount of calories and protein would be provided by 30g groundnut (or groundnut paste) plus 115g cassava; or 30g beans plus 220g sweet potato; or 25g groundnut plus 180g plaintain. In each case the Protein Calories per cent value of the mixes would be similar to that of the cereal, but as can be seen, the mixes would be less bulky and physically easier for a small child to eat than the cereal porridge. For small children the groundnut would need to be pounded into a paste, and the beans would need to be soaked and skinned, but the overall work necessary in the preparation of these foods would be no greater, and possibly less, than that needed to prepare the cereal.

Jim McDowell, *Appropriate Technology*, 1983.

Women and Winter Vegetables

Winter vegetable cultivation is one of the largest agricultural projects implemented in CARE's Women in Development Project (WDP). In this project, high-quality seeds and seedlings of carrots, cabbages, radishes, tomatoes, cauliflower and two types of local leafy vegetables, 'lal shak' and 'palong shak' are distributed by CARE and then cultivated by village women with supervision by CARE field staff.

The implementation of the winter vegetable project began in October 1980. The CARE extension staff, who are mostly women with in-house training in health, agriculture and community development, spend two or three days per week in each of the villages under their supervision. Each worker works in her assigned villages all year, but most of her time during October, November and December is devoted to winter vegetables. The extension workers encourage the village women to make a 'kitchen garden', a small vegetable garden near the household. The village women then decide which of the seven types of vegetable seeds/seedlings available in the programme they would like to cultivate. Field observation indicated that several major problems were commonly experienced. The most serious problem was lack of fencing before substantial damage had been done by domestic animals. As only a 'katcha' fence of bamboo, sticks, banana leaves, or straw woven between split bamboo would be put up, the lack of fencing was probably more the result of negligence than cost. However, for the poorest women, cost may be a major factor. It should also be noted that where poultry are relatively intensively raised, even a well-made fence is not sufficient to keep out chickens. Quite frequently, one would see chickens had flown over a 1m fence to eat the vegetables.

This problem is so serious that household vegetable gardening is not recommended in villages where poultry are intensively raised by the free-range methods that are almost exclusively used in Bangladesh villages. The most successful winter vegetable cultivation occurred in a Hindu village where the villagers kept no chickens or goats in their homesteads.

Another major problem inherent in winter vegetable cultivation as an intensive household activity was that of excessive shade. Many models for intensive household cultivation assume that the household would be rationally planned so that buildings and trees are placed to allow some areas to receive enough sunlight for gardening. In fact, a great number of household compounds are haphazardly laid out so that no sunny areas remain. It is unrealistic to assume that the household can be rearranged to permit such cultivation. In these cases, some of the common components of a 'maximum production from minimum land strategy', namely poultry, trees and gardens are competing. In some cases they are mutually exclusive, rather than complementary.

Some participants simply had no idea of how to grow winter vegetables. Quite often the seedlings were planted too close together, a common tendency when an inexperienced gardener plants in a small area.

One common curious practice is to tie up young cabbage leaves with strings to form a ball. The village women explained that they thought the leaves of the cabbage had to be bound up for the cabbage to form a head. Other women damaged their young seedlings by planting them with too much fresh cowdung or leaving them in direct sunlight without shading.

Other production problems were those not easily avoided even by experienced cultivators. Insect damage was sometimes severe. Extension workers advised the use of chemical insecticides if the household had access to them.

Otherwise, more traditional remedies were suggested such as sprinkling fine ash on the affected plants. Another significant problem was theft. Villagers sometimes relied on traditional measures, such as placing an amulet in the field. Extension workers asked all participants to encourage their neighbours to grow vegetables, the logic being that if everyone had a plot of their own, there would be less incentive to steal.

Implementation of the winter vegetable problem also occasionally suffered from CARE's lack of experience. Due to an unusually late monsoon rain and some delays in planting for the production of winter vegetable seedlings, CARE was late in distributing cabbage and cauliflower seedlings. CARE staff had not placed sufficient emphasis on the need to put up fencing right from the time of first transplanting the seedlings. Germination of the 'lal shak' and 'palong shak' seed, bought from local seed sources, was poor in some cases.

Sandra Laumark, *The ADAB News*, 1982.

The Food Factor

It is necessary to approach the improvement of food production with a 'beginner's mind' which is not preoccupied with Western agricultural practice to the extent that the many valuable features in traditional practice cannot be appreciated. Ecologicaly balanced mixed cropping makes very good sense in the tropics. Strengthening and developing what is already being done is likely to be more appropriate than any attempts at 'modernization'. E.F. Schumacher might have been talking about this very issue when he wrote '...there is a plentiful supply of know-how, but it is based on the implicit assumption that what is good for the rich must obviosly be good for the poor...this assumption is wrong'.

Jim McDowell, *Appropriate Technology*, 1983.

The New Age of Organic Farming

A quiet but intriguing scientific revolution is taking place in tropical agriculture; ideas from the canon of organic farming are gaining impetus from the energy crisis and growing environmental stresses — not for romantic or religious reasons, but simply because they promise results at lower costs. The era of cheap oil led commercial farmers to neglect the old, tried organic methods of waste recycling and crop rotation using legumes to fertilize the soil by 'fixing' atmospheric nitrogen. Now the cost of running machines is rising and so are the prices of oil-based inputs such as fertilizers and pesticides, and old organic methods, and some new adaptations of them, are again beginning to look economically attractive. The increasing problems of erosion, loss of soil fertility and shortage of water in many parts of the Third World, are pointing agronomists in the same direction. But the pendulum will not swing back the whole way to organics. The exciting prospect now is of a new synthesis and symbiosis of inorganic and organic approaches.

The quantities of organic material potentially available are vast: the most commonly quoted estimate is that all organic wastes in developing countries contain about 130 million tonnes of the three principal plant nutrients, nitrogen, phosphate and potash (NPK) — eight to ten times the amount of chemical fertilizers used. Very conservative estimates of the amount of human excreta available in India alone (based on 133g of faeces and 1200g of urine per day) suggest they might provide more than 3 million tonnes of nutrients a year.

Recycled organic waste acts as both a fertilizer, to boost yields, and as a soil conditioner; reducing erosion, increasing the water holding capacity of soil and facilitating the uptake of nutrients added in the form of mineral fertilizers. But organic inputs alone are not enough.

Even the Chinese supplement organic fertilizers with minerals (in the ratio of about three mineral to seven organic) which are needed to optimize the balance of nutrients in organic materials, to add P and K or to bring down the ratio of carbon to nitrogen.

Research has now shown beyond any reasonable doubt the benefits of maintaining soil cover in keeping down erosion and increasing water absorption and retention and while few organic gardeners belong to the 'no-digging' school proper, most believe in disturbing the soil as little as possible and many preach the value of leaving non-competing weeds to grow.

Ideas similar to these are proving especially appropriate to the tropics, where rainfall can do a great deal more destructive damage to the soil and where organic matter that is mixed in with soil breaks down very rapidly. Organic matter on the soil surface takes longer to break down, and dissipates energy. This prevents large raindrops from 'impacting' the soil surface which otherwise would prevent subsequent rainfall from penetrating, so that it runs off and carries much of the topsoil away with it. Soil cover also helps to reduce fluctuations in soil temperature and provides good conditions for soil organisms.

Paul Harrison, *New Scientist*, 1982.

Amaranth Rediscovered

Agricultural experts, academicians, scientists and environment specialists have expressed their concern over the predominant cultivation of 15 species of plant, out of a total half million of which about 300 are developed.

This ultimate concentration on a few specific crops has created a disaster in the context of world food supplies. Among the new neglected and under-developed crops that have considerable potential to provide the significant part of the food needs of the future is amaranth. Amaranth has potential to replace wheat or maize in some countries.

It is a broad leafed plant that yields cereal grain — the size of grains of sand. These seeds are rich in nutritional value; they have at least twice as much protein as wheat or maize. They taste like popcorn. Amaranth can be grown in either tropical or temperate regions.

Amaranth is attracting interest in the US and some Third World countries where malnutrition is endemic.

The second crop is the winged bean which has enormous potential. The winged bean has two great advantages. Firstly, all of it is edible, and secondly, it thrives in poor soil including backyards. The winged bean has higher nutritional value than the soya bean. It contains up to 40 per cent protein and mashed up beans make a superb infant weaning food.

Ap-tech Newsletter, 1983.

The Winged Bean — a Nutritious Alternative

The winged bean (*Psyphocarpus tetragonolobus*) is the Indonesian equivalent of the soya bean, and is traditionally cooked as a vegetable. But it has many other valuable properties, which no one exploited until a local technology organization, Yayaban Dian Deba (YDD) stepped in. The beans compare well with soya beans, the protein content of more than 39 per cent is roughly the same as that of soya bean. The plants are easy to grow and, being leguminous, they fix nitrogen and so enrich the soil. Apart from being cooked as vegetables they can, like the soya bean, be used to make a nutritious bean cake ('tempe'), or a curd know as tahu ('tofu'), or a sauce, similar to soy sauce, called 'kepac' (and pronounced kechap). YDD has introduced the winged bean to farmers as an alternative to traditional crops. One of its project areas on Gungung (mount) Kidul, also close to Yogyakarta, now produces 10-15 tonnes of winged beans per month. One year ago YDD set up its food technology department with the express purpose of providing the converted farmers with an outlet for their winged bean crops, and to exploit the bean further. By February 1983 the YDD kepac processing plant was in its trial stage and producing 400 litres of kecap per day.

Omar Sattaur, *New Scientist*, 1983.

American Agriculture: Energy Efficiency versus Labour Efficiency

Today, over 100 million people are starving to death all over the globe. Another 1.5 billion people, nearly one-third of the human race, go to bed malnourished each night. With worldwide population expected to double in the next several decades, demand for increased food production will be greater than ever before in history. American agriculture is already producing 20 per cent of the world's wheat and feed grains and exporting over half of it to countries around the planet. Certainly, looking at the statistics, one would be hard pressed to deny what everyone accepts as gospel: that American agricultural technology is extraordinarily efficient. Yet, the truth is that it's the most inefficient form of farming ever devised by human kind. One farmer with an ox and plow produces a more efficient yield per energy expended than the giant mechanized agrifarms of modern America. Hard to believe, but it's absolutely true.

A simple peasant farmer can usually produce about ten calories of energy for each calorie expended. Now, it's a fact that an Iowa farmer can produce up to 6,000 calories for every calorie of human labour expended, but his apparent efficiency turns out to be a grand illusion when all other energy expended in the process is calculated in.

147

To produce just one can of corn containing 270 calories, the farmer uses up 2,790 calories, much of which is made up of energy used to run the farm machinery and the energy contained in the synthetic fertilizers and pesticides applied to the crop. So for every calorie of energy produced, the American farmer is using up ten calories of energy in the process.

Can Western Agriculture be Sustained?

Quite a long time ago Schumacher and I did a little study on what would happen if total world agricultural production and food processing were based on European and American levels of energy use. It turned out that all known oil reserves would disappear off the face of the earth in 30 years. So far the West has thought in terms of substituting energy and chemicals for people, even though the highest productivity per ha usually comes from small farms where a lot of labour is used. The West has a totally unsustainable form of agriculture, where the amount of fertilizer used has to be continually increased simply to keep production at its present levels.

People have tended to think that the only function of agriculture is to produce cheap food. But it has a lot of secondary functions: to make sure soil remains in good condition, to keep the water table pure, to maintain genetic variety, to ensure the quality of food. We in the West have thought in terms of a battle with nature, but as Schumacher said, if we win that battle, we'll be on the losing side.

If we hook farmers in developing countries into a Western style of agriculture, then they are doomed, because it is not sustainable and the environment can't stand it. What we really need to do is to develop an approach to agriculture that minimizes external inputs into farming and really develop the biology of farming, instead of just throwing chemicals into the soil without any real understanding of what that is doing to soil structure.

The present, when there is an agricultural surplus in Western countries, is an ideal time to experiment with farming systems that may initially be less productive, but will be sustainable. They will lower costs, too — organic farmers I know spend only $25 per ha or so on fertilizers, instead of $250 per ha for conventional farming. At the moment Western farmers' incomes are caught in a cleft stick between rapidly rising costs and lowered prices due to overproduction. Instead of thinking of how to increase production, it would make economic sense now to concentrate on ways of lowering costs.

George McRobie, *Ceres*, 1983.

How Pumps Divide the Peasantry

The oasis of Atar, capital of the Adrar region of Mauritania, lies one day's journey to the North of Nouakchott. The principal source of revenue for the Atar oasis is agriculture.

Post-flood agriculture is practised by the poorer peasants, those who have very little land, whereas the richer ones prefer the certainty of irrigation.

For centuries, irrigation water has been raised from the underground water sheet by means of a very simple system, used throughout the world, and know here as 'chadouf'. A long pole works on a central spindle. At one end, a heavy stone serves as a counterweight; at the other, a wooden rod is fixed at one extremity to the pole by a piece of leather, and at the other to a water container, formerly a goatskin, but now, thanks to modernization, made out of the inner tubes of tyres. The peasant lowers the wooden rod, the container fills up and the counterweight helps him to bring it to the surface. The container then empties into a small storage basin.

The 'chadouf' cannot draw water from a depth greater than the length of the rod, usually about four metres. If the groundwater lies any deeper, it can no longer be reached by the container. The peasant must then haul

up the container himself, which is exhausting, or do without water and abandon his plantation. The trees will die, which means starvation for their owner.

The installation of motor pumps at Atar is recent and has revolutionized agriculture. This machine gives the peasant control over his irrigation water: it draws instantaneously and at any time, in almost unlimited quantities, all the water he needs and, in practice, at any depth, even if another motor pump has to be installed as a relay.

However, whereas the 'chadouf' costs little or nothing to set up, the motor pump is expensive to buy, maintain and use, since it requires oil and fuel. Like all agricultural machinery, it accelerates the change from subsistence to market agriculture, since the farmer must sell more to offset his water costs. But that is not the worst. Since everyone needs water at the same time, the agricultural calendar being the same for all, massive demands are made on the groundwater. Its level is dropping continuously; from 4m twenty years ago, it is now more than 20m deep. The greater the drought, the greater the need for water, and the less the underground sheet is replenished. Palm trees, whose deep-growing roots could find water several metres down, can no longer reach it at such a depth. This is why only irrigated palm trees can survive; the others are dying from drought, and the total of 400,000 palm trees in Atar in 1965 must by now have been reduced by half.

For the time being, in Atar, there is no check on the quantity of water drawn by each cultivator. In these circumstances, the drop in the groundwater level means the elimination of the small cultivators who have no motor pumps.

Is there a solution to this problem? It seems doubtful, since the community at Atar is controlled by the rich cultivators. Is it to their advantage if water is available to everyone, when they have difficulty in selling their own produce? Of course, surveys have been made to look for underground water to the north of the town, but the general phenomenon of desertification that has afflicted the whole Sudano-Sahelian region for several years does not allow for much optimism on this point.

Finally, if the motor pump breaks

down (obviously, while it is being used) it must be repaired or replaced immediately; otherwise the whole harvest is at risk.

There is a repair shop, where the State employs one workman to repair the machines free of charge, on condition that spare parts are provided by the owner — spare parts that can only be bought in Nouakchott. The peasant must go and get them, or have them sent, with the risk of buying the wrong part, at an exorbitant price. The system is misconceived and completely choked up. A visit to the workshop is depressing. In a place adequate for normal repairs, there is a huge pile of motor pumps; no accounting, not even a ledger to record the entry and exit of machines; nothing is done to salvage from motor pumps permanently out of order the parts necessary for the repair of other machines. In practice, this huge immobilization of capital in a poor population means that, once a motor pump breaks down, the owner buys another one, if he has enough money.

Ceres, 1980.

The Bamboo Tube-well

In the Kosi area of Bihar, India, tube-well irrigation was first promoted by government programmes intended to increase agricultural production. In the late 1950s, there was limited initial local response mainly among a few large landowners. However, when the droughts of the mid-1960s coincided with the appearance

of the new hybrid wheat seeds and the Intensive Agriculture Areas Programme, there was a new campaign to sink wells with direct government credit and a 50 per cent subsidy on investment costs. The combination of high food-grain prices, new varieties with greater yield potential and low-cost credit coincided with a substantial increase in tube-well investment. The windfall profits that a few farmers made during 1965-67 were frequently cited as the conclusive demonstration of the potential of tube-well irrigation.

Techniques chosen for public credit-supported programmes for tube-well investment failed to consider local physical or socioeconomic conditions. The official package consisted of 10 or 15cm diameter wells, iron casing with a brass screen, sunk to a depth of about 45m or more. A small number of peripatetic rigs controlled by the Minor Irrigation Directorate sank all the wells. These rigs employed a slow and expensive percussive drilling technique. Transport costs and delays multiplied as each rig had to move long distances by bullock cart to sink perhaps only one or two wells as individuals decided to invest. Wells were to be powered, where possible, with electric pumps connected to the slowly spreading rural electricity network. This involved another government agency, the Bihar State Electricity Board, which apart from problems of co-ordination, was faced with the linking up of a network of scattered pumpsets. Water was to be delivered through a system of concrete or brick channels.

The potential command of a 10cm well on levelled land was expected to be 6 to 8ha but the 5-hp pumpset could irrigate 12ha of irrigated dwarf wheat. Fragmentation of larger holdings as well as the small size of holdings in comparison with the potential capacity of such a tube-well system prevented most farmers from profitably investing in tube-well irrigation or benefiting from the purchase of water. Mortgage requirements for the package costing initially Rs.8,000 also limited credit-financed investment to farmers with at least 3.2ha.

My hypothesis is that the combination of the potential profitability of tube-well-irrigated farming, once high-yielding wheat had been introduced, and the embodiment of tube-well technology in a package that restricted the opportunities for profitable investment induced a process of local innovative activity. The invention of the bamboo tube-well was no chance occurrence or isolated act of inspiration but part of a process of induced technical and institutional innovation. This hypothesis can be substantiated in two ways: from interviews with farmers and contractors during the period in which they were actively engaged in experimentation to reduce tube-well investment costs, and by an investment appraisal of alternative choices of technique that confronted the potential investor with respect to the three major components of a tube-well system, the well, pumpset and delivery system.

Evidence from interviews and time series statistics on the sinking of different sizes and specifications of tube-well (Table 1) indicate that small-scale contractors and farmers began experiments to adapt tube-well technology to local conditions from 1965 onward. A few contractors and farmers with business contacts outside the region found that they could substantially reduce investment costs by importing their own materials, sinking shallower wells and using local contractors to install the wells. By privately installing wells, they also cut out the lengthy delays and other hidden costs of credit-financed investment.

Through experience, contractors found that the very high water table throughout most of the region and the deep deposits of stone-free sandy alluvia made it possible to sink wells to only 30 to 36m and still provide an assured water supply. In these conditions, expensive brass screens brought little advantage and did not necessarily lengthen the life of a well. Without fully realizing this, local cultivators began to install cheap iron screens because they preferred lower investment costs and a higher immediate return to a potentially more durable investment.

Local contractors also discovered that, using the simple 'sludger' drilling method previously developed to sink narrow diameter wells to be powered by handpumps, they could sink 7.5 and 10cm wells in the soft strata down to 45m at lower cost than government rigs. Apart from the blind pipe and an auger, this technique uses only local materials.

Local cultivators found that it was possible to further reduce the cost per ha of an irrigation system by sinking several wells, all of which could be powered by a single mobile diesel pumping set. In this way, again rejecting the offical choice of technique (electric power) they could overcome both the indivisibilities in pumping set investment and adapt the technology to take account of fragmented holdings and uneven land. Due to the existence of a high water table throughout most of the region, in all but the driest summers, pumps could operate at field level and power units could be moved from well to well with comparative ease in contrast to regions of Uttar Pradesh and the Punjab where pumpsets are placed in excavated pits.

Farmers also rejected another official choice of technique in preferring to construct only 'kutcha' earth channels rather than to install cement channels that represented an unprofitable investment where holdings were fragmented, terrain broken and few wells irrigated more than 2 or 3 ha.

The development of the bamboo tube-well was the culmination of these many attempts to reduce the cost of tube-well technology.

Modifications tried included coconut coir wrapped around steel and afterwards bamboo frames. Eventually one succceeded in sinking a well with bamboo casing instead of steel pipe, also using the coir and bamboo screen.

In assessing the impact of these innovations on employment and income distribution patterns in the Kosi region, one ought to take account of the consequence for landless labourers who, by 1971, comprised half of the population of this largely rural area, as well as for large and small farmers.

The techniques of assembly and sinking of bamboo tube-wells as well as the complementary work on land levelling and channel construction largely involved unskilled labour and a minimum of capital equipment. My estimate is that in 1972-73, 300,000 man-days of additional employment were created by the fabrication and sinking of at least 14,000 bamboo wells, and 100,000 man-days through subsequent earthworks. In addition, the maintenance of a stock of 40,000

Table 1
Irrigation Equipment in the Kosi Region of Bihar, India
(Number of units)

	1965/66	1969/70	1970/71	1972/73	1977/78
1. Tube-wells					
State tube-wells	—	—	—	—	193
6'' and 4'' all metal	180	2,300	2,900	3,500	5,211
3'' all metal	20	340	456		
Bamboo	—	330	1,438	19,500	50,187
Total	200	2,900	4,812	23,000	55,591
2. Private pumpsets:					
Diesel powered	336	na	2,693	na	22,087
Electric powered	63	na	435	na	1,148
Total	399	na	3,128	6,589	23,235
3. Ratio of private tube-wells to pumpsets	0.5	*	1.5	3.5	2.2

bamboo wells would generate 150,000 to 200,000 man-days of employment annually, according to whether one assumes an average life expectancy for wells of 4 or 3 years. However, this is considerably less than the additional employment generated by tube-well-irrigated farming which even in 1971-72 was estimated as at least 1.7 million man-days when there were less than 5,000 operational wells in the region. Approximately 30 per cent of incremental net product from more intensive cultivation went to agricultural labour. The low cost of bamboo tube-wells and the development of a market in pumpset services enabled many more farmers to introduce tube-well-irrigated farming profitably. It was the overall labour-using character of the package of innovations associated with tube-well-irrigated farming more than the labour-intensive nature of bamboo tube-well fabrication and sinking techniques that had the greatest impact on employment. However, any assump-

tion that the primary beneficiaries of low-cost technologies will be small farmers appears to rest on an oversimplified analysis.

First, the development of a low-cost well brought at least as much benefit to the larger farmers with their fragmented holdings. Most of those involved in innovative activity were large landholders seeking to find ways of more profitably exploiting the potential of tube-well irrigation. Even their larger plots often included land at different elevations that could not be irrigated from a single well without prohibitively expensive investment in land-levelling and channel construction. Larger farmers first recognized the possibilities of spreading the service of a pumpset over several wells. Among a random sample of 54 tube-well investors surveyed in 1971, there was one farmer with 11 wells and several others had two, three and more bamboo wells. As Table 2 shows, the distribution of the first 1,500 bamboo wells included few small farmers. It

Table 2
Distribution of Tube-well Investment by Holding Size
(Up to May 1971)

Holding size (ha)	Bamboo tube-wells	All-metal wells 3''	All-metal wells 4'' and 6''	Total
		(% of column total)		
Less than 1	—	—	—	—
1-1.9	0.4	4.3	0.6	0.8
2-3.5	11.4	21.9	6.3	9.1
4-7.5	42.1	26.7	24.2	29.3
8 and more	45.7	47.1	68.9	60.8
Total	100.0	100.0	100.0	100.0

was the provision of subsidized credit for bamboo tube-wells in 1972-73 and the development of the pumpset service market that enabled small farmers to sink tube-wells in large numbers. As the lower cost wells could be profitably installed on smaller plots, this also left more spare capacity and the sale of water, mostly to small farmers, also became more widespread. However, dependence on the purchase of water or hire of pumpset services is another reason why the small farmers will be less able than the larger farmers who own their own equipment to exploit fully the potential of tube-well irrigation. The sharing or marketing of pumpset services introduces into the operation of small well systems the problems of organization and distribution that plague larger tube-well, low lift and surface systems; the unit of control is no longer the unit of crop production decision-making. Since the usefulness of irrigation water depends critically on its timing, it is reasonable to assume that owners will always satisfy their own water requirements first. Potential water buyers have to make their own requirements consistent with those of

the seller. Also, they must expect to bear more of the costs of any breakdown of equipment or shrtage of fuel in terms of reduced yield due to untimely supply of water. The expected value of services will be higher for owner-users than for buyers. This analysis is supported by evidence for 1971 showing the those who hired pumpset services irrigated less frequently, and applied less supplementary water in growing high-yielding varieties of wheat.

Experience elsewhere on the operation of co-operatives for pump hire and government-managed deep tube-wells suggests that these alternatives to private sale of services are unlikely to overcome the problem of unequal access to scarce services. When time-specific water requirements of crops such as high-yielding wheat make water a constraining input, then the same more powerful members of the community will be able to ensure that they have first call on available services. There remains the problem of unequal access to other complementary inputs: fertilizer, better seed, pesticides, mechanical draught power for peak period operations. These are all part of the problem of the small farmer who faces multiple constraints in competing with the large and powerful farmers for economic resources. This is why the relaxation of a single investment constraint is not a sufficient condition for a social revolution.

The bamboo tube-well and associated innovations in the Kosi region are only one example of the adaptation of lift irrigation technology to highly specific local environment conditions and farming systems. But most other examples reported from the Indian subcontinent conform to a similar pattern: lift irrigation was first introduced in some government programme, rarely preceded by research and development into what would provide the most cost-effective package. Innovation and adaptation were largely left to farmers and small local contractors more sensitive to the needs of their potential customers.

Edward Clay, *Ceres*, 1980.

When the Hills Came Tumbling Down

Sukhana Lake, the showpiece of Chandigarh, was once 14 metres deep. It is now only 4m deep. Every year since 1958, when it was built at a cost of $1 million, the lake has been collecting sediment from the Shivalik hills. It has lost three-quarters of its storage capacity, with a yearly loss of 3-4 per cent.

In 1970 the government started dredging operations which have already cost nearly $3 million but have failed to improve the situation. Another $7 million have been earmarked for this purpose, while the total cost of the Sukhomajri project stretched over 16-14ha amounts to only $8.7 million.

A man called Mr. Mishra is well on his way to solving the problem permanently by simpler techniques and much less expenditure. When first acquainted with the problem in the early 1970s, Mr. Mishra decided to walk up and along the stream feeding the lake to detect the true cause of such heavy siltation. Soon he reached the top of the watershed, the source of all the trouble, namely Sukhomajri village.

Mr. Mishra found that the devastation was being caused by heavy grazing in the watershed area. Four hundred tonnes of top soil would go tumbling down the river every year from every hectare of the area. It takes 400 years to make one centimetre of top soil. With 6cm of top soil disappearing annually, Mishra calculated that the villages were losing 2,400 years of their ecological history in one monsoon.

Sukhomajri village is situated 35km from Chandigarh and lies at the point where Kansan, the main stream feeding the Sukhana 'Choe', starts. Choe is the local word for gorged river beds which widen every year due to the silt. These choe are filled with rocks and boulders and are mostly dry. During the monsoon they collect rainwater, rapidly causing floods downstream.

Mr. Mishra — now project direc-

155

tor of the Central Soil and Water Conservation Research Centre in Chandirgah — has found that no grant is ever given for catchment protection programmes anywhere in India. The catchment area of the Sukhana Lake covers twelve villages and does not belong to the Indian Government as it was not purchased along with the lake.

Mishra's first contact with the villagers began with a fight. He accused them of damaging the top soil and sedimenting the lake below. The villagers bluntly told him to go away. But Mishra adamantly stayed on and studied the various strategic points where he could implement some of his 'lab-techniques'.

He constructed a few 'brushwood dams', basically piles of firewood and twigs placed horizontally between two wooden pegs to check the erosion from the top of the gully-head. The villagers uprooted the pegs and took the firewood away for burning.

This cold war (which was not always very cold) went on for two years.

One day a villager came running to Mr. Mishra wailing that his land had disappeared in the gorge at the end of the village. This gorge, or gully, would widen with each monsoon like a a monster, gulping away the village land. Mr. Mishra explained to the villagers that their whole village would be devoured by the gorge unless they stopped indscriminate grazing. The villagers' resistance weakened and real work started in 1975. An earthen dam was first built at the base of the gorge. The monsoon arrived and rain water collected in the man-made reservoir. The villagers then urged Mr. Mishra to give them water for irrigating their fields in the village below. He promised to do so on the condition that they stop grazing in the slope above the reservoir. Soon the grazing stopped.

Acacia trees were planted on the slope of the reservoir. The vegetation bloomed and began to hold the soil. Two hundred trenches were dug along the contours of the sloping terrain at the head of the gorge to trap the silt and water coming down. These contour trenches helped to increase the moisture in the soil and support quick growth of trees and grasses. Bhabbar grass (*Eulapiosis binata*) was grown on the ridges of the trenches. This grass grows up to 5-6ft; it holds the top soil and can be cut every month for fodder and making ropes. A series of 'check dams' made of local stones were also constructed to check the sediment from falling into the gorge.

As a result of these measures, the erosion rate has dropped from the earlier flood of 400 tonnes per hectare per year to a trickle of 2.5 tonnes. The extension of the gorge has been checked and agricultural land saved. The dam also provides enough water to irrigate the wheat crop of the village during the dry months. The water for irrigation is conveyed from the dam to the fields through underground pipes so that even the farthest farmer can get the same amount of water as everyone else. The reservoir being at a height much above the village provides water through a simple gravity system.

The villagers have formed a 'water-users association'. Each farmer is charged half a dollar for one hour of water use and each gets an equal quota of water. This way the new reservoir has come to represent a water bank for the villagers, landless or otherwise. Landless farmers can sell their water quota. Because of this water every hectare of the pre-viously eroded land now provides productive employment to the village. 'It is probably for the first time that water is being used as a community asset, distributed equally to every family,' comments Mr. Mishra.

Rajiv Gupta, *Changing Villages*, 1982.

Mechanical Innovations in Africa

In Uganda, the introduction of a biological innovation in the form of a new crop (cotton grown for the market) was followed immediately by the introduction of a mechanical innovation (the ox-plough) to permit land to be planted to cotton without reducing the area of food crops. Within twenty years the plough had become the universal implement for primary tillage among the cattle-keeping people residing on the plains of eastern Uganda.

In southern Zambia, ox-drawn equipment was brought in by European settlers who were growing food crops for sale to the Copper Belt. The Tonga people of the area quicky took up cash cropping also, substituting maize for the traditional cereals and adopting the ox-drawn implements they saw in daily use on European-owned farms.

The common feature in both these cases was a disruption to traditional equilibrium subsistence systems that generated sufficient cash income for farmers to be able to purchase the mechanical equipment needed to consolidate and expand cash cropping. Interestingly, credit purchases played no major role in assisting the adoption of new equipment. This shows that, other things being equal, innovations which do more than cover their own costs will be adopted. This is very clearly demonstrated by evidence from Kenya. Traditionally, Kenyan farmers use mould-board ploughs for tillage and hand hoes for planting and weeding. The equipment innovations introduced there consist of a package of chisel shares, Indian 'desi' ploughs and A-shares mounted on a modified hoe tool-bar capable of performing all three critical operations: tillage, planting and weeding. Where the typical farmer employing two adult workers and cultivating 3ha of land adopts this package without any modification of traditional farming practices, the marginal return to an additional worker amounts to 17 quintals of maize per ha. If it is adopted together with modern bio-chemical inputs and improved farm management practices (optimal schedule of planting and weeding, plant protection, proper dosage and timing of fertilizer applications, etc.), the net return to labour rises to 69 quintals, or four times the level of return obtained when these same equipment innovations are used in conjunction with traditional cultivation methods. However, it is interesting to note that the cost of the equipment innovations (the equivalent of 13 quintals) is more than covered by the additional output generated in a single season even in the absence of improved fertilizer and other farming practices.

One of the best illustrations of how a well-designed and implemented R & D programme can provide the basis for wider dissemination of farm equipment innovations comes from Botswana, where R & D work by the Ministry of Agriculture resulted in the design of an animal-drawn, plough-mounted planter. This equipment increased output which had hitherto been limited because the vast majority of the country's farmers used traditional broadcasting methods. After extensive field trials, prototypes were supplied to the Mochudi Welders' Brigade for local manufacture of the planter. The Ministry then bought this equipment (and a cultivator for weeding) from the Brigade and, under the Government's Arid Land Development Programme, made the machines available to some 500 small farm households who benefited from an attractive subsidy/loan package. Under the same programme donkey draught power was introduced, which, in turn, stimulated the use of an improved harness locally made from used tyres by the Mochudi Farmers' Brigade.

There have been a number of attempts in recent years to introduce scaled-down tractors and other motor-powered tillage equipment which it has been claimed would be suitable for small-scale farming in eastern Africa. Perhaps in the hopes of securing as large a market as possible, however, manufacturers and designers appear to have misspecified their target group by a wide margin. For example, in Malawi, annual running costs of one such tillage device — the Snail — are five times the average annual farming income in one of the more prosperous agricultural parts of the country. Another motorized item of equipment, the 16hp Tinkabi tractor from Swaziland, is being used on a trial basis in both Zambia and Malawi. Although the economic and social impacts of a representative range of such equipment have yet to be evaluated, the Tinkabi could reduce labour inputs in Zambia considerably (relative to oxen-based cultivation). The Tinkabi is being sold for over US $5,000 in Zambia, where the average monthly rural household income is less than US $90. Similarly in Malawi it is far too much machinery for the average size of farm. On the supply side, ploughs and other items were originally imported into Uganda (largely from Europe, although American and Indian equipment was also tried) and distributed through what appears to have been a highly effective network of Asian and European rural traders. Some problems were encountered in obtaining spare parts and in having repairs done, but these were quickly overcome through the initiative of private dealers. Moreover, the distribution of farming implements to the rural areas was facilitated because the Government had invested heavily in basic road and rail transportation and storage infrastructure. An efficient distribution system, linked to a network of after-sale service facilities, therefore contributed to successful diffusion of innovations in the past; and there is no reason to expect that these factors will be any less important in future attempts to promote equipment innovations.

B.H. Kinsey and Iftikar Ahmed, *International Labour Review*, 1983.

Ethiopia's One-Ox Job

Of all the sights that typify the agricultural system of the Ethiopian highlands, that of a farmer using what appears to be his own pair of oxen to cultivate his fields is perhaps the most familiar. But the appearance is deceptive: the chances are that one or even both the oxen the farmer is using do not belong to him. International Livestock Centre of Africa's (ILCA) surveys show that more than half the highland farmers of Ethiopia own fewer than two oxen — at least 30 per cent have only one, while over 20 per cent possess none. These farmers face severe problems in cultivating sufficient land early enough in the cropping season to ensure timely planting and hence good yields of the subsistence crops they depend on for survival.

Farmers with only one ox have several solutions open to them at ploughing time. Their most likely one is to find a 'mekanajo' (an ox to pair with their own ox) by joining forces with a neighbour in a similar position: the two single oxen form a pair, which the farmers then take turns to use on alternate days. 'Mekanajo' arrangements involve no extra cash cost, but they have other serious drawbacks. Firstly, finding a partner whose farm is near enough to make the arrangements workable is often difficult. Secondly, because the two oxen are not accustomed to working as a pair, valuable time is lost in retraining them and working efficiency is lowered. Thirdly, farmers ploughing only on alternate days are frequently unable to sow their crops by the most favourable date.

Another solution is to rent a second ox 'minda' from a farmer who has more than two. Under this arrangement the ox is on permanent loan for the whole of the cropping season, so that the drawbacks of the 'mekanajo' system are avoided. But the cost of a 'minda' arrangement is substantial, averaging 200kg of grains (cereals and pulses) plus the maintenance of the ox for the rental period — a total equivalent to about 50 per cent of the purchase price of an ox.

The farmer with no oxen at all is even more severely disadvantaged. To cover his lower deficit over the medium term he must try to rent one ox on a 'minda' basis and pair it with another under a 'mekanajo' arrangement. If he cannot do so, he must fall back on short-term solutions that are even less satisfactory, such as giving two days of manual labour in exchange for each day a pair of oxen is borrowed, or renting oxen on a daily cash basis. In the latter case the cost is high: US $1.50 per day, or US $2.50 plus food if a handler is hired as well.

ILCA's surveys reveal that the number of oxen owned strongly influences the area cultivated: farmers with two or more oxen plough at least 2.7ha., those with one ox crop 1.0ha., whereas those with no oxen prepare an average of only 0.2ha. In addition, farmers with more oxen grow a greater proportion of cereals, which have a higher market value than pulses but are very labour-intensive and require more draught power for land preparation.

Underlying the problems of these small farmers is the tacit assumption that two oxen are needed for cultivation. This assumption, which has hindered progress for centuries, was first challenged in early 1982.

To find out if a single ox could be used instead of a pair, simple modifications to the traditional 'maresha' harness and yoke were made and tested on the Debre Zeit station. The traditional neck yoke designed for an oxen pair,was replaced by a simple 'inverted V'-type yoke and a swingle-tree, joined by two traces made of nylon ropes. A simple metal skid was attached under the shortened beam to overcome the tendency of the modified 'maresha' to penetrate the ground at an oblique rather than an acute angle. The tests showed that the single ox could cultivate 60 to 70 per cent of the area ploughed by a pair. Cultivation was slightly shallower than with the traditional 'maresha', but the desired depth could be recovered with extra passes. Oxen previously used in pairs needed up to two days retraining before they became accustomed to working as singles.

There are some 6 million draught oxen in Ethiopia, the highest population of work animals in sub-Saharan Africa.

Widespread use of the single ox could dramatically reduce the numbers of oxen as well as the breeding and relacement stock needed to support food crop production, thereby increasing the feed resources available for each working animal. Not only would grazing pressures be reduced, lowering the risk of environmental degradation, but the nutritional status of the remaining oxen would improve, enabling them to plough faster and more efficiently. In addition, more timely cultivation of larger areas of land would lead to increased food production and allow more balanced cereal/pulse rotations to be practised.

The single ox technology has two other major advantages which make the prospects for its uptake look encouraging. Firstly, it does not put subsistence crop yields at risk, and secondly, it requires minimum investment — the new yoke and harness can be made cheaply from local materials, while the modifications to the maresha can be carried out by the village blacksmith.

If uptake is successful and occurs on a large enough scale, the single ox technology will have far-reaching implications for the smallholder farmers of Ethiopia, who are among the world's poorest people.

International Agricultural Development, 1983.

Farm Power for Farm Women

No government programmes have attempted to experiment with and introduce animal drawn equipment for women, despite the fact that the Siscoma 'houe occidentale' package for rice (seeder with rice seed plates, single mould board plough and 3-tine weeder) is available in Senegal, and is designed for donkey traction which women are physically able to handle. This equipment is not readily available on the open market (though many men buy new or second-hand Siscoma implements from Senegalese farmers) and women would either need credit to purchase the equipment themselves or to form societies to buy and share implements.

Action Aid carried out trials on rainfed rice in 1980 using the 'houe occidentale' package drawn by two donkeys and operated by women. The experiment was successful technically and the women were enthusi-

astic about the advantages of saving both time and energy, and being able to cultivate a larger area.

Agricultural equipment and training in the use of draught animals have been available to men in a variety of programmes: ox-plough training, the irrigated rice projects (power tillers and pedal threshers), the Rural Development Project and the cotton project (Siscoma packages of animal-drawn implements). None of these are made available to women growing rainfed and swamp rice (not even the pedal threshers) while women cultivating groundnuts or cotton may sometimes be helped by husbands and other relatives but this is often in return for help on the men's fields (for example, cotton picking) or for a small cash payment. Women have no automatic right to use men's equipment.

Gambia Food Strategy Report, 1981.

2 POST-HARVEST SYSTEMS AND FOOD PROCESSING

Cereal Post-Harvest Systems

Demand in developing countries for improved harvesting and post-harvest systems is increasing, because:

— the introduction of new varieties and adoption of improved production methods lead to higher grain yields;
— the time-lag for harvesting and land preparation for the subsequent crop is reduced through more intensive multi-cropping;
— when more intensive multi-cropping is introduced, harvesting of one

crop takes place during the rainy season. Farmers therefore face unknown problems of handling moist grain in a humid atmosphere.

Furthermore, it appears that sufficient resistance to storage pests and diseases has not always been taken into account in the breeding programmes for high-yield cereals.

The grain harvest at small and medium-farmer's levels in the developing countries is still a very labour-intensive operation. It is a peak season of labour demand — the harvested crop has to be stacked in heaps prior to threshing in order to clear the field for the succeeding cropping season. The piles of unthreshed cereal crop are a major target for rodents and pests and, if wet, large and qualitative losses through mould and rotting can be expected.

Partial mechanization is being introduced in a number of countries. For example, the IRRI-reaper, based on a Chinese design, found a good response in many South-east Asian countries: in the Philippines and other countries of the region it is reported that acceptance of two-wheel tractors — as a possible basis for a reaper attachment — is increasing. A similar design is currently being tested in Egypt. Mini-combines are a technical solution normally beyond the reach of farmers, but could be introduced on a hire basis by entrepreneurs.

Displacement of rural labour by mechanization of cereal harvesting is, of course, a point of dispute. Fears have been expressed that landless labourers, particularly women, may lose an important source of livelihood. It has been argued, on the other hand, that the reduced harvesting time in intensive multiple cropping systems cannot effectively be utilized by the existing manual labour force. Thus, an important policy decision by governments is required, but each case has to be assessed according to the specific circumstances.

In many rural areas traditional methods of threshing can still be seen. For instance, animals used to trample on the unthreshed cereal spread out on a clean flat ground, or wooden sledges or steel disc rollers drawn by bullocks.

The farmers require a threshing machine which, in terms of size and investment costs, lies within their economic reach, is efficient in capacity and in separating grains from straw, in maintenance and in energy costs, can be manufactured and repaired locally, and gives a good and clean cereal product with little breakage.

Technical solutions are in hand in many countries, which fulfil most or all of these conditions. However, it appears that many of these threshing machines are still too big to be individually owned by small farmers. Multi-farm use on a co-operative basis or through entrepreneurs is already widely practised.

Cleaning and drying of grains are necessary for preserving them and limiting losses. Technical solutions for grain cleaning, e.g. winnowers, are available at various scales and technology. Simpler and yet fairly efficient winnowing equipment is also within reach of the smaller farmer. However, problems of drying are much more complex. Cereals can be safely stored if their moisture content is not more than 12 to 13 per cent. Yet, the relative moisture content of the surrounding atmosphere has an influence on the conditions of the grain stock, which has to be carefully watched. Furthermore, the cereal species and the availability of energy for drying have to be taken into account.

Various types of grain driers exist, requiring fuel energy on a scale suitable for co-operatives. For instance, fossil-fuel, electricity, solar cells and harvest residues such as straws, husks and coconut shells are used.

Investment and energy costs are still prohibitive factors for a wide dissemination: scaling down and simplifying driers to bring them within reach of small farmers are, in many cases, uneconomical because of the short time of seasonal use. However, the demand for drying at the farmer's level is considerable.

Cereal quantities for either subsistence or local market do not pass through co-operative or official marketing channels where larger driers can be installed and run economically. It is a common knowledge that farmers accept using commercial drying facilities only if they are sure that their grain, which is the basis of their families' subsistence will be returned to them.

In the Philippines traders and millers procure rice mainly during the season when they can benefit most from sun drying. They tend to avoid investment in driers, thus leaving the problem of excessive moisture in rainfed paddy to the farmer. This problem is aggravated where farmers, applying more intensive production techniques, suddenly have to handle larger quantities of paddy harvested during the rainy season.

It is an essential prerequisite that grains entering the store are clean, dry and not infested by pests and diseases. Minimizing grain losses during storage is a main objective of any post-harvest activity. This is most important where cereals have to be preserved for up to one year because harvest is once a year or where the risk of crop failure is high. In the case of reserve stocks the storage may even be for several years. With increasing surplus production for urban markets, grain storage has the function of levelling out seasonal price fluctuations.

Traditional storage types and technical solutions for the individual farmer have evolved in accordance with natural and socio-economic environment and normally are well adapted. To some extent they broadly satisfy most of the conditions for good grain storage.

In fact, quantitative losses have proved to be low in stores used by farmers for their own subsistence grains. A study in Bangladesh and India has shown that total physical losses are below 7 per cent in traditional stores. Simple and cheap improvements are possible, but these are only necessary if cereal production is rising beyond normal levels. The cement block silo with a capacity of 1.5 tonnes and the 'ferrumbo' type have been accepted by farmers in West Africa. They are particularly suitable for semi-arid and arid zones, and the investment costs are within reach of small family holdings in the savannah. These closed containers allow fumigation.

Open storage systems like the crib is mainly used in warmer humid regions. The free flow of air avoids wastage of the crop which is stored un-threshed.

163

An improved version of the crib has been developed. It can be built at low cost from locally available materials.

Another improvement necessary for both open and closed storage systems is better protection against attack by rodents. Introducing innovation into the traditional grain storage system is a long-drawn process. Systematic extension efforts would be necessary to have an impact on the target of a large number of farmers.

For example: a new storage pest, the Larger Grain Borer, *Prostephanus truncatus*, formerly known only in Central America, appeared a couple of years ago in eastern Africa. The pest mainly attacks maize through the cobs. Controlling this pest will require a change in farmers' storing methods which involves mainly keeping maize on the cob at home or in thatched houses.

Grain storage at village or co-operative level mainly serves as collection points for the marketable cereals. They have a buffer function and farmers can retain some control of 'their' grain until they are convinced of having sufficient supplies in stock to reach the next harvest. This type of store, mainly built as a shed, is promoted in many countries as a means of organizing marketing, improving storage and controlling pests. However, large losses occur since those in charge are most often not acquainted with or properly trained in handling cereals in such conditions. They lack equipment and pesticides to control pests, and many of the prerequisites for a good storage are not met. Fulfilling these conditions does not necessarily increase the establishing costs of the various types of local stores.

In Togo a model shed-store with a capacity of 50 tonnes has been developed. It is built solely with sand, cement and timber and is suitable for a vil-

Traditional storage: room for improvement? (*FAO, A. Tessore*)

164

lage or a co-operative. Experimental results with regard to storage quality and gas treatment have been encouraging.

At this level, grain is normally stored in bags. In larger grain silos the choice between bag or bulk handling, or a mix of both systems is a crucial decision . . .

The choice of storage system depends on a country's transportation infrastructure. Large grain silos are necesary to ensure the cereals supply of the growing urban populations in the developing counries, to create buffer or reserve stocks, especially in regions where the supply situation is notoriously precarious, and to facilitate cereals import or export.

In many developing countries high storage losses are estimated to occur in intermediate stores due to mould, rotting, insects and rodents. These losses are due, mainly, to lack of know-how in grain silo management and also to lack of incentives, for instance, in establishing quality standards and a relative pricing policy.

Peter Muller, *The Courier*, 1983.

Designing manually operated threshers

Wheat is a new crop to Bangladesh and farmers, used to paddy, have found it much harder to thresh. In December, 1979, the Comilla Co-operative Karkhana began development work on a human-powered wheat thresher in response to heavy demand from Bangladesh farmers. The idea was to introduce the principle of a rasp bar threshing drum and concave, used on European machines, as the threshing mechanism most suitable for wheat. The basic operation of the rasp bar threshing mechanism is as follows: wheat is fed into the machine between the rotating drum and the concave; the rubbing and beating actions of the rasp bars attached to the rotating drum and the parallel beater bars on the stationary concave cause the separation of the grain from the straw.

The smallest thresher designed to date (that is, when the Karkhana began development work) was the UK National Institute of Agricultural Engineering (NIAE) model, which had a 2.25hp engine. To design a manually-powered machine, several parameters had to be considered.

The power input. The NIAE thresher had an output of 250kg/hr. A manually-operated paddy thresher requires one man to work the pedal and has an output of 80-100kg/hr. However, as wheat requires more power to thresh than paddy, the following ideas were tested:

— Increasing the power input by designing the pedal so that more than one person could operate it — in the first instance this was for four people, giving a maximum of about 0.5hp;

— Designing the drum so that it also acted as a flywheel. When the wheat was fed in, the drum came under frictional load, the speed slowed down and the stored energy released assisted in threshing. Energy could be restored by pedalling the drum up to operating speed again before feeding in the next bundle;

— Further reducing the power whereby the operator held the bundle of wheat so that only the heads of grain were fed in. This became known as the 'hold on' method.

The drum speed. For a rasp bar threshing drum, the optimum speed of the rasp bar for maximum efficiency is approximately 25 metres per second (m/s) for wheat. The peripheral speed is governed by three factors; the diameter of the drum; the gear ratio; and the pedal speed.

The concave. This largely affects the amount of unthreshed grain. The parameters are: the length/angle of arc; number and position of beater bars; spacing of wire rods; clearance from the drum; and method of feeding, that is, holding on to the wheat bundle so that only the heads of grain are fed in, or passing the whole bundle through.

Three prototype machines were built and tested. For the first prototype, the Karkhana decided to use as many parts of existing paddy threshers as possible. Tests on rather poor-yielding 'Sonalika' wheat (200-600kg/acre, 15-16 per cent moisture content (m.c.)) showed that output varied from 6.8 to 27.2kg/hr and 10-38 per cent unthreshed grain. There were two to three people pedalling the machine and two feeding and passing the wheat, giving a maximum output of 6–7kg/man-hour. When tested with high-yielding 'Pyjam' paddy (1930kg/acre, 18 per cent m.c.) the output was increased to 60kg/hr using a total of three operators. In both cases, it took about 25–30 hr to thresh one acre. Thus, a minimum of three operators was required to thresh the wheat, whereas only one pedal operator is required to thresh paddy; however, one could expect a higher output with a better crop of wheat. There were also problems with heads breaking off and passing through unthreshed.

The Karkhana carried out modifications to the concave, increasing the number of beater bars to seven, which improved performance. Also, the feed entrance was enlarged. General observations showed that the drum speed was too low and stalling occurred if it was allowed to fall below 250rpm. However, the pedal speed was too high — 110 strokes/min at the desired drum speed.

There was great pressure from local farmers and businessmen who wanted to buy the threshing machine. When initial tests looked favourable, the Karkhana manufactured and sold eleven machines, which definitely proved the existence of a market for a manually-operated machine which was simple to understand and operate and within the price range of a middle-class farmer owning two to three acres in Bangladesh. The sale of these machines also gave the Karkhana a good opportunity to observe the machine in use in the field and to look out for any defects.

The main objection from users was that the thresher's output was not high enough and too many operators were required: output per man-hour was not competitive with traditional threshing by beating at 20kg/man-hour. Thus, the Karkhana was greatly encouraged by the market demand but saw a need to further improve the machine's performance.

In the following year, a completely new machine was designed with the focus on trying to obtain the ideal drum speed. This machine was difficult to pedal and took a long time to build up speed. The drum stalled very quickly under load because not enough force could be exerted at the pedal. The machine was only tested for a short time with paddy, at which point further tests were abandoned as its performance was worse than that of the first prototype.

After two years of development work, the Comilla Co-operative Karkhana had to conclude that human power is not sufficient to operate a rasp bar threshing mechanism for wheat. The output is neither economical nor competitive with traditional threshing.

Finally, the Karkhana fitted a small lhp electric motor to the first prototype. The wheat threshed was 'Sonalika' variety, approximately 600kg/acre and 14 per cent m.c. The drum was run at 900rpm (20.3m/s) and had an output of approximately 100kg/hr with two operators feeding alternately by the 'hold on' method. The motor was not overloaded. However, although a small electrically-powered machine would be very useful to farmers and could still be within the price range of middle-class farmers, its greatest limitation would be the availability of electricity in rural areas. Also, there are few small engines in Bangladesh and where they do exist they are both expensive to purchase and difficult to maintain.

A.R. Bose and J.A. Infield, *Appropriate Technology*, 1982.

The Carnage of the Green Revolution

The use of high-yielding wheat varieties is centred in the wealthier northern Indian states of Punjab, Haryana and Uttar Pradesh, where there is a shortage of agricultural labour. This shortage has attracted workers from poorer states such as Bihar, in India's north-east.

It has also opened a vast market for farm machinery, especially mechanical threshers to take over the labour-intensive job of separating the grain from the straw or husk. Today, there are thought to be some 8 million threshing machines in India, from the hand-operated to the latest mechanized models. According to a study by the Agricultural Ministry, another 50,000 are being added to that total yearly.

The increasing numbers of threshers have led to an increasing number of accidents involving these machines, up from only 500 in 1975 to 5,000 in 1980. During the past twelve years, about 10,000 farm workers have been incapacitated while threshing. During the most recent wheat harvest, some 1,000 labourers were maimed, compared to 900 last year.

In England recently a farm worker's arm was cut off by a baling machine. He carried his arm into a hospital and doctors sewed it back on. The state paid the bill. But India and other developing countries lack both the rural health system and the insurance programme to cope with such accidents.

Indian national and state laws require farmers to insure their workers against such accidents. But this is rarely done. The Biharis are largely illiterate, and farm owners can usually buy off a maimed worker with a few rupees 'compensation'.

Migrant workers have little political power.

Many workers are permanently crippled. Gaya Lala was once considered the most able man in his village in Bihar. Last year, a thresher took off his left arm. But he was lucky to survive; many of the victims die from loss of blood from wounds that would not otherwise have been fatal.

The demand for threshers has led village blacksmiths and small town artisans in India's wheat bowl to produce their own backyard varieties, which find a ready market. These substandard machines are responsible for 50 per cent of all thresher accidents, according to a study by Haryana Agricultural University. Few of these machines have the 90cm (35in) feeding system designed to take wheat into the machine and

Inadequately made machines have maimed thousands of farm workers.
(*Mark Edwards, Earthscan*)

keep hands and arms out. A study by the Punjab Agricultural Department found that 95 per cent of all thresher accidents happen while crops are being fed into the machines. By leaving off protective gear, the manufacturers save sheet metal. So the covers of the feeding areas on the backstreet machines are usually inadequate.

Human factors cause their share of the carnage. A recent field study by Punjab University found that fatigue was responsible for about 40 per cent of farm accidents. Landowners hire the labourers for only a short harvest season and are anxious to get the work over before the rains begin. The workers want to earn as much as they can, so are eager to work around the clock. To help them work, the landowners often supply them with drugs such as hemp or with alcohol. Poor lighting was cited as the cause of 9 per cent of all mishaps. Labourers are often required to work through the night to get the harvest in.

Alarmed at the rising number of Green Revolution casualties, the Agricultural Ministry has brought out a series of stringent laws aimed at banning the substandard threshers. But the badly built machines are in great demand and are likely to remain so, taking their toll among workers as they process India's harvests.

Radhakrishna Rao, *Earthscan Bulletin*, 1983.

The Flour Didn't Taste Right

In the arid regions of Senegal, where because of the soil and lack of rain, millet and sorghum are the only cereal crops that can be grown, women must pound grain twice a day. First they must separate the grain from the stalk, then they must mill the grain.

Here in the village of Morry Laye not far from the town of Kebemer, most of the women no longer have to perform these exhausting and arduous tasks. Nine months ago, after a harvest that was slightly better than that of previous years, the villagers decided to buy a millet mill. Of course, Morry Laye is not the only village with a mill, but this is a very special one, made (apart from the motor that powers it) by a village craftsman. The locally-manufactured millet mill is an improvement over the imported mills installed elsewhere, according to the Morry Laye villagers.

Traditionally, before the millet or sorghum is ground, it is either washed or moistened to trigger fermentation of the grain. This gives food a better taste, especially in the preparation of lakh (millet porridge) or tiere (couscous).

'You see, imported mills are different from ours' says one of the women in front of the hut housing the millet mill. 'You can't use wet grain in the imported mills. You must always use dry grain and this doesn't give the same taste. Even my husband complained about it — he doesn't like the imported mills. Some women even

169

have to resort to old-fashioned pounding methods because their husbands won't accept the new taste.'

Apart from the taste, there are plenty of other problems with the imported mills. For instance, the sifters may be too small and they may clog, with the result that it takes a tremendous amount of time to mill the grain, and fuel consumption is considerably increased.

'With our mill we have no problems. We use moistened millet and the flour comes out very white and fine. This did not happen with the imported mills, which gave us a kind of paste.' The women gives her calabash full of millet to the miller, who weighs it: three kilos. He starts up the motor and throws the millet into the mill. In less than three minutes, it is completely milled. A beautiful flour — fine and white — pours out.

When we ask 'How long did it take to pound three kilos of millet in the past?' the woman smiles and says: 'With my two daughters helping me, it used to take me two to three hours a day to pound our grain.' Now it takes three minutes. In addition to the time saved, there are many other advantages.

With the traditional system —

pounding, winnowing, washing and pounding again — as much as a third of the grain was often lost. Also, the grain was first separated from the stalk and the latter, although of great nutritional value, was fed to the livestock. With the mill, the stalk is finely milled and eaten with the flour. Local mills can also be used equally well for millet and sorghum and for peanuts and cassava. Different-sized sifters can also produce flour suitable for making couscous. The price of the mill, which costs between US $1,500 and US $2,000, includes installation, the miller's training, and the maintenance of the equipment for twelve months. This is less than the cost of an imported mill, and far better adapted to the villagers' real needs, particularly because there are no husbands complaining about the odd taste of the milled products.

A local craftsman has a vested interest in making good mills and training good millers: it is good publicity for him.

The first training course for rural craftsmen in the manufacture of millet mills was held in December 1982. The course was designed to standardize mill manufacture and to ensure that mill parts, screens, roots and blades are interchangeable.

Norbert Engel, *UNICEF News*, 1983.

New Presses for Old!

Everywhere one looks these days there are new devices being introduced into development projects: hand-crank grain mills, water hauling tools, pumps, oil presses, etc. Many

of these devices are indeed useful in many areas.

But ideas like these are certainly not efficient or culturally acceptable in all cases. Much time, money, and credibility can be lost when these devices are introduced on the assumption that there is a universal intermediate level of technology.

For instance, in Upper Volta I tried a model oil press developed by a local organization to test if it was any better than the traditional methods.

I took it to Dori and worked with a group of Mossi women to test its efficiency. First they showed me their traditional method of making peanut oil, and then we tried the new oil press.

The Mossi women produce oil in large quantities. Thirty kilograms of peanuts are needed to produce 7.5 litres of oil. Small quantity production was inefficient because of the time and effort involved.

Culturally, making the oil seems very much like a social event, with three to five women working together. The children play and share in making and eating 'Kuli Kuli' peanut cookies.

The preparation process is the same for both the traditional and the new methods. First the nuts are shelled with a decorticator. Then they are grilled over a fire. The skins of the grilled nuts are then removed by using a wood block and a mat made from millet stalks.

The next step with the traditional method is to grind the nuts into a paste at the local grain mill. Boiling water is then stirred into the paste. This requires little effort, although it does require fuel. The oil rises to the surface and is removed as the mixture is stirred.

The grain mill is not needed when using the oil press. The peanuts are instead chopped fine or pounded in the mortar. This takes much longer than taking the nuts to the mill, although the small mill fee is not paid.

I calculated that a half-litre of oil could be produced in half an hour with the traditional method. Two hours were needed to get the same amount from the oil press.

The particular oil press that we used had several problems, some of which could be solved. The small rounded plate that fits into the cylinder to press the nuts kept getting stuck at an angle inside the cylinder. This was because of uneven pressure caused either by uneven chopping of the nuts or by the fact that they were inadequately pounded.

The press is only large enough to take about two kilograms of nuts at a time. This does not produce enough oil to make it worth the time and effort.

The oil press is also made from imported materials by a skilled blacksmith in the capital city. It costs US$150, and is very heavy, hard to clean, and totally unfamiliar to the women.

The wooden mortar used in the traditional method is inexpensive, made from local materials by local craftsmen, easy to clean, and a familiar part of every household.

The most important thing is that the traditional method is actually faster and easier than the oil press.

The actual skimming of the oil is one of the easier steps of the whole production process. It would have made a lot more sense to develop a machine that removes the skin from the grilled nut. This is a slow, labour-intensive process that is still done with a mat and wooden board.

Another thing to consider is the by-products of oil production. Women make a bigger profit with the traditional method by selling the Kuli Kuli peanut cookies than they do from selling the oil itself.

The by-product from the oil press is a hard, cake-like mass of chopped nuts that can be used as animal feed, or pounded into powder for cooking sauces. The market saleability of this product has not been determined.

The oil press could be a viable piece of intermediate technology in another situation. It might be used for sunflower or sesame seed oil production. But it does not seem to be better than the traditional peanut oil production method in the Dori area.

Much better examples of intermediate technologies that reduce time and labour needs are the decorticator and grain mill. The decorticator has been in Dori for a few years, and has proven itself to be functionally efficient and highly acceptable. Shelling the peanuts by hand requires much more time and effort.

A simple decorticator machine is available in Ouagadougou. A small model costs $150, while a larger one costs $300.

The grain mill and motor is a much bigger investment, but it is becoming a familiar business in many villages. Its primary use is to grind millet, the staple crop of the Sahel. A separate part can be purchased to grind peanuts into paste.

If oil production is a viable project for this area, funding should be used to buy peanut decorticators or grain mills. These can be used on a co-operative basis in the villages as part of a credit and training programme.

Susan Corbett, *VITA News*, 1981.

 # *Banana Chips in Papua New Guinea*

The women in Situm village in Morobe Province had heard that a technology centre at the University in the nearest town could assist communities to start income generating projects. At their request, a team from the Technology Centre visited the village to see what could be done.

There was found to be an abundance of bananas in the area, with little use being made of these and much wastage resulting. Thus arose the idea of starting a village industry based on the processing of bananas into a tasty snack food. The Centre's food technologist suggested making the bananas into banana chips which are a popular snack food in many parts of Southeast Asia. This was thought to be a good idea and so training commenced in the method of processing which involves slicing, drying, deep frying and packaging.

The problem arose as to how to accomodate and finance a business based on the new skills. A small loan was given to help purchase/make equipment; the villagers themselves also contributed their labour to make an extension on the community store where production could be carried out. After two years, the women

Bananas are now big business in Papua New Guinea.

have become proficient in banana chip making and the business is flourishing. The men in the village have been very supportive and are actively involved in the community industry too. The outflow of young men to the towns has slowed down because they feel they now have useful work at home: this pleases the women.

There have been many difficulties which could not have been easily overcome without assistance from the Technology Centre. Apart from the obvious problems in learning the process itself, there was also a problem in that the kerosene stove was found to be very expensive. The Technology Centre showed the villagers how to build a 'lorena' wood burning stove to replace this and thus cut costs. The Health Authorities visited the 'factory' and were about to close it down because it did not meet with regulations. The Techno-

logy Centre intervened and taught the women how to comply by incorporating washing/drainage facilities in the 'factory'; taking care not to handle chips after frying; keeping animals and insects off finished foods: refraining from smoking on the premises, etc. Finally, the Technology Centre was needed to help establish a market for the product. First, chips were sold in snack-packs on the University Campus. These were a great success with staff and students. Second, local supermarkets were approached: trial orders sold out so quickly that store-keepers wanted to buy far more than the village industry could produce — even if the community started working at this full-time, which they prefer not to do.

The economics of the project have yet to be worked out. The setting of sale price is arbitrary, although the

173

villagers seem happy enough with the return they receive. There has yet been no repayment on the loan, although this is desired so that a revolving fund can be established to assist other groups establish similar village-based industries elsewhere.

Marilyn Carr, *Blacksmith, Baker, Roofing Sheet Maker*, 1984.

3 LIVESTOCK AND ANIMAL HEALTH CARE

Are Animals Appropriate?

The derivation of an appropriate food production technology for the poor must, of necessity, take into account the inescapable arithmetic of energy transfer in the biosystem and the fact that the efficiency of food production is dependent on the efficiency with which energy is used. The primary food producers are the green plants which use solar energy to convert elements from the atmosphere and the soil into plant tissue. The plants provide the food which grazing animals use to build up their body substance, and from which they derive the energy for life, movement, and warmth. In turn, the carnivorous animals obtain their food and energy needs from the flesh of other animals. At each step in this food chain a massive amount of energy is lost. Plants use only between 1 and 5 per cent of the solar energy available, and animals which eat the plants convert only about 10 per cent of their food into body substance.

It follows, therefore, that in any circumstances where it is necessary for man to obtain the maximum amount of food which the biosystem can provide, the direct use of plant sources, i.e. arable farming, offers the most efficient approach. Any system in which food has to be specially grown to feed animals, or in which animals are grazed on land which might otherwise be used for arable crop production is, in comparison, very inefficient and will result in a 90 per cent loss of productivity. Food production systems for the very poor must, therefore, make use of the most efficient part of the food chain by utilizing plant sources as much as possible.

In situations where climatic conditions restrict possibilities for arable farming, pastoral agriculture in which ruminant animals, such as cattle, sheep and goats, graze on non-arable land and convert plants which man cannot eat into food which he can eat, represents the best alternative. However, the organized production of non-ruminant animals, for instance, pigs or poultry, represents an inappropriate luxury which no poor community can afford. These animals require the same type of food as man, and thus compete directly for available food supplies. In addition, they are inefficient converters. If a family eats 10kg of maize it will obtain some 35,000 calories

174

and 900g of protein. If it feeds 10kg of maize to poultry or pigs it will obtain only 3,500 calories and 90g of protein — hardly a good deal by any standard. The use of pigs or poultry as food producers is only logical on a very small scale when the animals are finding food by foraging or eating household scraps. In such circumstances they can provide a very occasional, but tasty, supplement to the diet.

Jim McDowell, *Appropriate Technology*, 1983.

Small Animals for Small Farms

Major advantages of small animals over large animals for small farms are:

Small animals reach productive or sexual maturity more rapidly and at a younger age than do large animals. For example, sheep and goats can give first birth and commence lactation around eighteen months of age (less in some cases) compared with thirty two or more months for a cow. Some breeds of fowl reach maturity in six to eight weeks.

Large animals are usually selective consumers of feed and forage whereas small animals, at least some species, are less selective and consume whatever is available. For this reason, some animals can be raised on feeds and forages, including kitchen refuse, that would otherwise be wasted.

The space required for handling and feeding small animals is much less than that required for larger animals. This is particularly important in respect of feed production because the space required to produce feed for a few small animals is less than that required for one large one. Thus, it could become feasible for many small farmers, although not all.

Some animal species and even some animals of the same species have more efficient feed conversion ratios than others so that the former require less feed per kilogram of liveweight gain. For example, the ratio for fattening improved breeds of pigs on a balanced ration can be less than 4:1 whereas for cattle it is around 9:1. Thus, in most cases, animals with highly eficient feed conversion ratios would be best suited for a small farm with its limited feed resources.

The productivity of improved breeds of some small animals under good management can be similar or even superior to that of large animals when compared on an animal unit basis.

In general, there are fewer fertility problems in the case of small animals under good management than in the case of large animals.

Finally, the relatively small size and low cost of small animals makes them more freely available to low income households who have neither space nor capital for a large animal. In this respect, small animals might also be utilized by landless labourers.

A major constraint to the successful implementation of a programme of small animals for small farm projects may be the reluctance of the individuals involved to consume a product that is not part of their traditional diet. However, there are experiences that indicate that this problem can be overcome. Ghana has an extensive action campaign for promoting the consumption and rearing of rabbits and it is successful. Milch goat numbers increased rapidly in Europe and Japan during World War II, even though people preferred cow's milk. Rabbit production reached its peak in the United States during the same war and many traditional red-meat eaters found rabbit meat to be a delicious substitute.

It might be possible to develop a programme in some areas in which small farmers could raise small animals on a contract basis with industrial or marketing organizations. Arrangements similar to this is one of the reasons why the poultry industry has grown rapidly in certain countries.

D.L. Huss, *World Animal Review*, 1982.

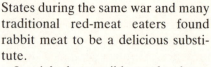

Poultry Production

The promotion of small-scale poultry production units in developing countries has often been suggested not only as a means of improving the diet of the rural population but also as a source of additional income for rural families. During the past two decades many developing countries have initiated schemes to foster poultry production in rural areas by providing such supporting services as the distribution of improved breeding stock among farmers, balanced poultry feed rations and massive vaccination programmes against the major poultry diseases. The performance of such schemes has been uneven, and results have been disappointing. Moreover, some governments, faced with growing domestic demand for poultry products, have been supporting large-scale intensive poultry production systems to the neglect of the smaller rural units.

A background paper recently prepared for a meeting of the FAO Intergovernmental Group on meat makes clear how modern large-scale poultry operations have become a major factor in making poultry production the fastest growing food subsector in developing countries. The study reports that in the higher-income and more urbanized countries of Latin America, North Africa and the Far East intensive commercial operations now account for between 70 and 90 per cent of total poultry production. In some Near East countries it has reached 90 to 100 per cent. Even in lower-income countries in Asia and Africa, large-scale operations account for between one-third and two-thirds of all poultry production.

'These enterprises,' the study notes, 'are heavily dependent on imported inputs, not only at the time of their establishment but, to a considerable extent, also during their operation. Hatchery, automatic feeding, slaughtering, processing and packaging equipment is generally imported and so is the equipment for the compound feedingstuff factories which supply these enterprises with feed.' In some cases, feed itself is imported directly from abroad, as are the required pharmaceutical supplies.

Ironically, although most poultry species originated in the tropics and subtropics, many developing countries depend on imported breeding or commercial stock. During the past five years, the European Community and the United States have more than doubled their exports of highly productive hybrid strains to developing countries. Most of these sales are by a small number of transnational companies.

Does large-scale intensive poultry production pose a threat to poultry development programmes in the smallholding sector? A Professor of Poultry Production at Ein Shams University in Cairo, does not think so. 'Rural poultry production and commercial or intensive poultry production should be able to coexist and not compete with each other,' he told the expert group that met in Rome in 1981. 'The target consumer of commercial or intensive production is mainly the urban population, while the main consumer of rural poultry production is the rural community itself. In fact, small units of rural poultry production may grow into an intensive unit when market conditions are suitable and the producer gets adequate training and experience.' He urged government extension services to seek out small producers who wanted to specialize

in poultry production and help them to form associations and marketing co-operatives.

There are a number of reasons, he says, why government schemes to assist small-scale producers have not had the desired impact. These include a lack of close collaboration between government authorities and farmers, the absence of organized marketing systems for poultry products, inadequate funding and credit and the difficulty of getting feedback information from farmers that would help to improve future programmes.

Gradually, however, a number of countries are beginning to make some progress in promoting poultry development in the smallholder sector. One successful model is the Fayoum co-operative in Egypt which distributes improved native birds to village housewives, supplies them with feed from its own mill and markets surplus eggs and birds. India's Special Poultry Production Programme, under the current Sixth Plan period ending this year increased its coverage among smallholders and landless labourers from 68 to 168 districts. Pakistan has had a co-ordinated national programme for the development of rural poultry since 1975 under which 800,000 birds of improved breeds have been distributed at the village level and more than 60 million birds vaccinated at cost. As poultry-rearing in Pakistan is carried out mainly by women, extension work has been conducted increasingly by female staff.

Yet even where such programmes have been effective, it is often found that government incentives have been benefitting the intensive commercial operations more than tra-

177

ditional producers in the countryside. Of particular benefit to the modern sector have been tax exemption, low interest credit, liberal policies regarding the import of equipment and feed, and subsidies on inputs or production. In Brazil, the second largest producer and leading exporter of poultry meat among developing countries, the industry has been able to borrow at interest rates about two-thirds below the commercial rate and to benefit from export performance premiums. In a number of petroleum-exporting countries subsidies on poultry and egg production have tended to benefit mainly the modern sector.

Ceres, 1983.

Milk and Money in India

Operation Flood is weaving a supply web between rural milk producers and urban consumers in India on a scale that has never before been approached in this vast and highly-populated sub-continent. Reaching the smallest towns and villages the scheme buys as little as a litre or two from the producer to deliver it, chilled and pasteurized, through automatic vending machines in the cities.

The Indian dairy scheme is designed to supply regular quantities of milk at reasonable prices to housewives in the four major metropolitan cities of Delhi, Bombay, Madras and Calcutta. A novel feature is that neither the Indian Government nor private sources were tapped for funding. Instead, an autonomous body, the National Dairy Development Board (NDDB), obtained surplus dairy products (skimmed milk powder and butter oil) free of charge from the European Economic Community and with the funds generated from reconstituting and selling these products set about establishing links between housewives in the cities and small farmers in the surrounding 'milkshed' areas.

Funds of Rs.116 crores (US$130 million) were raised for the first phase of Operation Flood which began in 1970. The second phase, which started in 1980, involves a further Rs.485 crores (US$540 million) generated in a similar manner.

Operation Flood is nation-wide in scope, but it is centred on the small town of Anand in Gujarat. This is also the site of the NDDB headquarters. The town has a history of successful dairy co-operatives spanning almost four decades. Amul, now a brand name for a range of Anand Milk Union Limited milk products, is a household name throughout India. Much of the credit for the present scheme's success must go, however, to its dynamic chairman Verghese Kurien, now in his early sixties, who has pioneered dairy co-operatives in India.

Key to Operation Flood is the organization of thousands of previously scattered and ill-paid rural producers, small farmers with one or two buffaloes, into highly efficient co-operatives. Milk, as Verghese Kurien never tires of pointing out, has two properties which distinguish it from other agricultural produce. It has a short life and there is a big seasonal variation in production (summer is a lean period, because fodder dries up). It is thus tailor-made for collection by co-operatives, such as that in Anand. It allows a farmer's wife (all over India the care of cattle and buffaloes, as well as control over the income they yield, is in the hands of women) to hand over supplies twice a day and receive cash payment at fixed rates and not according to the whim of the local wholesaler.

Anyone who visits the Anand collection centres can see orderly queues of women form soon after dawn and again at dusk. The women tip their containers into a pan, where an employee of the co-operative tests its fat content and notes this down in a register. The very next day, each 'producer' is paid according to the amount and quality of milk she has delivered. With the resources raised by selling milk in the cities, the NDDB has created eighteen 'Anands' in the country; by 1980, 1.3 million farmers had been drawn into the programme, with the help of 11,200 dairy co-operatives. For the first time in India — and probably any developing country of its size — the dairy 'industry' is being modernized and producers are being assured a fair return.

A somewhat startling feature of Operation Flood is that these unique rural organizations don't actually produce food for direct consumption by the farmers. On the contrary, the milk is consumed far away in the cities and in many cases, especially in Harijan or low-caste families, this means that the producers, by selling their milk, are themselves deprived of it. What the technocrats who man the NDDB believe is that milk co-operatives can help transfer incomes from the cities to the countryside. Every evening at collection centres in Gujarat, for instance, it is common to see women come straight out to the vegetable vendor who parks his cart alongside and buy their families' daily needs with the 'milk money'.

Darryl D'Monte, *Development and Communication*, 1983.

Progress — But for Whom?

Because, as the cliche puts it, everything is connected to everything else, even a conventional, modest, solid development project, can have the unforeseen effect of making vulnerable members of society, namely women, worse off than they were before. Any disturbance of the status quo can put them at risk. A good example comes from the 'white revolution' project in India.

On the face of it, the 'white revolution' scheme was inherently a good development. It involved the introduction of dairies in order to improve the production and distribution of pasteurized whole milk to urban areas. It

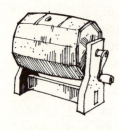

reduced waste, cut down disease and improved the nutrition of poor people in urban areas. Unfortunately, poor women in rural areas ended up worse off than before.

In Gujarat State, women of the poorer castes used to graze buffalo, seeking out marginal areas of grass. They would milk them, make butter and sell the fruits of their labours in a nearby town. They would keep the skimmed milk to feed their own families. Now all this has changed since the dairies handle the cattle, and the dairies are run by men. Thus the meagre but independent income the women used to earn has mostly disappeared. Poor families no longer get the benefits of the buttermilk, which they received free; instead they need cash to buy milk, but have lost their capacity to earn it. This development in Gujarat State is being parallelled in Maharashtra State in Mawat.

Development Forum, June 1980.

The Poor Man's Cow

In the eyes of many people familiar with developing countries, goats, with their vacuum-cleaning eating habits, can and do devastate land and lay waste to forests. At least one outraged person has blamed them for the decline of the Muslim Empire.

This is rubbish. Goats can survive on land that cannot support another species of livestock. Thus, 'goats have come to be associated with poor people, marginal land, pockets of poverty and hence are often found in areas where the last stages of erosion and decay have been reached. Goats then become the culprits, taking the blame for what is actually man's mismanagement of the land.

But attitudes towards the goat are changing, writes Wernick. 'The talk these days is of how the goat can combat hunger and provide, on a daily basis, the essentials of life for the family that cares for it.' Goat milk, for example, is very nutritious and can make a significant contribution toward providing the minimum nutritional requirements for a poor man's family, thus, in effect, becoming the poor man's cow. 'In developing countries, national and international programmes are now aimed at increasing the productivity of goats, improving the local breeds, introducing dual purpose goals where the animal previously had been raised only for meat (or not at all).'

'All these plans are well intentioned and some work,' Wernick writes. An African expert says that 'those that do

Goats are valuable—and manageable. (*Don Allen, UNICEF*)

are the simple ones. Complex and superficially attractive schemes hatched in laboratories and government offices are likely to founder on the facts of life in the tradition-bound societies without an infrastructure of technicians and veterinarians. It is cheaper and more effective to establish production capabilities of native goats in the traditional environment and to base future improvement of herds on these.'

World Development Forum, 1983.

When Camels are Better than Cattle

Over the centuries the camel has slowly been moving west and south from out of the Horn, a living indicator of a dying environment. The camel has now spread to the west of Lake Turkana, to the Nilotic people for whom the lake was named. The Turkana were originally cattle pastoralists when they arrived in Kenya in the eighteenth century. But their large numbers and herds soon caused severe environmental degradation. Their own oral traditions recount a land rich in rain, trees and grass just three generations ago. Today there is a desert. In the nineteenth century they began to raid the Rendille and Gabbra, stealing camels from them. Today the Turkana have large camel herds and without them they would not be able to survive in their increasingly hostile environment.

The same process is taking place further south in Kenya. The Samburu, northern cousins of the Maasai, live south of the Rendille in relatively heavy *Acacia* bush country. They are also blessed with several high rainfall mountain areas which provide good grazing for cattle — once the forests have been burned down, which is happening at an alarming rate. The Samburu have been compressed into about 60 per cent of their early twentieth century range by pressure from surrounding peoples and by land alienation for a national park and private ranches south of them, and their population has grown considerably over the past eighty years. The ancient response of migrating to better pastures is no longer open to them. Increased population necessitates larger herds to feed the people, which when coupled with smaller grazing lands inevitably leads to over-grazing, deforestation and desertification. As a result of a drop in the productivity of the land, which is then expressed in lower milk, meat and blood yields from livestock, the Samburu have recently become interested in acquiring camels. The price is very high, preventing the acquisition of as many camels as the Samburu would want.

The Pokot, who live south of the Turkana and west of the Samburu, are also turning to camels. Those fierce pastoralists have managed to build up quite substantial camel herds from incessant raiding on the Turkana. Camels are now even spreading south of Lake Baringo to the Maasai-speaking Njemps (Il Chamus). Camels cannot move any further south by natural means as the land is owned by farmers and ranchers. The land used to be occupied by the Maasai, but the British colonial government moved the Maasai south and opened the land to European settlement. Most of the land has now been transferred to indigenous Kenyan ownership since independence in 1963, but modern agricultural techniques and a private land tenure system still prevail. If it were not for that barrier of private land and fences in all probability camels would continue their southward migration down the dry Rift Valley to the Maasai and with them into Tanzania and eventually even further south. Kenya marks the most southerly extent of camel expansion in

Africa, yet they would do comparatively well in many parts of southern Africa.

People like the Samburu, Turkana and Pokot have a deep emotional attachment to cattle built up over centuries of interrelations and mutual dependence with them. A cow or bull is not simply a piece of property or a source of food for the Kenyan cattle pastoralist; each animal is like a part of the family, being named and cared for, and it can have an important social or ritual significance. What then is the great attraction of the camel for such cattle-loving people? Why do they desire it? In a trade a good milch camel will fetch between two and five cows and up to thirty sheep and goats — if a milch camel for trade could ever be found. The pastoralists with small camel herds are also constantly complaining that available camels are too few and too expensive. They want to increase their camel herd, but are unable to do so.

There are some excellent reasons why the camel is so sought after. Some the pastoralist realizes, others he does not appear to be conscious of. Indeed, anyone interested in combating desertification should be interested in the unique qualities of the camel, some of which have only recently become appreciated as a result of detailed research within the last decade. The camel is significantly superior to other livestock animals in terms of food production, its effect on the environment, and even in controlling human population growth. The camel should not be thought of as a specialized animal, adapted only to deserts. In fact, the camel is an extremely versatile animal, while it is the cow which is the more specialized in terms of its needs and potential uses.

The average female camel in northern Kenya produces from five to ten times more milk per lactation period than a cow. A camel will lactate for more than a year after giving birth while the cow usually ceases giving milk within nine months or less. During the rainy season when pasturage is good the camel will give an average of about 10 litres of milk a day, a cow will produce less than 5 litres. In the dry season the cow will practically dry up while the camel will continue throughout to give from 3 to 5 litres a day of milk — approximately that which a cow will give at the best of times. Thus, a camel will consistently provide a substantial quantity of milk for human consumption over the course of an entire year, while cow's milk production is relatively small and highly variable. Because of the very low milk producing capacity of the cow during dry periods, and their poor resistance to drought, the cattle pastoralist tries to have the largest herd that he can. There is no thought of culling unproductive animals to take pressure off the pasture; the expectation being that an unproductive animal may survive the next drought and then be used to trade for grain or for some calves to rebuild the herd. With everyone trying to maximize his herd the limited rangeland inevitably suffers.

The feeding habits of cattle and small stock are also much more destructive than those of camels. Cattle and sheep feed almost entirely on grass, as long as it is available. They also eat the green leaves of shrubs and herbs in the understory. The goat will devour almost anything within its reach, including *Acacia* seedlings, which seriously reduces tree reproduction. All three species

183

travel to and from grazing areas and water points in bunched up herds with their hard and sharp hooves kicking up clouds of dust as they scuff the earth. The cumulative results are the stripping of ground cover and the very destructive trampling of that barren land. Erosion gulleys that end up carrying away tons of top soil often start out as livestock trails.

Camels, on the other hand, have a very wide diet and eat the leaves of shrubs, trees and herbs, as well as grass. They do not overgraze any type of vegetation, and they can eat into the upper stories of vegetation that other animals cannot (except for tree climbing goats), thus lessening the pressure on the lower vegetation levels. Camels also disperse much more and travel farther than other livestock types while feeding, a trait which again lessens the effects of vegetation consumption. The dispersed movement pattern of camels also reduces the effects of trampling, although with their soft, flat hoofless feet little damage is caused anyway. In short, camels do not strip and kick up soil from the ground with the result that soil loss is minimized and trees have a much better chance to reproduce.

The camel is also much more efficient than the cow in converting vegetation into milk. Studies in northern Kenya show that a camel can produce one litre of milk for human consumption from about 2kg of vegetation dry matter. To produce an equivalent litre for human use a cow must consume more than 9kg of dry matter. The camel, then, is more than four times as efficient than cattle in converting its food to human food. The implications of this finding for the future of pastoral economies in semi-arid and arid lands cannot be underestimated.

Another environmental plus for the camel is the type of settlement pattern it permits for people. The more dispersed settlements and livestock are, the less the land is affected by tree and bush cutting for firewood and stock enclosures, and by livestock grazing and trampling. Camels are justly famous for their ability to go for long periods without drinking, and they can carry water long distances to settlements for human needs. Camel pastoralists can thus live in areas where there is good pasture but no water. Cattle pastoralists have no such option since their animals have to be watered at least every three days, making it necessary to live within a maximum radius of about 40km from a water source, although 15 to 20km is more common. This tends to concentrate cattle people in certain parts of the range, putting excess pressure on natural resources, while leaving other areas unused. Camel pastoralists can live up to 80km from water, allowing a more even distribution of settlements over the land.

In northern Kenya and most other semi-arid and arid areas rain falls unpredictably in patches over the landscape. It is rain that spurs plants to grow, so the pastoralists must be ready to go where the rain has fallen. Mobility is therefore essential, and it is that perpetual chase for patches of good grazing that makes the pastoralist a nomad. If a settlement stays in one area too long overgrazing results and the people create a surrounding circle of uprooted bushes and grotesque trees with lopped off limbs. Camel pastoralists like the Gabbra will move ten times in a year, but cattle people like the Samburu might stay in the same place for several years. It is a very great effort for cattle pastoralists to move, but camel people can pack everything on the backs of their beasts and be on the move with twenty-four hours notice.

The last environmental advantage of camel pastoralism over that of cattle pastoralism is one of the most important. Cattle pastoralists burn bush, forest and savannah to create grasslands, because cattle depend on grass. Camel pastoralists do not need to burn because camels do very well in a bush environment. Fire has undoubtedly done more to modify the earth's terrestrial habitats than any other single factor, and most fires have been and are anthropogenic. In highland areas where rainfall is high the creation of grassland plains might not be environmentally deleterious.

Daniel Stiles, *The Ecologist*, 1984.

Barefoot Vets

In most developing countries, animal health care is customarily provided by an inadequately-financed state veterinary hierarchy of professionals, from the chief veterinary officer to the field veterinarian supported by a range of para-professionals. No matter what changes occur in the political structure of a country, there is no way in which a group of marginal livestock farmers could exercise direct control over this service in order to ensure the level of clinical veterinary service they require. If the number of private veterinary surgeons were sufficient and the level of resources were adequate, then one solution would be for a group to employ its own vet; but for the foreseeable future, considering the real costs of training veterinary surgeons, it is unlikely that such a level of resources would be generally available in many countries.

An alternative animal health care package that positively discriminates in favour of marginal livestock farmers is required and could have the following characteristics: livestock owners would need to know each animal individually; it would concentrate on preventative rather than curative measures; endemic disease would be reduced by vaccina-

Where there is no vet

(*Basics, 1982*)

185

tion programmes and parasite control; where possible, levels of feeding would be improved; hygiene levels would be upgraded, especially for animals which are housed; improved breeding and selection of animals would be practised; simple first-aid measures would be available for injuries and the commonest diseases; an animal insurance scheme would be available to protect the farmer against loss. To implement such a package, the farmers would require a community-controlled animal health worker, who had received the necessary training in these skills and whose job it would be to transfer the skills to the livestock farmers. If such a package were generally adopted, the role of the veterinary surgeon would change from providing a 'fire-brigade' service to being one of adviser and trainer and providing a referral service. This example also raises an interesting point, similar to many other aspects of agricultural administration, that no matter how completely the community controls the animal health service at a local level, it would be able to do little about epidemics which strike the region. There should, therefore, be a strong demand from the community that epidemic disease be controlled regionally or nationally, a role for which the state veterinary service is perfectly designed.

Patrick Mulvany, *Appropriate Technology*, 1982.

IV

Technology for Development

Health, Water and Sanitation

HEALTH CONDITIONS vary greatly both within and between countries, but throughout the developing world they are substantially inferior to those in affluent countries. While life expectancy has increased in most developing countries since the 1950s, it remains low relative to developed countries and the rate of increase is declining. Low life expectancy reflects very high death rates among children under five years of age. In the poorest regions of low-income countries, half of all children die during the first year of life. For people who survive beyond the age of five, life expectancy is still less than in developed countries and disability, debility and temporary incapacity are often serious problems. It is estimated that one-tenth of the life of the average person in a developing country is seriously disrupted by ill-health.

Improving health is a popular priority in developing countries and health, along with education, has become a major indicator of development. Most countries now have publicly financed systems of health care and programmes of investment in sanitation, water supply, and health education. Government health facilities generally reach to the district or, in a few countries, even to the village level. In total as much as six to ten per cent of gross domestic product (GDP) is spent by the public sector and by private individuals on health care.

Despite the large expenditures on health, and the technical feasibility of dealing with many of the most common health problems, efforts to improve health have had only a modest impact on the vast majority of the population in most developing countries. This is attributable to three major factors. First, health activities have overemphasized sophisticated, hospital-based care, mainly in the urban centres, while neglecting preventive public health programmes and simple primary care provided at conveniently located facilities. Second, even where health facilities have been geographically accessible to the poor, inadequate training of staff, poor supervision, inappropriate services and lack of social acceptability have often compromised the quality of care they offer and limited their usefulness.

Third, the emphasis has been on curative medicine rather than preventive care through health education, and even when health education programmes are available, there are often infrastructural and socio-economic constraints limiting their impact. In many cases, these constraints relate to the lack of water and sanitation facilities near the home, to the special role of women in preventive health care, and to the time and cash constraints that women so often face. For example, as a recent World Bank publication explains: 'it is unreasonable for a health programme to exhort people to use purer water if women without a water supply in or near the home must carry more water ... Advice on better nutrition and health practices is wasted unless women are assured of resources to follow such advice and will not be prevented from doing so by their husbands and families.'[1]

The lack of a sanitary environment is a matter of particular concern. In the early 1980s, it was estimated that over half of the people in the Third World had no access to safe drinking water and three-quarters had no sanitation facilities: and the situation is getting worse. With an estimated eighty per cent of the world's disease linked to unsafe water and poor sanitation, it is obvious that the issues of health, water and sanitation need to be examined and tackled together.

Concern over these issues has brought about a radical change in thinking as to the ways and means of improving the access of the vast majority of people to basic health and sanitation facilities. Obviously, increased resources are essential and the two major international initiatives in this field (namely, WHO's goal of health for all by the year 2000, and the UN Water and Sanitation Decade's goal of water for all by 1990), have attempted to focus the attention of donors on this need.

Several of the general extracts in this chapter refer to the scale of the health, water and sanitation problem; the magnitude of the resources needed to overcome it; the disappointing response to promotional activities aimed at attracting resources; and, as a consequence, the increased need to identify systems, techniques and technologies which allow greater numbers of people to be reached at a lower cost per head. Frequent reference is also made to the need for systems and technologies which the users can participate in and help build or run and maintain themselves.

Do such systems exist? What do they entail and have they been successfully implemented in poor communities of the Third World? Most of the specific extracts in the chapter address these questions by showing there are viable technological packages available which provide practical, low-cost alternatives to the over-sophisticated techniques and technologies imported from the West. Some of these, such as the now famous Chinese health care model with its 'barefoot doctors' (part-time health workers chosen by and responsible to the community) and the various types of improved pit latrines, have been successfully applied on a very widespread basis.

Section 1 of the chapter covers primary health services. It starts with an extract by Susan Rifkin which summarizes the basic principles of the Chinese health care model. Subsequent extracts illustrate different aspects of adaptations of this model in South Asia and in Africa, which have been chosen to reflect the two major themes emerging from the health auxiliary literature.

[1] World Bank, *Recognizing the Invisible Woman in Development*, (Washington, 1979).

188

These are: the advantages and disadvantages of choosing and training traditional midwives or healers as auxiliaries; and the emphasis given to preventive care and low-cost, often home-made cures. As a contrast to these extracts, that by Ivan Illich illustrates the problems of modern medicine with its reliance on modern drugs.

The rural poor often need access to facilities dealing with specific health-related problems. These include simple surgery (e.g. to treat lacerations, abscesses and limb fractures); eye care (particularly cataract removal); and the provision of low-cost aids such as artificial limbs. By comparison with the literature on health auxiliaries, there are very few examples of these specialist schemes. However, the extracts on barefoot surgeons, eye camps and artificial foot making demonstrate the desirability and possibility of providing such facilities to rural communities on a low cost basis.

An important issue covered in the section is that of increasing the emphasis on the use of existing (improved) medical skills and knowledge which rely on locally available materials, rather than modern imported drugs. Extracts have been selected to demonstrate the range of traditional skills, the scope for applying modern scientific principles to improve traditional practices and cures, and the breakthrough being made in treatment of diarrhoeal diseases by the application of locally made oral rehydration mixtures. These contrast with the extract on fluoride levels in toothpaste which demonstrates how vested interests can lead to the promotion of inappropriate (often harmful) health products in developing countries.

Inextricably linked with the health problem is the challenge of providing safe water supplies and sanitation technologies which are capable of manufacture at the local level or, in some cases, by the users themselves. The second section of the chapter has extracts which describe a number of these technologies including hand-pumps, rainwater catchment schemes, ferrocement water storage systems, water filters, pour flush latrines and other types of improved rural and urban sanitation systems.

The availability of such technologies is a necessary condition for meeting the goals of health and water for all by the end of the twentieth century. It is, however, far from being a sufficient condition. Many of the extracts demonstrate that the problems associated with technology for water supply and sanitation are related more to software (organization, management, manpower and education) than to the hardware itself. The introduction of any improved technology is difficult. In the case of water supply and sanitation technologies, the problems are often compounded by the difficulties experienced by (predominantly) male engineers and extension agents trying to work with predominantly female end users.

A major constraint hampering the provision of adequate water supply and sanitation is the lack of awareness of the benefits that come from (and consequently the lack of social acceptance of) safe water points, and — even more so — of sanitary latrine and other arrangements for improving personal hygiene. The extracts reflect the particular socio-cultural difficulties involved in introducing pit latrines in many societies. Information and public education schemes aimed at tackling water and sanitation problems are often inadequate or totally lacking. Once improved systems have been introduced, they often fall into disuse because of the scarcity of qualified human resources and inadequate management and organization techniques, including a failure to capture community interest. It is estimated that an appalling thirty-five to fifty per cent of such systems in developing countries become inoperable five years after installation. Sanitation schemes also have a high failure rate. It is because of these figures that the extracts in this section concentrate more on the problems of acceptance, operation, maintenance and use of the technologies than on the hardware itself. Special note should be made of the role of women in projects of this type.

Some of the extracts included in other chapters have relevance to the health, water and sanitation theme. Readers are referred in particlar to the extracts on the production of low-cost drugs in Chapter VII, and the training of village health workers and village pump mechanics in Chapter X. In addition, the important issue of nutrition, which is closely linked to health and well-being, is referred to in many of the extracts in Chapter III.

1 HEALTH

Barefoot Health Care in China

In the gloom of the profound health crisis which affected all countries in the 1960s and 70s one ray of light was the Chinese experience in health. In the pre-1949 period, China was a country whose mortality figures attested to the depths of human poverty and suffering.

The first years of the Communist revolution drastically changed that picture. By 1956 China had virtually eradicated the most prevalent communicable diseases including smallpox, scarlet fever, diphtheria, and cholera; had built a strong preventive network in both urban and rural areas; and had extended and expanded health care to the majority of its 500 million people. This had all been done without large investments in building new hospitals, in training doctors or in buying sophisticated drugs or medical technologies. These achievements characterized what became known as the Chinese health care model.

The Chinese model was built on four health care principles. The first was the absolute commitment to providing some type of health care to everyone regardless of position, location or ability to pay. Private practice was abolished. Government facilities were consolidated. Services were decentralized. Priority was placed on health care in the rural areas where over eighty per cent of China's people lived. Mobile medical teams of specialists from urban areas were sent to serve and teach in remote areas with the purposes of both distributing health care and acquainting city doctors with the diseases most common to China's rural people.

The second principle was that prevention was to receive priority. Resources were not allocated to support large research institutions, send medical personnel abroad to receive advanced training or build more large, highly sophisticated curative centres. Rather, money, manpower and materials were mobilized to build a strong preventive network in both urban and rural areas. Mass inoculation campaigns were carried out. Anti-epidemic stations were set up and staffed with the equivalent of medical auxiliaries. Health education was emphasized. The thrust was to improve the living environment, housing and nutrition so people once cured did not merely return to the very conditions which caused their disease in the first place. Scarce health resources were not to be poured into a bottomless curative pit but were instead to be positively invested in ensuring that a healthy population could build the new nation.

The third principle was to unite Western and traditional Chinese medicine. The Chinese saw the value of both the 500,000 existing Chinese medical practitioners (as compared to the 20,000 Western trained doctors), and the medicines and treatments they utilized. Chinese medical care was affordable, accessible and acceptable to the majority of Chinese people. Rather than depending on a mass influx of foreign aid to provide Western medical facilities, or putting all scarce resources into the very expensive training of a relatively few Western-type medical personnel, the Chinese opted to use the existing resources in both medical traditions to deliver health care to the people.

The final principle and perhaps the one which most fired the imagination in other countries was that of community participation in health. This idea took several forms. The most wide-spread was the mass campaigns where all people in all production units participated in ridding the country of disease factors such as rats, flies, and bedbugs, sweeping the streets and cleaning them of rubbish, treating VD cases and digging irrigation canals to bury snails which carried schistosomiasis. These campaigns provided vehicles for mass health education and experience in eradication of disease. They also mobilized vast amounts of available manpower to address the tasks of disease control, sanitation improvement and increased agricultural production through irrigation works.

A second form of community participation was the creation of the now famous 'barefoot doctor' who was a part-time health worker chosen by and responsible to the community. The barefoot doctor not only provided an additional manpower resource but also a trusted contact person for health care of the community. The barefoot doctor was not so much an extension of the medical services as a confirmation that common people could act upon their own health problems. The barefoot doctors denied the monopoly of the medical professionals in delivery of health. They attested to the wisdom of laymen and their ability to take and carry out decision which made differences to the standard of living in poor rural communities.

The Chinese health care model was preventive, decentralized, rural-based and labour intensive. It defined health as not only the absence of disease but as a total improvement in the life of the individuals. It understood that health reflected existing social, economic and political structures. Most importantly, it believed all improvements in man were a result of political will. To the Chinese leadership, from Mao Tze-dong to Deng Xiao-ping, politics is always in command.

The impact of the Chinese health care model has been evident. Many of the ideas have been encapsulated in the WHO/UNICEF concept of Primary Health Care (PHC). The PHC document prepared for the Alma Ata Conference recognizes health as a reflection of socio-economic-political systems, highlights the need for community participation, urges the use of indigenous resources and commands respect for national self-reliance. PHC is seen as the way in which 'health for all by the year 2000' is to be ensured. It explicitly recognizes political will and community participation as necessary to meet this goal.

Susan Rifkin, *UNICEF News*, 1982.

 # Prevention is Better than Cure

The Gonoshasthaya Kendra project was set up near Dhaka in 1972 by Dr Zafrullah Chowdhury and a group of medical colleagues who had worked together treating casualties and refugees during the country's struggle for independence in 1971. Their first objective was to set up community-based health services for the 200,000 people living in the area.

From the outset the emphasis was on prevention. The team at Gonoshasthaya Kendra were committed to stopping their centre from becoming nothing more than a 'community disease centre'. They became increasingly aware that health care alone could do little to improve health without an attack on poverty. So the scope of the project was extended to include schemes for agricultural credit, literacy and vocational training. Both the wide range of community development work at Gonoshasthaya Kendra and the project's bold new initiative in establishing a modern drug factory are important because, by themselves, attempts to improve the supply of essential drugs could have little impact on health.

Health promotion is carried out by teams of paramedics who are given a year's training, partly at the base health centre and partly out in the villages. There are now over sixty paramedics who divide their work between the villages, four sub-centres and the main centre. The paramedics are able to handle the majority of common illness including diarrhoea and dysentery, scabies, upper respiratory tract infections, night blindness, worms, anaemia and 'body pain', which is mostly backache.

Difficult cases are referred to the four doctors at the main centre, who supervize the paramedics' work and carry out surgery. But most of the out-patients who come to the centre are seen by the paramedics who also carry out some straightforward operations. For instance, 85 per cent of tubectomies are performed by the paramedics, with a lower complication rate than the doctors. In the villages the paramedics make house-to-house visits to encourage disease prevention and keep an eye on the health of mothers and young children who are most at risk. Children are vaccinated free of charge and women of child-bearing age are immunized against tetanus. The paramedics carry vaccines in thermos flasks, in addition to a small number of basic drugs.

Drugs are only used when they are essential. Prevention comes first. For example, to help prevent diarrhoeal disease, women are encouraged to use tube-well water for cooking and drinking. They are also shown that instead of buying expensive anti-diarrhoeals, they can make a home-made rehydration solution, 'lobon-gur' from lobon (salt) and gur (molasses) — ingredients which are easy to get hold of locally. Rather than routinely handing out vitamin A capsules, the paramedics try to convince mothers to include plenty of green vegetables in the family diet. If medicines are necessary, the paramedic will wait while parents give children their medicine, to make sure they understand the correct dose ...

The traditional village midwives, the dais, have been incorporated into the health team as far as possible. They are trained to use more hygienic methods for deliveries, and are also involved in family planning.

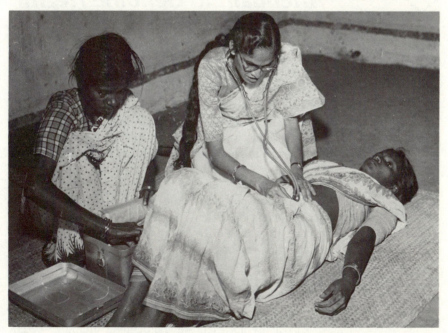

Village birth attendants bring, and save, life. (*UNICEF*)

The project can claim to have made an impact on health. For example, there has been a definite fall in the incidence of severe dehydration, and of skin diseases such as scabies. The birth rate in the project area is now about a third less than that for the rest of country, and whereas the infant mortality rate in the district is now about 120 deaths for every 1,000 babies born, the national rate is 140 deaths and higher still in some rural areas.

But the Gonoshasthaya Kendra team are the first to acknowledge that their health activities can only be seen as a qualified success. They have had difficulty encouraging active community participation in village health, so outsiders still play a key role as health promoters in the villages. Originally, health volunteers were recruited from the villages, but they showed little commitment and were replaced by the full-time paramedics. Many of the paramedics are young unmarried women from outside the area. As unmarried women in a traditional Muslim society, they cannot live alone so they have to be based at the centre, not in the villages. The fact that they are outsiders also means that many move on to live and work in other areas, leading to a fairly high drop-out rate.

D. Melrose, *Bitter Pills: Medicine and the Third World Poor*, 1982.

From Traditional Birth Attendant to Genuine Health Auxiliary

Clearly it is no longer the role of the traditional birth attendant to catch and cast out 'evil spirits'; she is being turned into a genuine health auxiliary. And an auxiliary who is all the more effective because . . . she has something that the midwife hasn't: the ability to offer deeply human services in the most remote places. The fact that she belongs to the village makes her a member of the family.

Because of the underdeveloped health facilities in African countries, because of the way their peoples cling to ancient beliefs and rituals, and because of the transformation the traditional birth attendants are undergoing, these practitioners — far from being doomed to extinction — are in the medium-term at any rate developing into a force that cannot be ignored and must be reckoned with.

Nevertheless, despite their numbers and their social and economic importance, the traditional birth attendants of Africa occupy a varied and ambiguous situation: extensively used in one place, illegal in another. While a dozen

African countries have programmes for improving the services offered by the birth attendants, no great effort has been made to give them any legal status.
World Health, 1983.

Is Medicine Healthy?

The year 1913 marks a watershed in the history of modern medicine. Around that year a patient began to have more than a fifty-fifty chance that a graduate of a medical school would provide him with a specifically effective treatment (if, of course, he was suffering from one of the standard diseases recognized by the medical science of the time). Many humans and herb doctors familiar with local diseases and remedies and trusted by their clients had always had equal or better results.

Since then medicine has gone on to define what constitutes disease and its treatment. The Westernized public learned to demand effective medical practice as defined by the progress of medical science. For the first time in history doctors could measure their efficiency against scales which they themselves had devised. This progress was due to a new persepective of the origins of some ancient scourges; water could be purified and infant mortality lowered; rat control could disarm the plague. treponemas could be made visible under the microscope and Salvarson could eliminate them with statistically defined risks of poisoning the patient; syphilis could be avoided, or recognized and cured by rather simple procedures; diabetes could be diagnosed and self-treatment with insulin could prolong the life of the patients. Paradoxically, the simpler the tools became, the more the medical profession insisted on a monopoly of their application, the longer became the training demanded before a medicine man was initiated into the legitimate use of the simplest tool, and the entire population felt dependent on the doctor. Hygiene turned from being a virtue into a professionally organized ritual at the altar of science.

Infant mortality was lowered, common forms of infection were prevented or treated, some forms of crisis intervention became quite effective. The spectacular decline in mortality and morbidity was due to changes in sanitation, agriculture, marketing, and general attitudes towards life. But though these changes were sometimes influenced by the attention that engineers paid to new facts discovered by medical science, they could only occasionally be ascribed to the intervention of doctors.

Indirectly, industrialization profited from the new effectiveness attributed to medicine; work attendance was raised, and with it the claim to efficiency on the job. The destructiveness of new tools was hidden from public view by new techniques providing spectacular treatments for those who fell victims to industrial violence such as the speed of cars, tension on the job, and poisons in the environment.

The sickening side effects of modern medicine became obvious after Word War II, but doctors needed time to diagnose drug-resistant microbes or genetic damage caused by natal X-rays as new epidemics. The claim made by George Bernard Shaw a generation earlier, that doctors had ceased to be healers and were assuming control over the patient's entire life, could still be regarded as a caricature. Only in the mid-fifties did it become evident that medicine had passed a second watershed and had itself created new kinds of disease.

Foremost among iatrogenic (doctor-induced) diseases was the pretence of doctors that they provided their clients with superior health. First, social planners and doctors became its victims. Soon this epidemic aberration spread to society at large. Then, during the last fifteen years, professional medicine became a major threat to health. Huge amounts of money were spent to stem immeasurable damage caused by medical treatments. The cost of healing was dwarfed by the cost of extending sick life; more people survived longer months with their lives hanging on a plastic tube, imprisoned in iron lungs, or hooked on to kidney machines. New sickness is defined and institutionalized; the cost of enabling people to survive in unhealthy cities and in sickening jobs sky-rocketed. The influence of the medical profession was extended over an increasing range of everyday occurrences in every man's life.

The exclusion of mothers, aunts, and other nonprofessionals from the care of their pregnant, abnormal, hurt, sick, or dying relatives and friends resulted in new demands for medical serivces at a much faster rate than the medical estiblishment could deliver. As the value of services rose, it became almost impossible for people to care. Simultaneously, more conditions were defined as needing treatment be creating new specializations or paraprofessions to keep the tools under the control of the guild.

At the time of the second watershed, preservation of the sick life of medically dependent people in an unhealthy environment became the principal business of the medical profession. Costly prevention and costly treatment became increasingly the privilege of those individuals who through previous consumption of medical services had established a claim to more of it. Access to specialists, prestige hospitals, and life-machines goes preferentially to those people who live in large cities, where the cost of basic disease prevention, as of water treatment and pollution control, is already exceptionally high. The higher the per capita cost of prevention, the higher, paradoxically, became the per capita cost of treatment. The prior consumption of costly prevention and treatment establishes a claim of even more extraordinary care. Like the modern school system, hospital-based health care fits the principle that those who have will receive even more and those who have not will be taken for the little that they have. In schooling this means that high consumers of education will get postdoctoral grants, while dropouts learn that they have failed. In medicine the same principle assures that suffering will increase with increased medical care; the rich will be given more treatment for iatrogenic diseases and the poor will just suffer from them.

After this second turning point, the unwanted hygiene by-products of medicine began to affect entire populations rather than just individual men. In rich countries medicine began to sustain the middle-aged until they became decrepit and needed more doctors and increasingly complex medical tools. In poor countries, thanks to modern medicine, a larger percentage of children began to survive into adolesence and more women survived more pregnancies. Populations increased beyond the capacities of their environments and the restraints and efficienies of their cultures to nurture them. Western doctors abused drugs for the treatment of diseases with which native populations had learned to live. As a result they bred new strains of disease with which modern treatment, natural immunity, and traditional culture could not cope. On a world-wide scale, but particularly in the USA, medical care concentrated on breeding a human stock that was fit only for domesticated life within an increasingly more costly, man-made, scientifically controlled environment. One of the main speakers at the 1970 AMA convention exhorted her pediatric colleagues to consider each newborn baby as a patient until the child could be certified as healthy. Hospital-born, formula-fed, antibiotic-stuffed children thus grew into adults who can breathe the air, eat the food, and survive the lifelessness of a modern city, who will breed and raise at almost any cost a generation even more dependent on medicine.

Ivan Illich, *Tools for Conviviality*, 1975.

Barefoot Surgeons

The manner in which Vietnam developed its nationwide surgical service during the war with the USA constitutes the most dramatic success story in village surgery that has yet emerged from anywhere in the world. The havoc created by the incessant bombing forced the Vietnamese to develop an unprecedented, decentralized medical infrastructure that could undertake operations at any place and time within the country. This medical achievement remains unmatched even in countries which have enjoyed peace.

The Vietnamese decentralized surgical skill by teaching surgery to all their medical personnel including specialists in child care, epidemiology and tuberculosis. Almost every doctor was, therefore, turned into a practitioner of general surgery. These doctors could operate inside a trench with planes flying overhead and a man pedalling away on a bicycle to generate electricity for the makeshift operating theatre. With just a few weeks of training, Vietnamese health workers also learned to perform operations to cure entropion, a frequent complication of the eye disease trachoma ...

Many operations can be carried

197

out by village-level 'barefoot surgeons' after suitable training. The doctor who directs one of the world's best rural health care projects near Bangladesh, has taught village women to perform tubectomies — operations to sterilize women. The success rate of these women has consistently proved to be as good as that of post-graduate doctors ...

Surgical costs can be cut dramatically by using simple equipment and techniques wherever possible. An operating table of antique design, which has an overhead shadowless lamp with a household bulb and a sucker made from a motor-car tyre pump, has been used for years in several hospitals, even in the UK.

Different strategies are needed for different types of surgical problems. Problems such as laceration, superficial burns, abscesses, ulcers and certain uncomplicated limb fractures which require surgical care can be handled at the village-level itself. But for more complicated problems like hernias, a common ailment in many parts of the developing world, a three-tier approach may be necessary in which each level of the health service — village, health centre, and district hospital — has a clear role to play. Hernias cannot be treated in a village, but village-level health workers can certainly diagnose them, especially amongst children, and recognize the danger signals before the hernias strangulate. In normal circumstances, hernia cases will have to be referred to district hospitals for treatment or operation. However, in the case of an advanced strangulated hernia, medical auxiliaries at the front line health centre should be able to stabilize the patient by injecting intravenous fluids before a surgeon from the district hospital can arrive or the patient can be carried there. On a rainy day, when no transport is possible, a medical auxiliary should even know how to operate in an emergency, with a reasonable chance of success. Otherwise the patient will simply die.

Rita Mukhopadhyay, *Appropriate Technology*, 1982.

The [book] that brought about an instantaneous and practical transformation in my life was *Unto this Last*. I later translated it into Gujarati, entitling it *Sarvodaya* (The Welfare of all) . . . I believe I discovered some of my deepest convictions reflected in this great book of Ruskin, and that is why it so captured me and made me transform my life . . . The teachings of *Unto this Last* I understood to be: (1) That the good of the individual is contained in the good of all. (2) That a lawyer's work has the same value as the barber's inasmuch as all have the same right of earning their livelihood from their work. (3) That a life of labour i.e., the life of the tiller of the soil and the handicraftsman is the life worth living. — Mahatma Gandhi

A Handful of Light

For three weeks people from near Bombay have been preparing for an Eye Camp. The schoolhouse, scrubbed and smelling of carbolic soap, has become a temporary hospital. Playgrounds are assembly points, classrooms clinics, and the schoolhall an operating theatre. Do-it-yourself experts have improvized a public address system and a bubbling contraption to sterilize the operating instruments.

Members of Lions, Rotary and Church clubs are cooking 2,000 rice meals a day. Every house in the township has hospitably provided bed space. In the courtyard of a temple, extra beds and sleeping mats are arranged incongruously around ancient statues.

Our jeep stops outside the improvized hospital. The doctors and nursing auxiliaries climb out and go inside. They start work at once; preliminary examinations and tests have been performed in advance.

The three Indian eye specialists and eight clinical assistants operate with practised precision learnt in many Eye Camps. The assistants, with rapid skill, do all the preliminaries and administer local anaesthetics. The surgeons' time is saved for the essential tasks of removing the lens of the eye and final stitching.

The principle of a cataract operation is simple. The lens of the eye, like the lens of a camera, collects and focuses light. When the lens becomes opaque from age, disease or injury, sight can usually be restored by removing the 'cataracted' lens and providing a spectacle lens of appropriate magnification as a substitute.

The surgeons move swiftly from one table to another as each patient reaches the critical point in the operation. Pausing only for a brief snack at mid-day, they have, by late afternoon, completed over 150 cataract operations. Most India eye surgeons have experience of Eye Camps and some have personally performed over 100,000 cataract operations during their working lives.

After nightfall, with electric torches, the doctors visit their patients, most of whom are lying bandaged and impassive, talking quietly to their relatives who care for and feed them. This routine continues for 8-12 days until the time comes when the bandages are finally removed and the patient tries on a pair of spectacles.

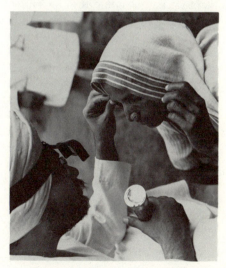

Travelling eye doctors help to prevent, and cure, blindness. (*WHO*)

That seems like the day of miracles. An old man looks out on the hills. A woman sees her children for the first time. A young child whose congenital cataract has just been removed stretches out his fingers to grasp a handful of light.

Last year, in villages and base hospitals throughout India, over 800,000 blind people had sight restored by cataract operations. But at least six million people remain blind in India for lack of cataract operations. Careful evaluation from thousands of operations performed shows that over 90 per cent of patients regain useful sight. ... Each week during the operating season, thousands of blind people will experience the miracle of sight restored by a five minute operation at a cost of less than $10.

Sir John Wilson, *Development Forum*, 1983.

Barefoot Technology — The Jaipur Foot

Medical rehabilitation work has been carried out over the past twenty years at the SMS hospital in Jaipur. The experience gained during this period clearly demonstrated that rehabilitation aids, such as artificial limbs, designed in the West, did not suit the lifestyles of rural people in particular. People in rural India usually walk barefoot and at home they sit cross-legged. The western style limb is thus unsuitable for them and many such patients who had been fitted with artificial limbs returned to crutches.

This became the starting point of a search for new alternatives which led to the evolution of the Jaipur foot. The Jaipur foot not only heralded innovation in design but also led to the use of technology which could be easily managed by the local craftsmen with locally available materials.

The Jaipur Foot is made of rubber and wood, both locally available. In order to make it sturdy and waterproof it is enclosed in rayon cord. The external surface of the foot has a layer of vulcanized rubber moulded in the shape of a normal foot. The first mould was made by a local craftsman who is now engaged in training other craftsmen in artificial foot-making.

The foot has become very popular with rural amputees who can walk over rough terrain or wade through mud and water without any discomfort. The foot can normally be used for three or four years without repair or replacement. It is inexpensive, costing about US$4.50. It takes forty-five minutes to fit the artificial foot from the time of initial measurements being taken.

Since 1975, more than 10,000 amputees have been fitted with the Jaipur Foot.

K.D. Trivedi, *Development Forum*, 1983.

Indigenous Healers

Indigenous medical systems show a wide range of ability in dealing with the problems of disease and illness. For the most part they have not acquired the technical knowledge to compete successfully with cosmopolitan medicine in the treatment of disease. However, historical studies in Africa show that certain groups did develop skills equal to or in advance of their contemporaries in Europe in the late 1800s and early 1900s, most of which was lost or suppressed during the colonial period. Caesarian sections, using banana wine for analgesic and antiseptic and cautery to control bleeding were done in Uganda in the kingdom of Bunyoro-Kitara in the mid to late 1800s. Surgical procedures ranging from trephining to abdominal operations, amputations and the setting of fractures were attempted successfully. Rauwolfia was used in Nigeria for the treatment of psychoses long before it was known in Europe. Potent herbs were known and used although dosages were inexact. Vaccination was widely practised. However the explosive increase in scientific activity and knowledge which began in Europe and America in the early part of this century hardly penetrated the colonized countries. Most present-day indigenous practitioners continue to use the herbal medications and procedures taught to them during extensive periods of apprenticeship. This is more true in the countryside than in the cities where entrepreneurs may practise a syncretic form of 'traditional' medicine blending a mixture of components from varying ethnic cultures with 'modern' trappings such as white coats and stethoscopes. However, even in the cities where there is a mingling of ethnic groups and contacts between people tend to be more superficial, indigenous healers still practise effectively. Even when the curer knows little more about the patient's background than would a busy hospital physician, the combination of personal charisma, herbal pharmacopeia and healing practices which are compatible with the patient's religious views and concepts of disease causation is effective.

S.S. Katz, *Approtech*, 1980

. . . the modern economist . . . is used to increasing the 'standard of living' by the amount of annual comsumption, assuming all the time that a man who consumes more is 'better off' than a man who consumes less. A Buddhist economist would consider this approach excessively irrational: since consumption is merely a means to human well-being, the aim should be to obtain the maximum of well-being with the minimum of consumption. — E. F. Schumacher

Oral Rehydration: a cure not a solution

Dehydration from diarrhoea can kill a child within hours unless the vital salts and liquids lost from the body are replaced. Until relatively recently, the most common method of replacing them was by intravenous transfusion introducing the fluid directly into a vein through the child's skin. This is costly and must be done in a hospital.

In the Bangladeshi refugee camps during the early 1970s, cholera and other diarrhoea-type diseases were a major problem. Obviously there were no facilities for intravenous therapy. So fluids were given by mouth and this method was found to be very effective. Since then, various groups and organizations, especially the Diarrhoeal Disease Control Programme of the World Health Organization (WHO), have been vigorously promoting the use of early oral rehydration therapy as a life-saver for children with diarrhoea.

The treatment sounds simple enough, but putting theory into practice is not so easy.

In many communities traditional attitudes to diarrhoea are barriers to using oral rehydration therapy. For example, diarrhoea is often seen not as an illness but a normal occurrence and, therefore, not something requiring any special kind of attention. It is rarely considered dangerous as a child may suffer many episodes of diarrhoea and become very weak but not actually die.

In many parts of the world, mothers stop giving any food or liquid, including breast milk, with the start of a bout of diarrhoea. Breast milk is not only an easily digested food, but if it is withheld the amount of milk a mother can produce decreases rapidly. Breast milk also protects children against other infections. Mothers often do not feed their children for several days after the diarrhoea has ended.

This is particularly critical for the long term health of the child as there is a dangerous link between malnutrition and diarrhoea. The badly-nourished child is more susceptible to diarrhoea and diarrhoea sometimes impairs the stomach's ability to absorb nutrients for up to two weeks after the bout has ended. So the vicious circle continues.

Mothers' attitudes can only be changed through persuasive health education. What to drink and how to make it are of minor importance until mothers firmly believe that their chldren should drink more when they have diarrhoea. A change of attitude can often be achieved through demonstration. If one group of mothers in the community can be convinced about the value of early oral rehydration therapy and see it working for their children, others will soon be persuaded to use it.

Traditional attitudes do not, of course, only exist in villages. Senior doctors, paediatricians and all those involved in training health staff have to be convinced that oral rehydration therapy can play an important role in the management of diarrhoea and that it has far more widespread application than intravenous therapy or often inappropriate drugs.

Once mothers have accepted that a child should drink when it has diarrhoea, it should be relatively easy to discuss what is the best drink under the circumstances. But again, this is not as simple as it sounds. If mothers can be persuaded to encourage their children to drink as soon as an attack of diarrhoea begins, they can use any fluid that is available in the home to do this. For example, tea, milk, fruit juices or rice water. However, if rehydration therapy does not start quickly enough and the child becomes dehydrated s/he must drink sugar/salt solution. This can also be given at home at first but the child must be taken to a hospital or clinic if there is no response and more fluid is lost. Mothers must be taught how to recognize the signs of dehydration.

WHO and UNICEF, the United Nations children's fund, are distributing sachets of oral rehydration salts (ORS) designed to be dissolved in one litre of clean drinking water. Some countries are producing their own sachets of ORS. One of the major problems with using the sachets is that it is often difficult to distribute them. Frequently, even where they are available, insufficient information is actually given to people about using them. This is unfortunate because if, for example, the ORS is mixed with too little fluid the child will refuse to drink the over-concentrated solution and treatment may be seen as unhelpful and discontinued.

There have been attempts in many places where sachets are not available to teach mothers and health workers how to make up simple sugar/salt solutions using locally available ingredients. The main problem here is to measure out even

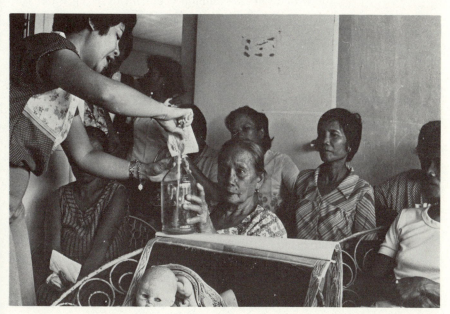

More women are learning how to prepare oral rehydration fluids. (*WHO*)

reasonably accurate amounts of unrefined sugar and salt. A simple formula for oral rehydration solution is to mix one level 5ml teaspoonful of salt plus eight level 5ml teaspoonsful of sugar in one litre of drinking water. A double ended plastic spoon has been developed to use for measuring out sugar with one end and salt with the other. The advantage of this spoon is that it measures out enough salt and sugar for a relatively small amount of water (200ml) so that the oral rehydration solution is more likely to be drunk quickly and not be left standing for hours to become contaminated. The main disadvantage associated with use of the plastic spoon is that mothers are not often given enough information on how to use it.

Sugar and salt can be measured out by hand, but measurement is often inaccurate. At least people do not have to depend on supplies of equipment from outside using this method. As with other measurement methods described, the success of this one depends on vigorous health education so that people understand not only how to prepare the fluid but why and when to give it.

Denise Ayres, *Waterlines*, 1982.

Fluoride — Too Much of a Good Thing?

The Fluoride Toothpaste Debate has raged in Kenya for almost a year. Both sides agree that fluoride levels of about one part per million (ppm) in drinking water can help prevent tooth decay. But in Kenya and other countries along the Rift Valley, the volcanic soil can naturally fluoridate water to concentrations of up to 45 ppm. Such high levels cause dental fluorosis cured only by acid etching or capping, and, over long periods, the spine may become calcified and cripple the victim.

Some 60 per cent of Kenyans suffer from fluorosis — the condition also appears in northern India and Thailand — yet virtually all the toothpaste sold here contains added fluoride. The Kenyan Dental Association (KDA) supports fluoride toothpastes, and is sponsored by one of its biggest distributors in Kenya, Colgate-Palmolive.

The debate became public when a young dental health lecturer at Nairobi University wrote letters to the newspapers. The issue was then taken up by the multinational companies as well as the KDA and Kenyan Medical Association. In November two UK-based multinationals, Boots and Beechams, launched non-fluoride toothpastes in Kenya. Following this, the Kenyan Director of Medical Services announced a ban on fluoride toothpaste advertizements. The government has maintained a low profile because it lacks funds to defluoridate water where levels are too high.

Research in the Philippines has recently suggested that in tropical climates where people drink much more water, a lower fluoride level of 0.5 ppm is more appropriate than internationally recommended levels. Western standards may have little bearing on Third World conditions.

Bryan Pearson,
Earthscan Bulletin,
1983

2 WATER AND SANITATION

Water, Sanitation, Health — For All?

In November 1980 the General Assembly of the United Nations formally declared the International Drinking Water Supply and Sanitation Decade (1981-90). Its official target is 'Clean Water and Adequate Sanitation for All by 1990'. But as the Decade began, over half the peoples of the Third World did not have safe water to drink. Three-quarters had no sanitation at all — not even a smelly bucket latrine.

And the situation is getting worse. About 100 million more Third World people were drinking dirty water in 1981 than in 1975. And 400 million more than in 1975 had no sanitation.

Diarrhoea, often caused by bad water, kills 25 million people in the Third World yearly. Some 16,000 small children die every day.

Eighty per cent of the world's disease is linked to unsafe water and poor sanitation. And illness can often drive a poor family into starvation.

The number of water taps per 1000 persons will become a better indicator of health than the number of hospital beds, prophesied by Dr Halfdan Mahler, WHO Director-General, at the UN General Assembly in November 1980.

To reach the Decade's targets will need $25 million a day in aid. At present, the world spends $240 million on cigarettes and $1400 million on arms every day.

In 1980, aid money spent on water was one-third of that required to reach the Decade's goal. Six months after the

Anil Agarwal et al.,
Water, Sanitation,
Health — for All?
1981.

Decade was launched, no government or aid agency had promised any extra funds.

Will the Decade bring significant improvements, or will it only raise unrealistic expectations?

If water were oil, we would have many, many times the financial commitment expressed to date for this Decade.
UNDP Administrator Bradford Morse

Is the World Water Decade Drying Up?

Are the United Nations, the international aid agencies and the national governments trying to quietly shelve the 'World Water and Sanitation Decade, 1981-90' after only two years?

In 1980 the then UN Secretary-General Kurt Waldheim said the official goal of 'Clean water and adequate sanitation for all by 1990' was 'eminently achievable'. He promised: 'The United Nations system will provide the overall framework, the technical support, the momentum and the promotional activities necessary for the programme's success'. Yet, less than three years later, senior World Health Organization (WHO) officials were saying they knew all the time that achievement of the Decade's goal was not possible.

What has gone wrong? In an official update on the Decade, WHO listed five major constraints: the absence of strong popular and official support; weak institutions; shortage of trained personnel; doubts about

technology and insufficient financial resources. These problems threaten to cripple the Decade. By the end of 1982 only 26 countries had set firm targets for 1990, and many were aiming at less than the 100 per cent target called for by the UN and its agencies.

Lack of money is the key obstacle. In 1980 the World Bank estimated that a global annual investment of $60 billion would be needed through the 1980s to provide every rural home with a latrine and a standpipe or handpump, and every urban home with a tap and sewerage connection. A cheaper option was to aim at only 80 per cent coverage using cheaper technologies, cutting the investment by half to $30 billion. As global spending on water and sanitation projects in 1978 had only been $7 billion, the second option was considered more realistic. So the 100 per cent aim of the Decade was virtually abandoned even before it had begun.

Since then the Decade has not attracted much more money for new projects. In 1981 only $10 billion went into new projects, which, allowing for inflation, meant that about the same number of additional water and sanitation services were provided as in 1978. The developing countries invested $8 billion of that $10 billion. Given their suffering economies, they are unlikely to increase that amount. World Bank loans for water and sanitation plummeted too.

In many countries the institutions which are supposed to be implementing the Decade's work are weak and lack trained staff. WHO has not devoted any more money for training, or recruited a single additional water engineer since the Decade began. Also governments must give greater priority to educating people about the benefits of clean water.

If resources are not forthcoming the Decade's aim of providing safe drinking water and adequate sanitation for all by 1990 and WHO's more ambitious goal of 'Health for All' by the year 2000 will become little more than cruel jokes.

John Madeley, *Modern Africa*, 1983.

Enough for Survival, enough for Health

Two thirds of our bodies' weight and nine tenths of its volume is water. That is why water is essential for life. People can survive for up to two months without food, but die within three days without water.

 A person needs about 5 litres of water each day for cooking and drinking.

 But the World Bank estimates that a further 25-45 litres are needed for each person to stay clean and healthy.

In many places the family's water must be fetched each day by women or children.

The most a woman can carry in comfort is 15 litres, each litre weighing one kilogram.

If she carries only enough water for her family (husband, mother, five children) to survive each day, she would need to fetch about 40 litres.

But to keep them all clean and healthy she would need to fetch 200 litres of water every day.

This is why the amount of water consumed depends largely on whether it has to be carried to the house.

Type of facility	Approximate consumption per person in 10-litre buckets
No tap or standpipe	1.2

Standpipe only	2.5

Single household tap	4.7

Several household taps	16.5

Meanwhile people in the industrialised world use
- 22 litres each time a toilet is flushed
- 150.000 litres to produce a ton of steel
- 750.000 litres to produce a ton of newsprint

A Woman's Lot

In the Sahel region of Africa, women make as many as four trips a day to fetch and carry water over long distances. This drudgery keeps them from performing other domestic functions. The heavy jars that they balance on their heads may eventually cause pelvic disorders and complications at childbirth. In finding out why women's homes were so far from the water source, it was found that men usually make family decisions and thus control how accessible water is to women. As husbands usually decide where the family home will be placed and water for domestic use is not one of their priorities, the distance to the water source can be as much as 12 kilometres.

New Internationalist, Septemeber 1981.

World Development Forum

A Pump for All Countries

Between 70 per cent and 98 per cent of rural households in Bangladesh use hand-pumps and tube-wells as their primary source of drinking water, depending upon their location and the accessibility of other water sources. More than 70 per cent of the rural population now has access to hand-pumps within 700 feet from their place of residence, while about 40 per cent of them live within 400 feet of such a water source. But very few people use tube-wells to meet all their water needs, mainly because of their relatively easy access to other sources of water, such as rivers, canals, ponds and open tanks.

The country now has a population well in excess of 90 million people — about 90 per cent of whom live in rural areas — and is growing at the rate of 2.4 per cent annually. Just to maintain the level of safe drinking water available, at least 25,000 new shallow hand-pumps need to be sunk annually on account of the population increase and the replacement needs of ageing pumps.

Handpumps are not a new phenomenon in this part of the world. What is, however, new and has become essential for the country's fast growing population to meet their drinking water needs, is the government's UNICEF-assisted rural water supply programme launched soon after liberation. Under this programme, begun in May 1973, more than 350,000 shallow hand-pumps and tube-wells have been sunk, bringing the total number in the public sector to more than 500,000. As a result, one hand-pump is now available for about 175 people, compared to about 400 in 1971. The programme aims at making available one tube-well for about 160 people by 1985 and for about 120 by 1990. To achieve this target, more than 50,000 new hand-pumps would have to be installed each year.

The suction-mode hand-pumps are sunk completely by hand, using the sludger method. The section hand-pumps, known as Bangladesh New No.6, were re-designed, marginally changing from the traditional pumps being used in the country, by outside consultants and further modified by UNICEF Bangladesh experts in keeping with local conditions. Its manufacture, transport, installation and maintenance need such minimal expense that it is now being imported by many developing nations, including Pakistan, Nepal, Burma, Vietnam and Papua New Guinea.

The maintenance of these hand-pumps substantially improved after the training of 190,000 tube-well caretakers in the rural areas in a three-year long programme. The training is very simple, lasting for only one day. The caretaker is shown how to dismantle and reassemble the pump, explained the usage of its parts and is also given a pair of very simple tools with which to dismantle, inspect and carry out the necessary repair work. That takes half a day. The other half is devoted to elementary health education; the impact of the availability of safe water on the people's health in preventing illness.

The results of the training have been impressive. Surveys undertaken in 1975 and in 1983 show that the percentage of operable hand-pumps and tube-wells increased from 75 to 90. Experts believe that with greater effort, this percentage can be increased to 95. A five point increase might appear insignificant, but in Bangladesh it can mean the supply of safe water for a few million additional people each day.

Another rising concern is the gradual depletion of the ground water levels due to large-scale extraction of ground water for irrigation. Ground water levels in such areas in the country (about 15 per cent) fall below seven metres, and make shallow hand-pumps inoperable. Force-mode pumps are needed to pump out ground water below such depth and there are only about 6,000 such pumps in operation because of the high cost involved — about US $2,000 apiece, seven times more than conventional shallow hand-pumps. This problem is likely to be aggravated

further in the near future as the government's policy is to increasingly use larger mechanized deep tube-wells to bring agricultural land under irrigation to raise food production. This will gradually, but nevertheless substantially, reduce the ground water levels.

The large scale extraction of ground water has caused dropping of ground water levels surprisingly quickly and when a balance between extraction and recharge has been achieved, as much as 50 per cent of the country may require force-mode pumps. This suggests that Bangladesh will have to replace, in the not too distant future, the suction-mode pumps by about 750,000 force-mode deepset pumps at cost of around US $1.5 billion, plus a recurrent cost of $100 million for replacement units annually. The magnitude of the task can be vizualised when one considers that the annual budget for the rural and sanitation programme is under $20 million!

In addition to the enormous costs

involved for the imported variety of these deepset pumps now in use in Bangladesh, they are also posing complicated and expensive maintenance problems. Almost 50 per cent of these pumps have become inoperable because of lack of maintenance funds.

The search for an alternative, and a cheaper variety at that, complete with simple mechanism — both for operation and maintenance — has been on for quite some time and a new pump has now been evolved in Bangladesh through close co-operation between the World Bank, UNICEF Bangladesh and a development engineer who runs the local Agricultural Workshop and Training School. It is called the 'Tara' Pump, all its parts below ground are plastic and it has a 'direct drive' operation — all of which minimizes maintenance. The pump can be used even by children without difficulty and has a capacity to pump up 50 litres of water a minute. It can be installed as quickly as the Bangladesh New No. 6, using the sludger method. Its routine maintenance can be done without any hand tools by the caretaker with the help of one unskilled assistant. All its parts, including Roboscreen-type pvc filters, are fabricated locally. Its cost is unbelievably low: less than $300, a seventh of that of the imported variety.

The implications of the successful application of this new pump are electrifying. It would be a boon not only for Bangladesh but for all developing countries.

For Bangladesh, using this Tara Pump instead of deploying the equivalent imported model would mean a saving of more than $1.2 billion. In addition, there would be annual savings of nearly $85 million in annual replacement costs. And the Tara could lead to the possibility of providing safe water to virtually all the rural people of this country. The problems that the International Drinking Water Supply and Sanitation Decade had posed before the planners in Bangladesh, and which seemed insurmountable, would not be so overwhelming.

By 1985, the final design of the Tara Pump will have been completed and experts involved in the project believe that up to 20,000 units could be produced annually, rising to about 70,000 units annually by 1990.

Sayed Kamaluddin, *UNICEF News*, 1983.

Clean Water in Kola

Kola is located in one of the drier parts of Kenya, with a mean annual rainfall of approximately 700mm, and like many areas of the Third World it has too little water for most of the year and then too much during the rainy season. Before the village self-help group (The Utooni Development Project) started work, the most reliable source of water was an unprotected spring which was exposed to pollution from the hundreds of people and animals that went there daily for their water.

211

Now life in Kola is changing. The spring has been protected and piped to a storage well where a hand-pump has been installed on a concrete platform so that clean water can be drawn off easily. Rainwater falling on the household roofs is being collected and stored for domestic use in large ferrocement water jars built next to the buildings in each compound. More rainwater is being stored in a catchment dam near the village and on one of the farms a shallow well has been dug and the water used to irrigate a small tree nursery. Alongside these efforts aimed at providing cleaner water in greater quantities and more conveniently, the group has been investigating designs for pit latrines and has built some in the village.

These changes have not come about by chance. The Utooni Development Project has worked hard to find out what can be done and to enlist the assistance of accessible organizations that can provide information and training on the technology options. Its success relies for the most part on the co-operation of all members of the group which in practice means that almost every household in the village contributes labour and money to the construction projects. Confidence in the village's collective ability coupled with a strong sense of needs and an awareness of what can be done about them has resulted in the organization and implementation of a continuing programme of works.

Progress is slow, inhibited by the shortage of labour as much as by general poverty: time spent building means time not spent on agriculture. This is a serious problem, especially in marginal areas. Despite this, there is a high degree of co-operation in the village in that others will look after a worker's crops while he/she is engaged on a project. Where projects are entirely built with communal labour each person works according to a schedule drawn up by the project committee.

The construction priorities are decided by the Utooni Development Project as a whole and the involvement of every member in the decision-making process is a major factor in the success of the projects. Good plans have, in the past, not always been the success anticipated owing to the non-involvement of the 'clients' in the planning and construction phases of the project. By attending training courses and then using the new skills in their own villages, the project members have gained far more than merely the end product. From being a muddy swamp with brown water, the spring in Kola now provides clean water for most families at a point conveniently near the market centre. A small earth dam was built across the shallow valley below the spring source to create a reservoir for water. A pipe conveys the water through the dam wall into a filtration trench which leads the water to a storage well.

The filtration trench, filled with sand and gravel, is 30m long and any seepage helps to irrigate a grove of banana trees planted along both sides. The discharge from the trench first enters a sedimentation chamber where any remaining silt can settle out before the water flows into the storage well. This is lined with concrete rings and the water is drawn off using a simple hand-pump installed on a concrete platform set over the well. This is constructed so that any overflow or spilled water is led away from the supply so that the risk of pollution is reduced. A stock-proof fence surrounds the spring source and the reservoir to prevent straying animals and humans from contaminating the water. Finally, facilities for watering animals, clothes washing and personal bathing are being discussed and if the water supply proves adequate provisions will be made. Although the pump and concrete rings were supplied by the Machakos Integrated Development Project, all the work was carried out by Utooni Development Project members.

Across the valley of a small rocky catchment area near the village a concrete arch dam has been constructed to store rainwater run-off that normally rushes quickly out along the watercourse.

By building the arch barrier at a carefully selected site where rock abutments on either side can take the load, water is held back for use in the dry season. Standing 4m high and about 30m long the dam wall is curved in both planes, across the valley floor and upwards, for added strength to resist the pressure of water trapped behind it. Steel reinforcing rod was set into the bedrock of the floor and sides of the watercourse and rocks and stones were carefully cemented together encasing the steel. A simple spillway for flood water was provided by a lowered section in the dam crest slightly to one side of the centre and a controlled outlet pipe set into the dam allows water to be drawn off. Maintenance requirements are minimal and any necessary work can be carried out

towards the end of the dry season when the water level is lowest.

Included in their programme of self-help development of small water supplies, the Utooni Development Project is planning to build three sand dams at selected sites in watercourses around the village.

These structures are designed to trap a deep layer of coarse sand particles so that run-off water can be stored beneath the sand surface in the void spaces where it is less vulnerable to pollution and loss by evaporation. A trench is dug down to the bedrock across the watercourse and into the banks and an impervious barrier is built from rock and stone cemented together. The dam wall is usually constructed in stages to a height of about 0.5m above the natural river bed. No spillway is required as the dam is designed to pass flood water over the crest without endangering it. In theory water passing over the dam carries the fine particles with it while coarser particles settle behind. This is desirable so that the storage space between the particles is kept to a maximum.

As the depth of sand behind the dam increases, the height of the wall is progressively raised so that the crest is kept about 0.5m above the surface. Depending on the local geology the water stored in the sand dam will to some extent infiltrate and replenish ground water reserves and consequently water may be obtained from wells dug either above or below the dam. Alternatively, a pipe may be built into the dam wall to tap the subsurface water, in which case it is usual to screen the intake to prevent it from clogging.

Ferrocement water jars are popular in Kola and the Utooni Development Project is now planning for three jars per household in the village. Approximately 150 jars are already in service and an estimated 350 more are needed. The jars are designed to store roof rainwater which would normally run to waste. They are intended to provide water for domestic consumption by members of each household.

Constructed of wire reinforced mortar (ferrocement) to a design adapted by the UNICEF Village Technology Unit, the jars made in Kola hold between 5 and 10cu m of water each and are built on site. A base of compacted laterite 'murram' is first prepared on firm ground to give the jar a stable foundation. A precast cement base made in the village workshop is next placed on the 'murram' and a specially prepared bag is then stuffed with leaves and grass to make a mould on top of the base. Chicken wire is next wrapped around the mould to provide a reinforcing matrix for the plaster and a pipe is positioned for the outlet. Barbed wire or plain fencing wire have been used as alternatives.

A 1:2 mixture of cement to sand mortar is plastered over the wire making a neat rim at the top and special care is taken that the walls are properly joined to the base. The structure is then covered with a polythene sheet and left for a day to set. The contents of the bag are then removed and the bag extracted. The inside of the jar is plastered to completely encase the wire as otherwise this may corrode and weaken the structure. Finally, a lid is cemented to the top of the jar incorporating a small gauze-screened inlet through which the water must pass. Rainwater is collected from the roof by

Ferrocement jars provide a cheaper way of storing rainwater. (*UNICEF*)

214

means of either proprietary guttering or by improvizing zinc sheets or recycled tin cans. A moveable down-pipe delivers water to the jar. After completion the jar must be cured for a month before it is used. This is done by partially filling the jar with water and covering it with a sheet of polythene to keep it damp.

The water running off the roofs is clean compared to the surface water supplied traditionally used in the area — especially after the first 'flush' of a downpour has been allowed to run to waste. The fine gauze screen over the inlet prevents leaves, beetles and other detritus being washed into the jar and reduces the likelihood of the water becoming a breeding ground for mosquitos. Finally, the natural acidity of most rainwater combined with the beneficial effect of simply storing water under hygienic conditions will do much to ensure that the water is safer than the usual alternatives.

The Utooni Development Project has also constructed some of these jars near the tree nurseries and experimental seed plots. This means that water is available as needed and repeated journeys for water are not necessary. These jars are filled by hand by a team of villagers on a rota basis. The basic design is the same but the lid of the jar has a moveable cover as it has no feed pipe. The workshop craftsmen are experimenting with the use of local fibres to strengthen the lids and bases.

Hilary Byrne, *Waterlines*, 1983.

Water Lilies and Broken Pumps

Ban Nong Suang and Ban Som are two typical small villages in northeast Thailand that were selected early in 1973 for field testing of a unique design of water filtration system constructed almost entirely from locally available materials. The filters were designed to be simple to maintain by the villagers without skilled supervision. They were cheap and could be easily copied by neighbouring villages should they so desire.

At the time that the field testing of the new filters was being planned, the people of the two villages had already taken steps to improve their supplies. They had built concrete storage tnks with the help of the local public works department as a first step, encouraged by the experience of neighbouring villages that had supplies installed already under a community water supply programme. These neighbouring supplies, built under a foreign aid programme, were designed as a package deal, and would be a familiar sight in any town in Britain. The designs allowed for coagulation by the addition of aluminium sulphate to settle the suspended solids, rapid filtration through mechanically washed sand filters, and finally chlorination before being pumped up to delivery storage tanks. At 900 Baht/head the cost of these systems was well outside the unaided reach of the villages to construct themselves.

215

The older and cheaper slow sand filters, whcih have played such an important part in improving water supplied all over the world, are also inappropriate for this part of Thailand because the expensive sand would have to be imported from the coast, and the filters cannot handle the turbid waters without becoming quickly blocked. both systems require skilled operation, which is clearly lacking in most rural areas.

Accordingly, a revolutionary type of cheap water filter using local materials was developed by Richard Frankel at the Asian Institute of Technology. This filter consists of two stages. The first is made up of the fibres from chopped coconut husks to filter out the bulk of the solids from the water; the second is a layer of burnt rice husks to 'polish' the water free from colour, tastes and odours — leaving the filtered water clear and sparkling.

The coconut fibres provide an indestructible mat on which the suspended solids and most bacteria are deposited. The burnt rice husks, 90 per cent silicon dioxide and 10 per cent activated carbon, act in a similar way to the rapid sand filters of conventional water works. The filter media are changed once every two to four months, depending on the condition of the raw water.

The system also requires a pump to raise the water from the source to the upper filter of coconut husk. This water then passes through the secondary filter of burnt rice husks and into a storage tank for collection by the villagers. Results from the filters installed at Ban Nong Suang and Ban Som were very encouraging: 90 per cent of suspended solids and 90 per cent of bacteria were removed.

Although this filter does not reach the standards set by the World Health Organization, it is a great improvement on no treatment at all and could prove a major element in helping to reduce water related diseases from rural villages. Thus the tales of its abandonment in the two villages provide an important note for caution.

In the village of Ban Som, nine large concrete water tanks had already been built by 1973 near the site of a proposed temple, to collect the rain water from the roof. They were built with the assistance of the Public Health Department, and the villagers contributed 200 Baht plus their own labour. The temple was not built, however, because of lack of funds. The existing dry season supply, a muddy stream 50m from the tanks, was badly polluted. Besides being the water supply for the villagers it was also open to the water buffalo and other animals.

The village head man volunteered to act as the plant operator during the field testing programme. The new water filter, constructed on top of the exsiting water tanks, was filled by a motor pump from the nearby stream. The new works were opened with great ceremony; dignitaries and officials came from many kilometres to be present. One of the villagers took two glasses of the treated water to the senior monk, saying that one was rainwater and the other filtered stream water. The monk, after slowly testing and comparing each sample, declared that one of them was in fact rainwater, to the jubilation and hilarity of the villagers. The water supply was successful and gladly accepted, costing each family about 8 Baht/month, mainly for fuel and repairs to the motor pump.

During the field testing, which lasted for eighteen months, the motor pump broke down seven times and eventually had to be replaced. When the field assistants finally stopped visiting the village, the broken pump remained un-repaired. The village head man was obliged to be away from the village for long periods, and because of the prestige the scheme had brought him was reluctant to delegate his authority.

In July 1977, the motor pump had been removed and the water tanks were empty. The villagers were back to using their old dry season source, the polluted surface stream.

Ban Nong Suang is a realtively new village built around a monastery and the villagers had also made concrete tanks as the first step to the construction of a water purificatio works. The existing dry season supply was a large pond with water highly charged with suspended clay and organic particles. This undrinkable water was used for washing, while cooking and drinking water was carried from a source 8km away by the women. The senior monk and teacher had requested assistance and welcomed the installation of the experimental filter.

Because the water had such a high turbidity, it was first dosed with aluminium sulphate to settle the suspended particles. This water then passed through two separate filters containing coconut fibre and finally through the burnt rice husks into the storage tank, where it was collected from faucets by the villagers. The filters managed to clean the water with ease and suggested that a whole series of filters could perhaps filter the water to standards acceptable by WHO.

In July 1977, the water pump had been broken for over three months and despite a small garage being in the village, there was no organization to effect repairs. The villagers, however, had a new water source alongside the old pond. The new source, a smaller catchment pond originally of similar turbidity had been flooded during the wet season several years ago. This flood water carried in a species of water lily that quickly colonized the pond. By its ability to prevent wind turbulence on the water, or by its encouraging a different ecology of micro-organisms, this water lily reduced the turbidity of the pond water to a level not much higher than that of the filtered water. This lucky circumstance was not missed by the villagers, who then refused to pay for the water pump to be repaired.

The failure of the water treatment systems at Ban Nong Suang and Ban Som is a failure not only of mechanical equipment but also of organization within the vilages. The Buddhist monks, with their minds engaged on Nirvanal matters, seem unable to make this organization. In Ban Som, the responsibility was vested with the head man, who could not carry his fellow villager with him to contribute regular upkeep and who was unwilling to release his own personal control of the project. In Ban Nong Suang, the monks were also unable to maintain an effective repair organization.

In relatively wealthy Britain, conventional water supply authorities collect water rates and take complete responsibility for the system; the poorer countries of the world do not have the cash or the skilled man-power. It has been suggested that this maintenenace can be carried out most cheaply by travelling techni-cians, barefoot or otherwise, but this does not answer the question of who will pay for the service and who will organize the collection of water rates in the villages. In addition, unless the villagers can understand the full ben-efits of a wholesome water supply, they will not be prepared to pay for it out of their meagre earnings.

Simon Watt, *New Scientist,* 1977

 # Pour-Flush Latrines in India

Use of pour-flush watersea latrines with a single leaching pit for disposal of human excreta was initiated in 1943 by the All India Institute of Hygiene and Public Health, Calcutta. Earlier their use was only in the rural areas. During 1958, the offset double pit system was introduced. Since 1967, the system has been adopted in the urban areas on a large scale, and nearly 400,000 units of this type have been constructed in India in different geological, hydrogeologi-cal and physical conditions and many of them are located in the densely popu-lated areas. A pour-flush latrine consists of a squatting pan of a special design (having a steep gradient at the bottom and particular depth) and a trap having 20mm waterseal. One person's excreta can be flushed by pouring nearly 2 litres of water. The excreta is discharged into the leaching pits constructed in the house compound or where it is not possible to do so for lack of space, into the pits constructed under foot-path or street. The squatting pan is connected to the leaching pit through a pipe or covered drain. These pits are lined with honeycomb brickwork or open jointed stones, so as to allow the liquid in the pits to percolate and gases to be absorbed into the soil; and at the same time preventing the pit from collapsing. The sludge gets digested and settles down gradually.

The pits are used alternately. When one pit is filled, it is stopped being used and the excreta is diverted to the second pit. The filled up pit is sealed and left unused; and in about 24 months the contents become rich organic humus, innocuous, free of pathogen and smell. When convenient, it is emptied and contents could be used as fertilizer. It is then ready to be put back into use when the second pit becomes full in its turn.

The size of leach pits depends on a number of factors such as: soil charac-teristics, sub-soil water level, interval of cleaning, number of users and people's eating habits. The studies in India have indicated that under dry con-ditions per year sludge contribution is about $0.045m^3$. The dimensions of the leaching pits for a different number of years are given below (in mm).

Pits can be circular, rectangular or a combination. Circular pits are more stable.

Users	Diameter (d)	Height (h)	Thickness (t)
5	900	1,100	50
10	1,100	1,400	60
15	1,250	1,600	75

Leach pit configuration and materials used for its construction can readily be varied to suit the site conditions. In case of space constraints, an oval pit with a partition wall or pits of smaller diameter and greater depth could be provided in unsaturated zones where the ground water-table is more than 2 metres below the bottom of pit and soil is of less than 1.0mm effective size, leach pits could be located even at 3 metres away from wells or hand-pumps used for domestic purposes. Where the water-table is shallow and soil contains coarse sand or gravel, pollution travel can be checked by providing a 500mm thick envelope of fine sand (average size not more than 0.2mm) all round the pits and sealing the bottom with clay on a polythene sheet.

The distance between leaching pits and house foundation depends on the depth of leach pits, foundation depth of the building and its structural condition. However, for an average condition a distance of about 1m is enough. Maintenance of pour-flush latrines is very easy and simple. Day-to-day maintenance consists only of washing the latrine floor and cleaning the pan. No other maintenance cost is needed. The cost of cleaning the pits can be covered by sale of humus obtained from the pits. The squatting pan and trap can be of ceramic, fibreglass reinforced polyester plastic (GRP) or cement. The cost of conversion of an Indian bucket/dry latrine into a pour-flush one comes to Rs.660, 850 and 1,000 for 5, 10 and 15 users respectively.

World Water, 1983-4

Zimbabwean VIP

Zimbabwe is taking rural sanitation seriously by the importance it is giving to the design and development of Ventilated Improved Pit Latrines (VIPs). An ambitious programme has been launched, aiming to build 60,000 VIPs in the next three years.

Several improvements on the basic hole in the ground make this safe, comfortable and very cheap. The pit is off-set (i.e. not directly under the squatting plate), which gives added stability and reduces fear of falling into the hole, and a ventilation pipe extracts smells from the inside of the latrine so that flies are attracted to the top rather than to the latrine itself. The upward flow of air in the ventilation pipe is encouraged by its warmth from the fermentation process and the suction of wind passing the top. The pipe is painted black because it absorbs heat, thus warming the air which then rises. A fly trap at its top prevents flies from entering and leaving it. Flies in the ventilation pit are attracted to light, and go up to get out rather than through the squatting plate.

A Tanzanian innovation, a mosquito trap over the hole in the squatting plate, adds to the comfort at very little cost. Mosquitoes are attracted by human smells rather than light, so tend to fly up to the hole which should be kept open.

There are a number of different VIP designs. The one which has been found best suited to rural conditions has a spiral superstructure, without a door, and is made of mud, wattle and thatch at a cost of under US $10 each.

Ideas and Action,
1982.

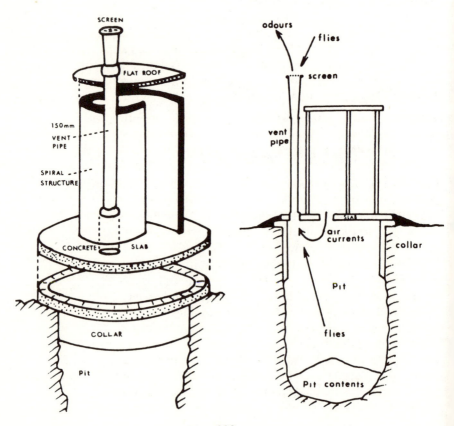

Community Participation as Token Jargon

The extraordinarily high failure rate of many sanitation programmes bears testimony to the pitfalls of centrally defined technical solutions which are presented in catchy jargon. When community participation is involved, it is usually not financed (on the assumption that communal labour is free and that community participation means a low-cost approach).

In one African country a primary school sanitation programme intended to be implemented by local communities has been a resounding failure. The technical design, drawn up before field experience, was based on an outsider's perception of what was needed and did not take into account village or the school-children's needs, or even local ideas on the provision of toilets. Even so, villagers and children were expected to contribute labour without being paid. Two years after the deadline for the first phase of the programme, in which latrines were to have been built in 60 per cent of the primary schools, only 8 per cent of the schools actually had them. Most of these were in complete disrepair after some time and even when in good order were only used by a fraction of the pupils.

Ideas and Action, 1983.

Comparable results are available from a great many other sanitation programmes with a similar approach which leads to the adoption of technologies that are inappropriate, neglects the need for local management, and may even be socially disruptive in unforeseen ways. In the Indian sub-continent, for example, the promotion of biogas systems among wealthier animal-owning villagers actually deprives the rural poor (who have no animals) of access to an important, and formerly free, domestic fuel. Moreover, excreta-disposal is in many cultures a very private affair and regulated by strongly held beliefs. A commonly held belief in Honduras, that women should not use the same latrine as men, effectively limits even the use of household latrines to female members of the family. In a number of African societies customary taboos prohibit fathers and their daughters-in-law from sharing the same facilities. There are also societies in which it is considered primitive to defecate on top of the faeces of other people. Development formulas with only token local involvement inevitably fail to recognize these facts.

V

Technology for Development

Biomass and Renewable Energy

ALTHOUGH NOT a basic need in its own right, energy is required by the rural and urban poor in developing countries to help them meet almost all other basic needs and to obtain adequate levels of income and standards of living. Principal energy needs are for irrigation; crop and food processing; manufacturing of building materials, household utensils, farm equipment, transport devices, etc.; cooking; domestic water supplies; lighting; warmth; running of community services; and transportation of goods and people.

Traditionally, rural families and communities have met these needs from locally available materials — primarily fuelwood supplemented by cow dung and crop residues. Although 'commercial' energy sources (coal, oil, electricity) have become more popular in recent years, these are used almost exclusively by the wealthiest 20 per cent of people in poor countries. Thus traditional sources continue to provide from 50 to 75 per cent of total energy needs of developing countries — and as much as 90 per cent in Africa.

In recent decades, the pressure on traditional energy supplies has increased dramatically given the rate of population growth, the accompanying demand for energy for domestic and rural industry uses, and the inability of the masses of the poor to gain access to non-traditional energy supplies.[1] On the supply side, the increased demand for agricultural land, uncontrolled exploitation of forests, and shifts in land-use patterns caused by a trend to larger scale agribusiness, have all tended to reduce biomass stocks and supplies. More recent increases in the price of liquid fuels and electrical power, and decreases in their local availability, have tended to shift industrial and domestic consumers away from liquid (oil) to biomass fuels, putting further pressure on biomass supplies.

---

[1] Rural electrification is a good example of unequal access. It is typical for poor countries to spend 15 to 20 per cent of their budgets on electrification, of which a considerable fraction often goes to rural electrification. Yet after decades of effort, most villages still do not have electricity, and in those that do, most people canot afford to use it. In Latin America only 2 per cent of the electricity generated is used in rural areas, where half of the people live; in India 10 per cent of the electricity is used in rural areas, where 80 per cent of the people live. In many countries, it is common to find a village which has electricity coming to its gate but which has only half a dozen houses with connections out of a total of a hundred or more. Similarly, heavily subsidized electricity allows the relatively better off to establish highly profitable industries, such as rice mills, which force traditional industries, based on human energy or biomass, out of business.

These changes have far-reaching ecological and socio-economic effects. Increasing demands for biomass have removed forest or other biomass cover and led to severe erosion and/or flooding in many areas, often reducing agricultural output. Reductions in irrigated levels which stem from rising energy costs and energy shortages reduce crop yields, and hence the supply of biomass in the form of agricultural residues. For rural industries, increased energy costs mean reduced profits and often closure. For urban households, increased energy costs mean higher prices for cooking fuels, reducing cash incomes for food and other essentials.[2] In rural areas, wood shortages mean that households, and that usually means the women, need extra time to collect fuel, reducing the time available for agriculture, child care, education and other more productive activities. Alternatively, households undertake less cooking, with adverse effects on nutrition; or they may use lower grade fuels such as agricultural residues, instead of wood. While this may reduce the pressure on wood supplies, it will also in many cases reduce the quantities of residues available for fodder and soil conditioners, which in turn reduces crop yields and biomass fuel supplies even further. Supplies of biomass for building and industrial raw materials are also affected.

With more than one billion people already affected in varying degrees by fuelwood/biomass shortages and with that number likely to increase to more than 2.5 billion, all possible ways need to be explored of tackling the problem. Basically, solutions can be sought either on the demand side — improved woodburning stoves to reduce the amount of waste heat in cooking, and improved steam engines, starling engines, briquetting, producer gas and biogas plants to make more efficient use of agricultural residues in providing energy for domestic and industrial uses; or on the supply side — planting trees and shrubs and developing new alternative sources of energy based on solar, wind and water power.

The first section of this chapter looks at the demand for, and supply of, biomass (fuelwood, dung, agricultural residues) and the efficiency with which they are used. The first extracts examine the deforestation problem — the one looking at the general dimensions and the other pinpointing the particular contribution of the tobacco industry to the problem. These are followed by extracts on the experience with wood-stove programmes. The extract by Foley and Moss gives a comprehensive picture of developments and makes the important point that even successful stove programmes should not necessarily be expected to make a major contribution to reducing the rate of deforestation at the national level. An important point covered both in this extract and in the Anton Soedjarwo story is the shift in emphasis in stoves programmes from owner-built to artisan-built stoves — a shift which is thought to be necessary if sufficient stoves are to be produced and brought into use within a realistic time-frame.

Given that improved stoves will do little to solve the fuelwood problem on their own, the rest of the section looks at the issues of increasing the biomass

---

[2] In parts of Africa fuel now accounts for up to 20 per cent of a family's income.

supply by planting more trees, and of increasing the efficiency with which agricultural residues can be utilized by domestic and industrial users. Two important extracts are those by the National Academy of Sciences, which introduces the concept of 'barefoot' foresters to persuade small farmers to plant trees for the future; and by Vandana Shiva who points out that not all forestry projects are appropriate. Particularly interesting are the extracts on briquetting of coir dust in Sri Lanka and the briquetting of papyrus in Eastern Africa.

The second section of this chapter looks at attempts to develop technologies which provide viable sources of renewable energy as an alternative to oil and diesel fuels. Implicit in the new renewable energy debate is the assumption that alternatives are, or will soon become, much cheaper than liquid fuels, thus enabling more rural families and industries to benefit from energy supplies. The idea of limitless supplies of renewable energy from the sun, wind and water is certainly an appealing one, and the renewable energy business is booming.[3] The potential is well illustrated in the extract by Barbara Ward, which gives a good introduction to the subject. An interesting point is that the fall in cost of solar photovoltaic cells predicted by Barbara Ward in the late 1970s is far from being realized — perhaps a warning that the promised benefits of solar energy for the rural poor may not be very easy to come by. Further warnings about blindly rushing ahead with renewable energy technologies are sounded in the extracts by Gerald Foley and Jeremy Rifkin.

Warnings apart, there have been some successful experiments with new renewable energy devices as alternatives to diesel and electricity for small power applications. One of the more successful is micro-hydro which has been used to provide rural electrification for lighting, heating and rural industries (sawmills, grinding mills, etc.) in many countries. Extracts given here illustrate its use in Papua New Guinea and Colombia. Some specific examples of wind power application are given in the extract by Christopher Flavin. An example of the use of solar energy for water pumping is also included — indicating the cost problem involved in disseminating solar-powered technologies to small farmers or poor rural families.

[3] An OECD survey found that solar energy is the most popular research area for Appropriate Technology Groups worldwide — particularly so in developed countries. Research on technologies using wind and water power are also very popular. (Nicolas Jéquier and Gerard Blanc, *The World of Appropriate Technology: a Quantitative Analysis*, OECD, Paris, 1983.)

As the world's resources of non-renewable fuels — coal, oil and natural gas — are exceedingly unevenly distributed over the globe and undoubtedly limited in quantity, it is clear that their exploitation at an ever-increasing rate is an act of violence against nature which must almost inevitably lead to violence between men. — E. F. Schumacher

There's No Smoke Without Fuelwood

Even though 90 per cent of Africa's primary energy needs are met by fuelwood, there appears to have been criminal neglect on the part of people, governments and donor agencies in ensuring the security of future supplies. In a continent of crises, the shortage of fuelwood supplies anticipated before the end of this century threatens the very social and economic fabric of African life.

More than 90 per cent of research expenditure in the 1970s went towards the development of forests for industrial purposes — wood pulp, veneer, etc — and yet by contrast, research into woodfuel, charcoal and other wood-based energy products, which account for more than 80 per cent of total wood consumption in the developing world, has been minimal.

While 4 million hectares of forest are lost in the arid or semi-arid parts of the world each year, 2.7 million ha of which are in Africa, only 1.4 per cent of aid to the arid Sahel is for forestry.

To add insult to injury, donor countries are actually contributing to deforestation through road-building and industrial-forestry programmes. The installation of an energy farm to produce fuelwood is a relatively easy matter, yet few donor agencies have taken up the challenge.

Niger's forests and woodland will be exhausted within twenty three years, and in Upper Volta, the demand for wood already exceeds the regeneration capacity. If the present rate of overcutting in the Sahel persists until the year 2000, native woodlands will then be able to supply only 20 per cent of the regional fuelwood demand.

The governments of the affected countries are equally culpable. In 1977, the UN Conference on Desertification (UNCOD) gave guidelines for measures against desertification, which included the protection of existing trees and the establishment of woodlots as sustainable sources of fuelwood.

However, little of the plan called for has yet been accomplished. Desertification is low priority for politicians because the people who live in arid zones have little political power. Governments want aid spent in cities, where results benefit the nation's elite. In short, forests offer no quick political advantage, and they lack prestige.

The need for fuelwood was re-emphasized at the 1981 Nairobi conference on New & Renewable Energy Sources, which called for evaluation of forest resources, research into and development of improved efficiency in stoves and cooking utensils, improvement of reprocessing of fuels, e.g. twigs, branches and dry leaves, development of substitutes, improvement of forestry management and increase of or establishment of reafforestation programmes with selected and tested species.

Another handicap is the shortage of foresters. In Upper Volta, for example one office is responsible for administering 10,920km² of forests, and yet it has only one motor scooter and a ration of thirty litres of petrol a month.

Perhaps the greatest impediment to change is the notion which resides in the collective consciousness of nomadic and semi-nomadic peoples that forests are there for you to 'help yourself'. Since time immemorial, people have roamed the continent using wood that nature kindly replaced by the time they came around again. But now, their habits persist even though the hard facts of population growth and urbanization mean that no replacement will be made before their return.

The removal of trees causes soil erosion and falling water tables, and ultimately, the blowing wind removes the last moisture from the soil. A graphic example is in the North Kordofan province of Sudan, where the desert is galloping along at a rate of about 100km in eighteen years. Farmers on semi-arid plains can be affected by deforestation taking place hundreds of kilometres away on watersheds of rivers upon which they depend for irrigation water.

Deforestation creates a vicious circle of poverty. A diminishing supply of woodfuel for cooking means that less food is cooked properly and lack of cooked food is adversely affecting nutrition. Women in Upper Volta may have to walk for 4-6 hours three times a week to gather enough wood to cook the evening meals. In Tanza-

nia, it may take 300 man-days (usually woman-days) per year of work to provide wood for the average household.

Animals that rely on trees for fodder have less to eat, and therefore they provide less protein, thus adding to the human protein deficit. As wood becomes increasingly scarce, animal dung, which itself is available in reduced quantities, is increasingly used for fuel, which in turn means that the soil receives less of its natural fertilizer, which results in poorer food crops.

Another aspect is that less wood means less boiled water, which in turn means more health problems.

An FAO fuelwood expert estimates that the diversion of dung from the soil to the fire in Asia, Africa and the Near East results in a production drop of some 20 million tonnes a year — enough food to feed 100 million people.

The burgeoning African urban population fares little better. Urban dwellers spend as much as a quarter

SOON TO BE MADE INTO A MAJOR NOVEL

Nick

(Punch)

of their incomes on woodfuel or charcoal. Charcoal is favoured in towns because it is cleaner and is cheaper to transport over the ever-increasing distances from wood sources. However, it takes ten tons of wood to make one ton of charcoal, and yet charcoal gives out only about three times the heat.

Town dwellers are exerting tremendous pressure on surrounding woodlands, and the price of fuelwood is rocketing. It increased fivefold between 1970 and 1979. In the Sudanese capital of Khartoum, where the population increased by 150 per cent between 1962 and 1973 and by a further 200 per cent between 1973 and 1980, only isolated woodlots survive within 90km of a city which was surrounded by dense acacia woodland in 1955.

The competition for wood is such that in Sudan, forest rangers have engaged in Wild West-style shootouts with armed crews loading trucks with illegally cut wood to be converted to charcoal and sold in the cities. In Kenya in 1981, around 20 million tonnes of wood were consumed — the equivalent of 7 million tonnes of oil, which is worth about $485 million. It is not difficult to see the advantage of forestry management for a country which spends over 60 per cent of its income on oil imports. In Nigeria, woodfuel plantations were started as far back at 1912 and have been helping to supplement the 'free forests' and to fight desertification, but pressure on the land is reducing the fuel plantations.

Just to supply the shortfall by the year 2000 in African countries south of the Sahara would require the establishment of plantations which, if collected together, would form a forest belt 6,000km long — across the Sahel from Senegal to Ethiopia — and 34km deep.

H. Harington, *African Business*, 1983.

Cigarettes Can Damage Your Health, and Someone Else's Land

The small Nepalese village of Madhubasa is fighting for its life. Lying at the foot of the Himalayas, between two rivers, the villagers' land is in danger of being eaten away by ever more frequent flash floods. The reason for the flooding is that the Himalayan slopes above the village have become almost denuded of trees; when the heavy rains come, water rushes down the hills, spills over the river banks and threatens to tear the heart out of the village.

Many of the Himalayan trees, whose disappearance threatens Madhubasa are axed for fuel to dry (cure) tobacco in a nearby cigarette factory.

Over 100 countries grow tobacco, most of them developing countries. In 1981, tobacco production was 5.66 million tonnes. About half the curing — the process of making tobacco fit for cigarette production — worldwide is done in flues, or barns, and most barns require a lot of wood. A small amount of curing is done by using fuels other than wood (in some parts of Zimbabwe, for example, coal is used) but probably around 2.5 million tonnes of tobacco are flue cured with wood. Between two to three hectares of trees are required to cure one tonne of tobacco. In Tanzania, as in most parts of Africa, the curing of tobacco is done by peasant farmers on their own premises. Each farmer who grows tobacco has his own curing barn and has to provide his own trees — which he may have to buy, if he cannot grow or collect them.

In other parts of the world, Malaysia for example, it is more common for tobacco to be collected from the farmers and dried centrally — which is less demanding on the timber.

Assuming, however, that as a global average, one hectare of trees is required to cure one tonne of tobacco then it is clear that the 2.5 million tonnes of tobacco that is flue cured each year using wood requires 2.5 million hectares of trees — annually.

Some 18–20 million hectares of trees are being axed each year for all purposes. One tree in every eight is being axed simply to dry tobacco.

As much of the world's tobacco grows in semi-arid areas, the disappearance of trees in these areas can hasten the process of desertification.

Tobacco appears to offer a good return to farmers. But, when the trees go the means of curing tobacco goes too.

One alternative to wood is solar units. These are becoming available, but tobacco companies are understood not to like them because they marginally change the flavour of the end product. But it is irresponsible to destroy good trees and land when an alternative is available.

Perhaps cigarette packets should carry the notice 'WARNING: Cigarettes can damage your health, and someone else's land'.

John Madeley, *Development Forum*, 1983.

Stove Programmes — Experience To Date

There has been an interest in the promotion of improved cooking stoves in developing countries since the late 1940s. In India, a mud stove called the 'Magan Chula' was introduced in 1947. After the country's independence, further efforts to promote improved stoves were stimulated by the publication in 1953 of Raju's paper 'Smokeless Kitchens for the Millions'. The major objectives related to the health and convenience benefits from using the stoves, particularly the elimination of smoke. During the early 1960s, Singer produced a report for the Government of Indonesia in which he made detailed recommendations on the introduction of stoves for cutting down on the country's wood consumption. He designed several new types, all of which were equipped with chimneys and showed improved efficiency in laboratory tests.

None of these early programmes appears to have had any substantial impact, and few stoves were built. The work, nevertheless, provided a design foundation on which many of the efforts to develop and disseminate improved stoves during the middle and late 1970s have been based.

One of the first of these programmes was in Guatemala. This was begun as part of the reconstruction work in the rural areas after the 1976 earthquake. It relied on the promotion of a heavy mud and sand stove, called the 'Lorena'. This stove is built in the form of a block, generally about waist high and a metre square, from which the pot holes and internal flue passages are carved. By now it is estimated that about 6,000 of these stoves have been built.

The Ban ak Suuf programe in Senegal started in 1980 and is based on a mud and sand version of the Lorena stove. By June 1982, a total of about 5,000 had been built. A variety of different types are being promoted. One of the most popular is a simple chimneyless stove called the 'Louga'. It consists of a thick circular surround enclosing the fire with a front opening for feeding in the fuel.

Upper Volta has been the scene of several stove programmes. Some have relied on stoves of the Ban ak Suuf type. Others have been based on designs made of bricks and cement. One of the most successful of these is the 'Nouna' stove. Over 1,800 of these had been built by 1982. The total number of stoves built in Upper Volta is now over 5,000.

A variety of programmes is now under way in Kenya. The most promising are in the urban areas where improved 'jiko' designs, based on the Thai bucket, seem to be starting to attract a significant amount of commercial interest. It is possible that self-sustaining local manufacture of these stoves will occur in the near future.

In Indonesia, Sri Lanka and Nepal, stove programmes have been running for the last three or four years. Interest is now concentrating on the production and distribution of ceramic liners for mudstoves. Sales of stove liners have now reached about 200 per month in Sri Lanka. In Nepal, a programme for the distribution of 15,000 liners over the next couple of years is underway.

In Sri Lanka, a ceramic charcoal stove is also being promoted in the urban areas. These are being linked with an effort to encourage the use of charcoal in cities, as a means of using up the large quantities of wood being produced as a by-product from the Mahaweli dam scheme and other land clearance projects. No special attention to energy saving is being paid in this programme; its purpose is to promote the use of charcoal among people who have not previously used it. Up to the present, about 20,000 of these stoves have been sold.

More programmes of one sort or another are under way in at least a dozen other countries. For the most part they are still at an early stage, with the number of stoves introduced so far being relatively small. In addition, design and research work is being carried out in various institutes around the world. These include Eindhoven University in the Netherlands, the Aprovecho Institute in Oregon, the Intermediate Technology Development Group (ITDG) in the UK, the Tata Energy Research Group in India, the Ceylon Institute of Science and Industrial Research (CISIR), the Bellerive Institution in Geneva, the Yayasan Dian Desa Organization in Indonesia, and the Research Centre for Applied Science and Technology (RECAST) in Nepal.

Many of the claims made about the effect of stoves in reducing the rate of deforestation are over-optimistic. The impact of stove programmes on national or regional rates of deforestation, even if they are highly successful, is likely to be extremely small. This is because the collection of fuelwood for local domestic use is rarely a major cause of forest depletion. In most areas, the pressure on woodlands and forests comes primarily from the need to obtain land for agricultural production. Firewood is a by-product from this process. Introduction of more efficient stoves may enable people to make better use of fuel resources as they become scarcer, but it will not have a significant effect on the deforestation which is taking place.

Hopes have also been expressed regarding the possible influence of adopting more efficient charcoal stoves in urban areas. Here again, however, the impact of even large and successful programmes is likely to be less than generally expected. Much depends on whether urban supplies are being obtained from land which is being cleared for agriculture anyway, or from areas which would otherwise remain untouched. Only in the latter case will reductions in urban demand have a direct impact on deforestation rates.

Changes in the price of wood and charcoal are likely to be at least as important as stove efficiency in dictating woodfuel demand. They will influence the extent to which people switch to and from other fuels such as kerosene or bottled gas. They also determine whether it is economic to plant trees to supply urban woodfuel needs, as is currently happening in parts of India and elsewhere. Establishing the long-term effect of introducing improved stoves is therefore a complex procedure, and can only be done though a detailed analysis of local conditions.

A Lorena Stove

The reasons why people acquire stoves of their own accord, whether these are new designs being promoted by stove programmes, or traditional versions which are already available, are not confined to energy saving. As the quality of their housing increases, people tend to upgrade their cooking arrangements. One of the obvious reasons for wanting a stove is that it can take smoke out of the cooking area. While there are sometimes reasons why smoke is required, working in a smoke-filled atmosphere is hazardous to health and extremely unpleasant. When people are very poor they may have no alternative. But as they become more affluent, removal of smoke from the kitchen generally moves upwards in their list of priorities.

Stoves also decrease the mess which comes from an open fire. If they are large, they allow a place for putting pots and cooking utensils down, allowing them to be kept away from the dirt on the floor, and out of the reach of children and domestic animals. They have important safety advantages over an open fire, reducing the number of burns that are suffered, particlarly among children. Stoves can also save time in cooking. This is frequently cited by users as a reason for being pleased with a new stove, though it is not necessarily related to improvements in fuel efficiency. For younger women, in particular, a new stove may also be attractive because it enables them to cook standing up. Other women have said that the improved appearance given to the kitchen by the stove is one of the reasons why they like it.

The promotion of new stoves is a process of intervention by external agencies into what is happening at a local level. It is motivated by an outside perception of what people need, and its intention is to help or persuade people to do what they would not do if left to themselves. From the point of view of programme promoters, a variety of dissemination approaches are possible, depending on the cost of the stove and the local economic context. In circumstances where fuel is commercially traded, a new stove can be promoted through normal market channels. For this type of programme to be successful, the new stove must be able to stand the rigours of market competition on its own merits, and be attractive to those who are expected to buy it.

The role of the promoters in such cases is to design and test a stove which is well suited to local needs and preferences, and introduce it into the market. The financial inputs from the programme will be directed towards the development of an appropriate stove and the creation of the manufacturing and marketing infrastructure through which it can be disseminated. It may also be desirable to provide finance to help cover the entrepreneurial risks of the early stages of manufacture.

In places where there is little or no market for woodfuel, and stoves are not bought, the problems of introducing a new stove are much greater. One approach is to rely on zero-cost, or very low-cost stoves made from locally obtainable materials. To avoid labour costs, these must either be built by their owners, or made on a co-operative basis.

The basic problem with this approach is that the stoves will inevitably be of a fairly low quality. They must rely on available local materials such as mud, sand and clay, whether or not these meet the necessary standards of strength and durability. The stoves will have to be built by people who are not trained artisans, hence their dimensions and assembly cannot be expected to be very accurate. Above all, they are not likely to be particularly durable. This all means that the benefits in energy saving will generally be small, and may disappear with time. Justifying the investment in such a programe is difficult unless there are other benefits from the stoves besides energy saving.

Providing subsidies is a way of stimulating the adoption of stoves where they are not commercially viable in their own right. In some cases, a temporary subsidy may be all that is required to overcome initial consumer resistance. Permanent subsidies are, however, much more problematic. Their effect is limited by the amount of funding available to the programme. They may also prevent the emergence of local stove-making entrepreneurs.

If stoves are to be acquired by a significant proportion of a country's population, it is virtually impossible for this to be done on the basis of subsidized distribution to everyone. Even if this could happen, the question of replacement of stoves after they wore out would remain unresolved. It is essential therefore that subsidies are seen as a short-term expedient.

It is important to take into account the fact that energy saving may not be the principal reason why people want a new stove. In some cases, energy saving may only be obtainable at the expense of other desirable features, or at a financial price which people are unwilling to pay. If stoves require extra work in operating controls, or in cutting up fuel into small pieces, this also can be a disincentive to potential users.

There is no doubt that under the appropriate conditions improved stoves can bring numerous benefits. When they are properly designed to suit their real context, they can make it easier and more convenient to economize on fuel thus saving time in tending the fire. They can make cooking quicker and more convenient. They can eliminate or cut down the nuisance and health hazards of smoke.

They reduce the amount of mess and dirt caused by an open fire. They can increase the comfort of the dwelling by controlling the emission of heat during cooking. They can make a major contribution to domestic safety, particularly that of children. In short, improved stoves mark an advance in living conditions, and have a valuable role to play in the domestic economy.

Gerald Foley and Patricia Moss, *Improved Cooking Stoves in Developing Countries,* 1983.

 # *Owner-Built Birth Control*

In many ways cooking is like making love, both activities represent the more personal and private sides of our lives, and the technologies we put to use in both realms reflect decisions made at the household, not community, level.

Now we can easily imagine the problems and barriers the family planning promoter faces when he or she attempts to change the long established norms of lovemaking. New birth control technologies, the male vasectomy for example, will have little chance of success in areas where such technology runs against established social and cultural norms — in this case the Third World male machismo. Likewise we can imagine the problems our family planning promoters might face, if instead of simply supplying the preferred birth control device or service, he or she attempted to teach the local acceptor to make their own. What a comedy this would be! And how miniscule the effect on the population explosion. Imagine, millions and millions of potential acceptors studying condom and IUD manufacturing as a prerequisite to birth control.

As ridiculous as this sounds, this is exactly the approach we (and many other stove promoters) have taken, in our search for a solution to deforestation. By embarking on fuel efficient stove programmes which require local vllagers to 'build their own', we have resigned ourselves to years and years of work, which by all indications will result in but a small improvement in the problem. Imagine 30 million wood burning cooks in Indonesia all studying combustion theory and stove construction as a prerequisite to acquiring a fuel efficient stove! How many training programmes, how many field workers, how many years will it take?

After five years of work in just this vein, we are becoming convinced that the promotion of mass-produced stove technologies (like mass-produced birth control devices) would represent a far more viable and rational approach to the problem of deforestation.

Anton Soedjarwo in Marcus Kaufman, *From Lorena to a Mountain of Fire*, 1983.

For Every Child a Tree

In Gujarat State, western India, school children are helping to plant trees as part of a re-afforestation plan launched by the State's Forest Department. The children are turning their schoolgrounds into tree nurseries where they care for the seedlings until they are mature enough to be replanted elsewhere.

The Gujarat experiment could serve as a model for other parts of the Third World affected by deforestation. Millions of seeds are distributed free of cost to interested schools. Boys and girls plant these in mud-filled plastic bags, they care for

them after the seedlings sprout and then return them to the Forest Department. The school is paid fifteen paise (half a US cent) per sapling and the saplings are replanted all over the State. The school uses the money earned in this way to help the poorest pupils. One or two teachers in each school are trained in nursery techniques and they teach the children the crucial importance of trees in preserving the environment by binding and improving the soil, reducing pollution, providing both fodder and firewood, and so on. By enlisting the help of school children, the Forest Department is guaranteed a cheap and steady supply of seedlings without having to create a huge bureaucratic machine.

In the village of Kelia Vasana, an hour's drive from Gujarat's biggest city of Ahmedabad, over 400 children from the local school have grown thousands of fast-growing trees. Already, fifteen hectares have been planted with more than 5,000 trees. The involvement of schools is part of a much wider policy of 'community forestry' in Gujarat. This involves the growing of trees to meet the small-scale needs of the people. Private groups, city dwellers and, above all, farmers are all being encouraged to grow trees on available land.

To some, Gujarat's policy may sound like one of those fine ideas that are doomed to failure when put into practice. But as befits the birthplace of the Gandhian movement, the State has shown that community participation is not a dead letter and already more than 3,000 villages have turned 12,6000 ha into woodland. The driving force behind this participation comes from enlightened self-interest rather than concern for the environment. Gujarat villagers have started to plant trees because they can see quick returns.

T.N. Nirian, *Appropriate Technology*, 1979.

Barefoot Foresters

Today, international aid agencies and foresters in the Third World are receptive to new notions about the purposes and practices of forestry. In essence, they recognize the modern necessity of taking forestry outside the forests — of involving people throughout the countryside in growing trees to meet their own requirements as well as to protect the land off which they and their livestock live.

Firewood production is particularly appropriate to this philosophy. It is less dependent on silvicultural expertise than sawtimber is and therefore can be done by nonprofessionals who learn the basic techniques for their own use. Firewood can best be produced like a farm crop without government intervention.

However, although the cultivation of firewood species does not demand continuous professional supervision, a forest service may be needed to provide seed or planting stock and advice for getting the trees established. Further, silvicultural practices (such as weeding and pest control) can greatly increase yields. What is sorely needed is the greater involvement of trained forestry experts in firewood production at all levels from the village woodlot to the national forest.

Trees for firewood can be planted in 'nonforest' areas: along roadsides, in shelterbelts, on farms, on unused land, and in schoolyards, cemeteries, churchyards, market squares, parks and home gardens. Fuelwood trees can be cultivated in small woodlots, even as individual specimens around a house or village. In some areas, such as Java and the People's Republic of China, home gardens already supply a good share of family firewood needs. Correct spacing of planted trees is important to production, but it is unnecessary to assure geometric precision as is required where mechanized equipment must pass between them.

Rural areas can probably supply their own fuelwood from small, local plantings, but urban areas can best be supplied from concentrated large plantations, strategically located and possibly government administered.

Firewood plantations, if carefully managed and protected from fire, animals, and 'poachers', can be self-renewing. They are usually managed on rotations of about ten years (much less in some moist tropical regions). The timing varies with the quality of the soil, species used, temperature, moisture available, and intensity of cultivation. Rotations of less than five years seem feasible in many areas, especially for those species that regenerate by sprouts (coppice). What is needed is a change in priorities in the use of trained foresters and agronimists. Forests for fuel can be treated just as one more farm crop. This makes firewood production more suitable for developing countries with few foresters. It seems possible that agronomists, rather than foresters, will be responsible for much of the small-scale firwood production in the future.

The existing forests are too important and too vulnerable to be abandoned by foresters in favour of village woodlots in the farm lands. In addition to making fuelwood production an agricultural responsibility, the suggestion has been made that what is needed are 'barefoot foresters' to persuade small farmers (whose economic horizons usually extend only to the next harvest) to plant trees for the future, to teach how to do it, and to introduce cook stoves that conserve firewood. Such extension services are particularly important in the case of individual and village-level fuelwood projects. Even with finances available, poor management and inadequate extension work are often critical bottlenecks.

National Academy of Sciences, *Firewood Crops*, 1980.

Fuel-efficient Stoves

In 1740, the threat of a fuelwood shortage around Philadelphia moved Benjamin Franklin to design his 'Pennsylvania Fire-place', a cast-iron heating and cooking stove that greatly reduced heat lost up the chimney. In Franklin's words: 'By the help of this saving invention our wood may grow as fast as we consume it, and our posterity may warm themselves at a moderate rate, without being obliged to fetch their fuel over the Atlantic'.

National Academy of Sciences, *Firewood Crops*, 1980.

Eucalyptus —
An Inappropriate Choice

In 1980, the State Government of Karnataka submitted a 'proposal' to the World Bank asking for funds to back a five-year programme of social forestry. The stated aim of the programme is both to combat ecological degradation and meet the basic firewood timber needs of the rural population. Reading the proposal, however, it seems that the State Government has learned few lessons from the failure of other social forestry schemes elsewhere in India. Indeed, the authorities seem set on repeating just about every mistake in the book.

Under the programme, the State Government intends to forest 110,000 hectares of privately-owned land — some 60 per cent of the total area in the scheme. This despite clear evidence from other social forestry programmes that it is only communally-owned forests which can possibly bring the desired benefits to the local population — and despite the State Government's purported commitment to making 'the main strategy' of the scheme 'the utilization of hitherto unutilized' communal land.

In fact, the area given over to private forest farms is likely to be even higher than the State Government has estimated. Thus, whilst a similar World Bank project in Gujarat aimed to forest only 1000 hectares of privately-owned agricultural land, in one district alone 10,000 farmers have already taken to growing eucalyptus in place of foodcrops. And, although the Gujarat project clearly identified 'marginal' land as the target for farm forestry, large areas of irrigated food-producing land have already gone under eucalyptus cultivation. There is no reason to believe that the experience of Gujarat will not be repeated in Karnataka. Inevitably those who will lose out are the very people whom the social forestry programme is intended to help.

The problem is further compounded by the choice of species which farmers are encouraged to plant. For whilst the pattern of land-ownership determines who accrues the benefits of forestry, the nature of those benefits depends on the type of trees planted. It is quite clear from the proposal that the main species to be encouraged in the project will be eucalyptus. No mention is made of such traditional farm trees as honge, neem or mango. That emphasis on eucalyptus is hard to justify. Eucalyptus fulfills fewer ecological functions than the traditional trees; it cannot be browsed by animals; it allows no under-growth and thus provides no fodder; it produces no fruit, nuts or other food; it is not favoured as a fuel for cooking because it burns too fast; and, above all, it is too expensive for most villagers to buy. If other social forestry schemes are anything to go by, then it is more than likely that the harvested eucalyptus will be sent straight for sale to the pulp and rayon industries. Without safeguards to ensure that eucalyptus does not cover all the land under the social forestry project it is impossible to guarantee even a minimum supply of basic forest products for the rural population.

Such safeguards are also essential if employment is to be protected. On the basis of our study it is observed that the labour displaced through eucalyptus plantations is far more than the employment generated. For each hectare of land lost from food crops to eucalyptus, there is a loss of some 250 man-days of employment per year. Assuming that the present trend towards eucalyptus monocultures continues, then by the end of the fifth year of the project the loss of employment will be 137.5 million man-days.

The State Government's social forestry programme also aims to increase agricultural productivity by improving soil and water conservation and by releasing cowdung, at present used for fuel, for use as fertilizer. Assuming that 50 per cent of the firewood to be grown would be sufficient to release the total amount of dung being used for fuel, 0.4275 millions of tons of extra food grain are expected to be produced over a period of thirty five years. That figure, however, is misleading. Firstly, the rural people of this particular region do not use much cowdung for fuel. Secondly, when the land at present under foodgrain is put under eucalyptus, there will be a direct loss of some 5.77 million tons of foodgrains, oilseeds, etc. That loss has not been taken into account in the project proposal.

Vandana Shiva et al., *The Ecologist*, 1982.

The Tree Wall

A very simple way to protect newly-planted trees, especially fruit trees, e.g. mango, paw-paw, pear trees, against abuse by goats and cattle has been successfully

PROTECTIVE STRUCTURES
FOR NEWLY PLANTED TREES

(GATE, 1983)

adopted for a long time in the transitional forest and savannah areas of middle and northern Ghana. The trees are simply protected by a circular mud structure which has ventilation openings in the wall.

The 'tree wall' is about 750mm high in the initial stage. Later on, when the tree develops, the structure is enlarged and made about 1m to 1.2m high. By that time the tree trunk has developed sufficient strength to withstand rough treatment. The advantage of this system is also that the ground immediately around the newly planted tree can be kept fairly moist even in very dry and hot conditions. The diameter of the structure depends on the tree species and on how early branches and leaves develop.

Gate, 1983

Coconut Power

A huge reserve of 'free' energy has been found piled up around Sri Lanka in the form of 'coir dust' — the dust left over when the coir, or outer husk of the coconut, is removed to make ropes and mats. Some four million tonnes have been produced over the past thirty years and now the fibre mills are churning out another 150,000 tonnes each year. Until recently, no-one knew how to dispose of this 'nuisance waste'.

Ceylong Tobacco, a private company, is pioneering the production of coir-dust briquettes. The coir dust passes through machines (essentially mincing presses) and comes out in solidified lengths of about 10cm thick. These can then be broken up and used to fire boilers and furnaces.

Coir dust briquettes give off, weight for weight, about half as much heat as diesel oil, but cost less than one-quarter, thus representing considerable savings. They are also much easier to handle than wood. Trials have shown they are ideal for small-scale industries such as brick-making, tile works, laundries, bakeries and potteries. But they are also good for larger industries, such as tobacco-drying plants which consume vast numbers of trees each year throughout the tobacco-growing tropics. Briquettes also promise savings in tea-drying, one of Sri Lanka's largest users of firewood, and any industry which uses fuel oil for the generation of hot air and steam.

Mallika
Wanigasundara,
Mazingira, 1983.

Papyrus: A New Fuel for the Third World

Papyrus is the largest of the sedges, a group of plants closely related to the grasses. The stems or 'culms' grow up to 5m and are topped by a large feathery crown called the 'umbel', which is the flower-bearing structure or inflorescence of the plant. However, flowers are either absent or largely inconspicuous and the many thin green rays that form the bulk of the umbel are the principal photosynthetic organs of the plant.

Most papyrus is now concentrated in east and central Africa. Some one million hectares of the Sudd are covered with papyrus. It forms a virtual monoculture, excluding all potential competitors apart from an occasional climbing hibiscus.

Papyrus grows in two types of location: in the Sudd the floating mats occur at the edge of many shallow lakes, or even fill them completely. In the second type, primarily in southern Uganda and Rwanda, the swamps cover the floors of valleys which carry the rivers that for the most part feed Lake Victoria. In both situations the papyrus lies on deposits of peat built up over many centuries from decomposing papyrus and other vegetation.

We have learned a lot more about the ecology and physiology of the papyrus recently, and one of the most striking discoveries is that it is among the most productive plants in the world. The harvestable standing biomass reaches a maximum of 32 tonnes per hectare (t/ha) at Lake Naivasha in Kenya, although harvestable biomass varies from place to place. (Biomass is the total moss of growing material including both living cells and dead components, such as wood.) By comparison, grass grown on the finest English pasture yields around 10t/ha. Papyrus will regrow rapidly after it has been harvested and can probably regain its original biomass within nine months to a year in most cases. This would mean that the annual production of the most productive swamps is about 30t/ha, a value that compares very favourably with forests as a source of biomass.

Papyrus is clearly productive enough to be a source of fuel, but when it is simply air-dried it is not dense enough to provide the concentrated heat required for cooking, for instance. The breakthrough has come with the idea of compressing the air-dried papyrus to about one-twentieth of its original volume into a briquette using a machine similar to those used in developed countries to produce briquettes from waste straw and wood chppings. The resulting fuel produces little smoke on burning and it has a low ash content — an admirable substitute for the rapidly disappearing charcoal which is the main source of energy in much of urban Africa.

Initial attempts to produce briquettes from papyrus are being carried out in Rwanda; the briquettes will supply its capital, Kigali. Large areas of swamp (more than 5000ha) lie within 40km of Kigali, which reduces the cost of transport, one of the main determinants of cost of fuels. The papyrus will be cut by hand as the swamps are accessible for at least seven months of the year, when the water levels are low. The papyrus will then be dried and compressed into briquettes at the edge of the swamp.

Large-scale exploitation of these swamps clearly has some dangers; not least that posed by removing large amounts of biomass from the ecosystem with a tight circle of nutrients similar to that of tropical rainforests. At first, however, the briquetting factory will need only relatively small areas; we have calculated that about 60ha should supply the equivalent of 20 per cent of Kigali's requirements for charcoal.

We still have to determine just how much can be cropped year to year from these swamps and the effect that continued harvesting will have on their ecology. For instance, production might fall after some years due to lack of nutrients. And opening up the canopy by removing papyrus will allow more light to penetrate and might increase the proportion of other plant species in the swamp.

Michael Jones, *New Scientist*, 1983.

2 RENEWABLE ENERGY

Digging Mankind out of the Nuclear Rut

No doubt, when social historians look back on our day, they will be astonished at our almost obsessive concern with sufficient supplies of energy. Our planet is, after all, one vast system of energy. The sun's rays that fall just on the roads of North America contain more energy that all the fossil fuel used each year in the whole world. The winds that rage and whisper round the planet are a vast energy reserve caused by the uneven solar heating of blazing tropics and arctic poles. The World Meteorological Organization (WMO) estimates that if placed on the earth's most steadily windy sites, windmills totalling 20 million megawatts of electric generating capacity might produce electricity at a commercial price. (At present, the world's total electric generating capacity is under 2 million megawatts.) In the Southern continents, less than 6 per cent of the power that could be produced by large hydroelectric schemes has been developed, and this figure leaves out the multitude of small hydro sites, so numerous that once developed, they could exceed an electrical output of many thousand megawatts.

Nor should we forget the energy locked up in plants and wastes. Indeed, in some developng lands, 90 per cent of the energy is derived from wood, often with disastrous risks of deforestation. At the other end of the scale of sophistication, the US Naval Undersea Centre at San Diego has an ocean-farm project cultivating Californian kelp. The hope is that the solar energy captured by the plant on an ocean farm of, say, 470 square miles could theoretically be converted into as much natural gas as is consumed in America at present. All in all, the fear of running out of energy must be said to have a social, not a rational base. Modern citizens have become so accustomed to fossil fuels and so transfixed by the necessity of building vast electricity systems that they simply do not see that their whole life is surrounded by a multiplicity and variety of energy reserves which not only exceed all present sources but have a further advantage: they are not exhausted by use. A ton of oil burnt is a ton lost. A ton of kelp will be growing again next year. Even more reliably, the sun will rise and release an annual 1.5 quadrillion megawatt hours of solar energy to the plant's outer atmosphere. There can be no running out of such resources.

But can they be harnessed? A tornado is a fine exhibit of energy unleashed but is hardly a useful one. The fundamental question with all the renewable sources of energy is how to develop the technologies for using and storing them, and to do so at reasonable cost. Admittedly, this issue of 'reasonable cost' begs a number of questions. It fails to take into consideration the huge and now discounted investment in nuclear-weapons research upon which fission power is based. It neglects a whole range of environmental and social costs — from strip mining coal to radioactive-waste disposal. And over a twenty five year period, the whole energy picture was totally distorted by the fact that oil and natural gas, both wasting resources, fell in price while their scarcity was actually increasing.

The truth is that all of us, consumers, technicians, engineers, scientists, have a tendency to see what we are used to seeing, and a twenty five year energy 'binge' based on fossil fuels has conditioned us to looking for a very narrow range of solutions. Perhaps the first need of all is for citizens in developed societies to open the eyes of their imagination and conceive of energy patterns which will come in new shapes, forms, and sizes. If they do, they will find it already clear that the technologies are available, will become cheaper, and could even lead to a more humane and civilized mode of existence.

We can begin with the most abundant of all sources — solar energy. The first point is that at least one-third of most developed nations' energy budgets is required for temperatures of well under 100°C — in other words, for hot water and space heating. Sunlight can provide useful heat for both of these needs at exactly the right temperature — and there are no distribution costs, since the sun can fall on each building. We should also remember that the sun already does much of our space heating without any cost or equipment at all. Picture the problem if every house, store, and factory had to begin heating itself from a temperature of minus 240°C.

But for particular uses and locations, the diffuseness and irregularity of the sun's radiance have to be overcome. Much can be done by the almost forgotten arts of good siting and good design, so that, for instance, a southern orientation is combined with overhanging roofs which allow the low winter sun to shine through windows and exclude the excessive heat from the higher summer sun. Thick walls of masonry are also heat traps in winter and exclude

241

unwanted heat in summer. The traditional architecture of the Middle East, besides being beautiful, has a singular technical appropriateness for dealing with high daytime temperatures and chilly nights.

But the sun traps can be more specific. Instead of just relying on generally well constructed buildings, one can add a 'thermal storage wall'. A thick block of masonry is placed behind a 'window' of glass and traps the solar heat coming through. A simple series of vents circulates the heat round the house, and when the weather changes, the vents are reversed to expel the hot air and cool the house down.

Solar collection units can be small and easily added to places that are already built. The simplest form is a 'flat-plate collector' — a glass-covered panel which allows the sunlight to pass through the glass and be trapped as heat inside what is basically a boxlike apparatus whose collecting powers have been much enhanced by the use of new materials. This heat is then passed on to the house's hot-water system, or can supplement its space heating.

In the near future the cost of these solar panels will be significantly reduced. Many of them are made almost by hand by small firms. Once the market is large enough, the big manufacturers will begin much larger production runs and turn out these units at very much lower cost. The big manufacturers are also conducting research into concentrating solar collectors so that higher temperatures can be reached. These would greatly widen the solar collectors' possible range of use.

Possibly the most interesting and promising of all solar devices is the photovoltaic cell, a spin-off from space exploration, which directly turns the sun's radiance into electricity which can then be used for all manner of domestic and commercial purposes. Such cells, incorporated into the roof, could give homes and factories in hot climates virtually their own independent energy source and still leave over heat for space and water heating — one reason why some prophets see the tropics as likely to become as thoroughly transformed by solar power as was the chilly north by coal and steam. A major breakthrough is needed in the price of photovoltaic cells, but this appears to be on the way, as production costs are falling dramatically year by year and may soon be near the 'competitive' price of US$500 per peak kilowatt: in 1958, unit costs per peak kilowatt were $200,000; in 1977 they were $10,000; by the early eighties they should reach $500 and by the late eighties even fall as low as $50-$150.

Another possibility that is arousing increasing interest is that of using solar power as an integral part of new or existing generating systems, so that traditional fuels — coal, oil, gas — can be saved and reserved simply for peak periods (or for cloudy days). Two forms look promising. One is a 'power tower'. A boiler is fitted into a tower, and computer-guided reflectors follow the sun round the sky focusing the beams on the boiler, which then produces steam for driving the turbines. A 200-foot tower with 5,500 mirrors focusing on it is now operating in Albuquerque and this will serve as a testbed for a 10-megawatt solar power tower in California, which by 1981 will become the first solar power station in the United States to deliver electricity to the grid.

The other area of research is 'solar farms'. Parabolic reflectors, spread over a wide land area, concentrate sunlight onto pipes containing salts or gases. As they heat up — to as much as 600°C — their energy is transferred to storage tanks, and it can be drawn on when needed to produce steam for electricity generation. The limitation on both the solar farm and the power tower is that they require large sites and are perhaps most suitable for desert areas, where they are not too likely to be next door to busy consumer centres. (On the other hand, they may well require no more distant and isolated sites than those thought safe for nuclear power stations, given citizens' growing unwillingness to have nuclear power plants too near their own particular neighbourhoods.) On this point of land requirements, we should note that the small-scale solar collectors which fit on rooftops and walls do not compete for land with agriculture, urban development, or recreation, since they are placed in areas already in use.

When we turn to wind power, once again the problem lies with the technologies for effectively using the inexhaustible cost-free but diffused wind power round our planet. Techniques can vary from the simplicity of the fleets of windmills in central Crete, where over ten thousand of them pump up the farmers' irrigation water by catching the wind in triangles of white sailcloth, all the way to the completely new and complex concept which now interests the US Department of Energy. The inventor has designed a circular tower and uses the wind to induce a mini-tornado inside. Then the difference in pressure between the outside air and the inner turbulence drives a turbine. This kind of tower could clearly operate on a quite small and dispersed scale or, like solar towers, be used to supplement grid systems.

Conventional wind turbines are also being thought of for windier sites, and a great variety of new techniques — from traditional propeller types to contrivances that look like vast eggbeaters turned upside down — are beginning to leave the researchers' drawing boards. One such upside-down eggbeater — a so-called 'Darreius rotor' — is already feeding electricity into the grid in Canada. Since the larger the scale, the greater the capture of energy, it seems likely that the WMO picture of a wind-driven world, although much exaggerated, could have some elements of reality by 2025.

When we turn to falling water as a source of energy — for hydroelectricity — the big reserves are in the developing world, since the industrialized nations have used up most of their best sites, and there is a limiting factor here — in both developed and developing lands. The reservoirs of big dams cannot be permitted to flood irreplaceable natural ecosystems or invaluable farmland. Where larger new water-power resources may be available in the industrialized nations is on the coast. There are a number of sites such as the river Severn in Britain or the Bay of Fundy in North America where the rise and fall of tides is on a sufficient scale to justify the kind of system the French have been operating at La Rance, where reversible-blade turbines allow power to be generated both when the tide ebbs and when it flows. The Russians, too, have an operating tidal power station — although at pilot-project level — and are now looking at several sites at which viable tidal power stations may be possible.

Quite new experiments are being considered for use in areas where the seas are normally stormy. It is suggested that the waves' power can be used to generate electricity, and several research groups in Britain, Sweden and Japan are experimenting with different ways of tapping this source. There is a particularly promising project in Mauritius. Yet another proposal is to use the difference in temperature between deep and surface water as a power generator. In the tropics, the sun-warmed surface water is ten or more degrees centigrade warmer than water 1,000 to 15,000 feet below it. By the early 1980s, a 25 megawatt prototype should be operating off the Florida coast, with the possibility of a commercial demonstration plant by 1985.

All these renewable energy sources — sun, wind, waves, tides — can be used as substitutes for fossil fuels and contribute to the production of electricity. But they can only produce power intermittently — the wind, for certain, 'bloweth where (and when) it listeth' — and this may well not be at a time of peak load or urgent domestic demand. During sunless, windless, waveless winter days in, say northwestern Europe, all the renewable-energy technologies will not be producing very much. Yet consumer demand for energy will be very high. Oil — as gasoline or as a heating fuel — and natural gas are easily and cheaply stored. Fuel can be drawn from the store when its heat or its power is required. But you cannot simply turn on the sun or the wind. So methods of storing their energy must be found if they are to meet a sizable part of an industrial nation's energy needs.

Storage, however, is a problem with all grid systems, coal-based, oil-based, nuclear, which supply electricity. That we may be surprised by this fact simply reflects the degree to which we have become accustomed to the idea of having a very large number of generating stations which come on stream only at times of peak load — over the breakfast eggs or Sunday lunch or when the advertisement break in a highly popular television series sends half the population to the kitchen for a cup of coffee. These 'surges' are only containable today within a system of continuous and expensive overcapacity. The trick is to get types of storage that are cheap enough to take its place. Otherwise it will have to be maintained (at high expense) for the dank, still days when wind and sun power will be insufficient for peak use. A variety of ways are being explored. Surplus electricity can be used to compress air, charge up a large flywheel, or pump water up to a high reservoir. In each case, the 'energy store' can be drawn on when needed as we now draw on our extra generating capacity. Another possibility is to generate hydrogen from water, store it, and then use it when needed to fuel a conventional power plant or generate electricity directly in a fuel cell. Then, at the consumer end, the surges can be minimized by encouraging small domestic energy stores. For example, a greatly reduced electricity tariff for off-peak hours can encourage the use of hot water tanks that heat up 'off-peak' and then supply the house's hot water needs for the whole day. Such tanks would, incidentally, have to be larger than conventional ones and, of course, well insulated.

Now, with rising prices and shrinking reserves, all the new possibilities — sun, wind, water — can begin to enter the public imagination, and there are at last signs that the mood is changing. The range of new experiment is not negligible and there is evidence of growing momentum. For instance, the US Department of Energy's solar budget for 1979 is between $400 million and $500 million. The Swedish Government will be funding over three years a major programme of research into biomass conversion, wind power and solar power. The French spent some $32 million on solar energy in 1978. Apparently, they see solar technology as a major potential industry with excellent export prospects. The Japanese Government has a billion-dollar research programme, 'Operation Sunshine', to develop energy-income technologies that could lessen their extreme dependence on imported fuels. As all these technologies begin to take hold, the effects may go much further than the provision of energy. They may well imply profound and beneficent changes in the whole developed way of life; give choice, variety, and safe options to the central process of energy use; and begin to dig mankind out of the nuclear rut.

Barbara Ward, *Progress for a Small Planet*, 1979.

A Questioning View

An economic transition from subsistence is the aim of development strategies everywhere. When it occurs many of the problems of fuel supply become resolvable. With rising rural incomes, the depletion of free-good wood resources opens the way to commercial subsititues. A typical transitional sequence is, first, the commercialization of firewood, next a more extensive use of charcoal: then comes the employment of kerosene, benzine, bottled gas, or even coal. When economic conditions improve sufficiently the use of electricity becomes universal. . . .

Another feature of the economic transition from subsistence is a widening of the area of economic activity and a much greater range of end-uses for energy. Mechanized agriculture, water pumping, food processing, small manufacturing and

workshop enterprises, services, and the provision of social amenities all use energy in different ways. In the European economic transition out of the Middle Ages, this was the time of windmills, water-wheels, tidal mills, and eventually, sweeping all before it, coal. It is within this modern transitional sector that the greatest number of possibilities for the use of renewable energy technologies occurs. Identifying exactly where initiatives should be made, however, requires care. There is a superficial attraction in enumerating the tasks that might be powered by renewables and then endeavoring to supply the hardware to accomplish this. All over the developing world however, can be found the derelict remnants of projects based upon such an oversimplified analysis.

Technology cannot be abstracted

245

from the priorities and constraints of the society it is supposed to serve. Often a lack of energy occurs because of the absence of effective demand rather than a scarcity of supply. The shortage of finance, technical skills, and access to markets for their products prevents people from mobilizing the energy resources already available. The low level of energy use, being a symptom rather than a cause of the poverty in which people live, requires a simultaneous creation of demand and supply to relieve it. Conditions will not be improved by simply widening the technical opportunities for supply.

Effective energy demand will manifest itself by a widespread and diversified pattern of energy consumption. Its absence is evidence that the developmental process is at an early stage and that almost certainly the highest priorities are not yet in new forms of energy. Since renewable technologies are generally less versatile, more expensive, or more demanding in repair and maintenance than conventional technologies — compare for example the availability of mechanics able to repair a windmill with those who can tackle a diesel engine — the chances of their successful introduction as a cutting edge of development are less than in the case of conventional technologies.

On the other hand, where conventional technologies are already in use there may be opportunities for supplementing them or even replacing them with renewable technologies. Thus windmills may be able to yield useful economies where diesel pumping is an established fact. Solar collectors or crop driers may be competition with oil-fired water heating or crop drying. Alcohol production may be able to supplement existing petrol consumption in countries with a well-developed technical capacity, provided the environmental problems, as well as competition with food crops can be satisfactorily resolved. In summary, where there are existing uses of high-cost commercial energy, renewable energy technologies may prove economically competitive and technically feasible. . .

New and renewable energy sources have acquired a special symbolic significance within the developed countries, which prevents them being viewed with the economic and technical detachment used when evaluating motor cars, dishwashers, and railway locomotives.

One result is that too much attention is paid to new and renewable technologies as elements in the energy strategies of developing countries. The visiting energy experts, or the delegation from an international aid agency, are driven two hundred kilometres to see the country's two working biogas pits or a solar pump, as if these were items of major importance. Growing a few trees, harnessing a pair of bullocks, or training a single motor mechanic in tuning motor vehicles for fuel economy would have a far more beneficial effect than most of the renewable energy gadgets now being promoted for use in developing countries.

The projection of Western 'post-materialist' aspirations upon societies which have not yet achieved the minimum material requirements of dignified human existence is insidious and dangerous. At its worst it leads to a diversion of developing country resources into ill-conceived programmes of renewable energy hardware. The energy problems of developing countries are too serious for any romantic preconceptions about certain technologies to distort the way in which the problems are analyzed or prejudice the solutions to be attempted.

Gerald Foley, *Ambio*, 1980.

A Renaissance for Wind Pumps

The technology that opened the American West in the mid-nineteenth century may turn out to be a lifesaver for many semi-arid parts of the world during the late twentieth century. Diesel and electric pumps have become prohibitively costly during the last decade, and a cheaper way must be found to bring up the water so desperately needed for irrigation, livestock watering, and general household use, particularly in developing countries. Fortunately, wind pumps are a well-established technology suitable in areas where wind speeds average as little as eight miles per hour, and they are especially appropriate to the modest water needs of homes or small farms. Wind power is also well matched to the task of providing water, since during a brief windless period the user can simply draw on water pumped into a storage tank when the wind was ample. Storing water is far cheaper than storing electricity.

Approximately one million mechanical wind pumps are in use today; most of them are located in Argentina, Australia and the United States, where they mainly provide water for household use and for livestock. With a pump connected to a well in the ground, the user does not have to depend on a utility company for either water or the energy to raise it. No good estimates are available on the amount of energy supplied by these windmills, but the figure is probably not large. Mechanical wind machines generally have an energy capacity of less than half a kilowatt, so at best the world's wind pumps supply a few hundred thousand kilowatts of power — less than the capacity of a single large thermal power plant. Nevertheless, mechanical wind pumps play a crucial role in the many areas where they are the only economical way to draw water. Coal-fired electricity does not reach parts of the Australian outback, for instance, and getting it there would be prohibitively expensive.

Most mechanical wind machines use anywhere from four to twenty blades to capture the wind's energy, which is then transferred by a drive shaft to a pumping mechanism below. The machines can be made of wood, cloth, metal, plastic, or a combination of these materials, with the particular mix depending on the availability of materials locally and their cost. The most common wind pump in use today is the American multibladed fan-type machine. This horizontal axis design, little changed since its invention in the nineteenth century, is extremely rugged and will operate effectively at average wind speeds of less than ten miles per hour. Most of the parts of these machines, including the blades, are made of metal, and the diameter varies from two to several metres. The cost of wind pumps now sold commercially runs from around $4,000 to over $10,000.

The market for these conventional wind pumps is fairly large and well established. Along with manufacturers based in Argentina, Australia and the United States, a healthy wind-machine industry can be found in New Zealand, the Philippines, South Africa and West Germany. Although sales declined somewhat during the fifties and sixties, particularly in the United States, many remote areas never gained access to cheap fuel supplies and continued to rely on wind power. The wind-pump industry has been particularly strong in Australia and South Africa, where the machines are standard equipment on farms and where there is an infrastructure that includes a market for spare parts and repair services.

To realize the full potential of this technology in the Third World, new designs for wind pumps are needed — designs that are more appropriate to people's needs and to the often harsh conditions under which the machines must operate. A great deal of research has gone into this problem in recent years, mainly by private appropriate technology organizations supported by national governments and international aid agencies. Some of the most promising programmes are those of the Intermediate Technology Development Group (ITDG) in Great Britain, the Steering Committee on Wind Energy for Developing Countries in The Netherlands, and Volunteers in Technical Assistance (VITA) in the United States. Wind-pump experts have considered designs from all over the world and then adapted them to a variety of local conditions and needs.

The sailwing or Cretan windmill, first developed in the Greek Islands, but now used in several Mediterra-

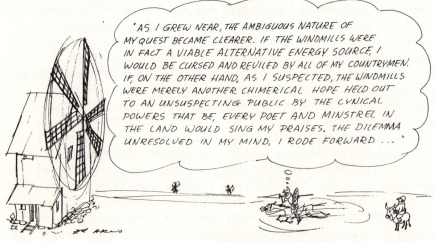

(*Ed Arno* © 1983, *New Yorker Magazine*)

nean countries, is one traditional design that has received a great deal of attention. This horizontal-axis windmill consists of a rotor with several metal or wooden spokes that have cloth sails stretched between them; the sails are, in effect, the blades of this machine. Cretan windmills are typically from three to six metres in diameter and their sails can be furled or removed during periods of high wind. Although the cloth must be replaced every two to three years, the rest of the machine can last ten to fifteen years if it is well constructed. Thousands of these windmills still provide irrigation water for farms in the arid Mediterranean region. A similar machine is used widely in Thailand, with the sails constructed of bamboo mats, a plentiful local material. Several thousand are in use, mainly for rice-paddy irrigation.

Researchers have been particularly drawn to the sailwing windmill because the design lends itself to local manufacture out of a variety of indigenous materials. Improved versions of this traditional design have been built in Colombia, Ethiopia, Gambia and India to meet the needs of local farmers. The Omo River project in Ethiopia, for example, significantly improved the sailwing's efficiency by adapting the Greek design and by using a double-acting pump. Similar machines have been tested in Canada, Malaysia and Sri Lanka recently. India's National Aeronautical Laboratory is one of several institutions in that country doing wind-energy research, and it has built a ten-metre diameter sailwing windmill that reportedly is more efficient than previous models.

Another innovative design based on traditional windmills is the Savonius rotor. This is a vertical-axis machine typically made of two oil-drum halves mounted around a perpendicular shaft so as to catch the wind. The machine is relatively simple and inexpensive to make but has the disadvantage of being heavy and inefficient. In addition, Savonius rotors have no built-in protection from high winds, so they must be attached to very sturdy towers, which adds to the cost. One partial solution to this problem is to use a wire frame and cloth sails instead of oil drums. Despite these drawbacks, the Savonius rotor does have a considerable number of proponents, and examples of it can be seen at development projects in several countries.

Many windmill development programmes in recent years have relied on a variety of materials such as thatch and wood that are both cheap and locally available. This contrasts with earlier projects that emphasized the need to import the latest technologies from industrial countries. The advantage of this new approach is that the projects often directly involve and benefit the rural poor who are most in need of an inexpensive nearby source of energy. Wind pumps can be considered a prime example of what E.F. Schumacher called an 'intermediate technology'.

Researchers and government planners are now turning their attention from individual prototype projects to the goal of making wind machines an integral part of development in the Third World and to building an indigenous manufacturing capability. To accomplish this, they must address a number of problems encountered in some early projects. Improving the ruggedness and lengthening the life of the machines is the first priority, since a windmill that lasts only a few years or that falls over in a major storm is rarely economical. Most wind-machine experts agree that improved engineering and the development of somewhat more modern production facilities will give windmills a longer life and will allow the quality control needed to turn out a consistently good product. Larger scale production can also in many cases lower the cost of wind pumps. Fortunately, a number of the designs developed recently are well suited to production in rural workshops or village factories. Some of the parts can be imported if necessary, but most of the manufacturing is well within the range of rural industries.

Las Gaviotas, a rural development institute in Colombia funded by the United Nations and the Colombian Government has taken a leadership role in this area. Researchers there spent six years carefully designing a reliable and inexpensive fan-type wind pump that functions in low winds and that is well suited to providing a family's water supply or to small-scale irrigation. A medium-sized production facility has been built to turn out 1,400 windmills a year, and the government is placing the wind machines in rural areas throughout the country. The Las Gaviotas factory takes advantage of plentiful local labour to produce an inexpensive machine that is also technologically one of the best anywhere. This programme is unique in both its scale and approach, and it has been widely acclaimed by experts throughout the world.

Christopher Flavin, *Worldwatch Paper*, 1981.

Generating the Links between Engineers and Rural Villages in Papua New Guinea

In 1975 the University of Technology, Lae, Papua New Guinea, in conjunction with the Appropriate Technology Development Unit (ATDU), based at the University, installed a small hydro-electric system near Baindoang, a remote rural village. Wind power and solar power had been considered but were not seen to have any great potential, although the latter may be feasible in the future. Since liquid fuel must be flown to Baindoang, diesel generation is not economical over time despite its lower initial costs. Hydro-

electricity was chosen as the power source because of the heavy continuous rainfall and high head available. Local initiative, essential for success, led to the choice of Baindoang: the headmaster of the local school heard a radio broadcast about a University seminar on rural electrification and invited the University to investigate possible sites in his area.

Baindoang is at an altitude of about 1500m and has a generally mild climate with fairly cold nights. It is a subsistence farming area with a small amount of coffee grown as a cash crop. Access is by either small aircraft or an extremely difficult two-day trek through very rugged mountainous terrain from Lae, the nearest town. The hydro-electric system is near the primary school and airstrip area, the focal point of the surrounding community of seven villages. The main village of Baindoang is 2km away at an altitude of 1800m and in the near future it will be linked to the system by a high-voltage transmission line.

Since most rural villages have no hope of a government-operated electricity supply, arrangements must be created which permit self-sufficiency. Initial costs must be kept to a minimum, management and operation must be provided by the people themselves, and maintenance arrangements handled on a co-operative basis.

The objective of the project was to install a scheme which was not associated with a mission, government or business interest, thereby allowing villages to achieve more self-sufficiency. It was essential, therefore, that the local community be involved in planning and installing the system in order to be prepared for independence from expatriate involvement, apart from occasional technical assistance. Thus outside involvement was restricted to technical advice and aid. The community, through the leadership of the village council of elders, decided and managed the non-technical details of the project. The council also established priorities for the use of electrical power: a two-way radio, a health aid post, the school, the teachers' houses, the trade store, coffee processing equipment, and a freezer. The elders also

encouraged the participation of local carpenters and an electrician.

The installation was managed by two communities. The first, consisting of University and ATDU staff, assessed various technical possibilities. A staff member then visited Baindoang and presented various alternatives to the council of village elders which, with some of the school teachers, formed the second committee. In this way, for example, it was decided that a hot water supply, shower block and washing facilities should be built. The system was installed in 1978. The community provided and managed the unskilled labour, which often consisted of women and children digging trenches and carrying materials. The University and ATDU provided skilled and semi-skilled labour; University students were encouraged to participate (which they did with great enthusiasm) by assisting with manual labour, house wiring, and design work. (Course credits were the reward for their efforts.)

The hydro-electric scheme is a high head (180m), small flow (8l/sec), system driving a pelton wheel giving a measured 6.8kW of electrical power. A single phase, 240-V, 30-amp, AC alternator is directly coupled to the turbine. The customary, expensive, and complicated mechanical speed governor was omitted in favour of an electronic load controller having no moving parts and requiring little maintenance. This device ensures that the electrical load on the generator is held constant, keeping the generator at constant speed. It was developed by the University and costs about US$200. The load controller enables electricity to be diverted from a 'continuous load' (a water heater) to a 'temporary load' (a radio broadcast or coffee processing).

The total steady electrical load is approximately 4.5kW, consisting of a 3kW heater and a 1.5kW lighting load which run continuously twenty four hours a day. A large hot water tank is used for showers and general domestic use at the communal facilities. No thermal insulation is used on the tank so the water never exceeds 60°C and the heat loss is available for drying clothes. Variation in the lighting load does not significantly alter the machine running speed but the load controller is required when an intermittent load such as a freezer or power tool is used.

In the event of a loss of a system load, a simple current sensing device will switch on a large water heater in the tail race of the turbine to prevent overspeed and damage to the generator. This method is simple, reliable, and inexpensive.

Answering the modest power needs of Baindoang required some bureaucratic intervention. After ensuring that initial electrical requirements were met, a few bush houses were wired up. Several safety problems, in particular the danger of fire in grass roofs and electric shock should a roof leak, led to the establishment of a standard for wiring bush houses by the Papua New Guinea Electricity Commission Authority.

Since stimulating self-sufficiency was a considerable influence on the engineers of this hydro-electric project, simplicity and reliability were essential to its design. The turbine and generator set was purchased from an overseas manufacturer known for reliability. All other

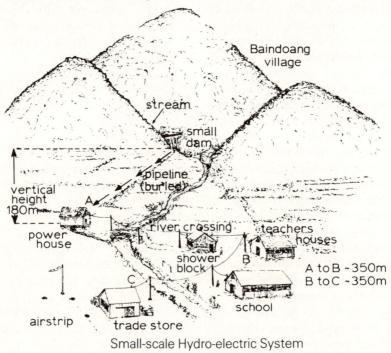

Small-scale Hydro-electric System

252

equipment was purchased in Papua New Guinea to minimize later problems of replacement. For future projects, pelton wheels will be manufactured locally. This will reduce the costs for high head schemes by up to several thousand dollars. A maintenance facility is being set up to service this and a number of other systems in the area.

The actual cost of the project is difficult to establish. Overhead costs such as staff time, airfares, and experimental facilities were included in the initial research project and the student field trip budget. The large amount of 'free' labour involved could not be assessed.

Travel and subsistence expenses were reduced by using the local labour force. Use of local materials avoided expensive airfreight charges. The hardware cost of the scheme was 10,000 kina ($14,000), including the community washing block and all electric wiring. Funding came from various aid organizations and University sources, plus a 10 per cent contribution from the village itself. Contributions from the future users are considered essential for any project of this sort; a sense of ownership is implanted as is active involvement in the project.

The transmission line to Baindoang, the next stage of development, will be expensive but the potential reward is high. The proposed 2km line, which will carry up to 3kW, and the wiring of up to forty houses will cost about $10,000. The scheme is supported by the government to discover low-cost methods for future systems and the socio-economic impact of rural electrification.

Peter Greenwood and C.W. Perrett, in Robert Mitchell, *Experiences in Appropriate Technology*, 1980.

Light in the Valley

The el Dormilon community in Central Columbia consists of some twenty small subsistence farming families living in the valley of the river Dormilon — with the only economic activities being the extraction of timber from the forest and of aggregate and ballast from the river, cattle farming and coffee growing. Besides suffering from a lack of infrastructure, such as roads, the area is showing signs of erosion where cattle farming has been started on the steep slopes cleared of trees.

A house-to-house survey was carried out to establish the perceived need for power and the potential to pay jointly for a micro hydro-electric plant. It became clear that while there was enthusiasm for the idea of electricity for lighting and cooking, the poorer members of the community would find it hard to afford the payments for a hydro plant used only to give domestic power. Some income-earning use for the power was needed, to take

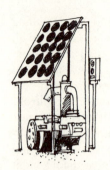

John Burton and
Ray Holland,
*Appropriate
Technology*, 1983.

advantage of the power which would be produced twenty four hours a day. Possible industrial applications included coffee hulling and sugar cane crushing but the one most often cited by the villagers as being of most use to them was timber sawing. Timber which was then sold at the roadside as logs, for processing elsewhere, could be sawn into planks, and there were numerous possibilities for making packing cases, broomsticks and even furniture, as well as building materials.

A study of the value that could be added by the establishment of a small sawmill and the repayment cost for the whole plant, showed that the cost of the sawmill and hydro plant would be amply covered by the income from the sawmill, and indeed there should be a substantial surplus. The electric generator and transmission lines would be paid for by the villages donating a fixed tariff, equivalent to their saving in kerosene and candles.

Further savings (in time if not necessarily in cash) could eventually accrue through the use of electric cooking. Because of the small size of the plant, limited by the flow of the nearby river and the available head, it was not possible for each household to have a one or two kW electric cooker and for them all to cook simultaneously (as they would normally do). The solution, which has been implemented by installing fifteen units, is to use heat storage cookers. They use a very low wattage and store energy as heat in a cast-iron block for twenty four hours per day, which can be extracted when needed. The heat stored in the cast-iron block should be sufficient for all the household cooking needs, but experience will decide how best to use these cookers and some modification may prove necessary to adapt them to Colombian cooking habits. Once all cooking is done electrically there should be a total saving in fuelwood collecting time to the community as a whole of some 4,000 hours a year.

Poverty is a great enemy to human happiness: it certainly destroys liberty, and it make some virtues impossible and others extremely difficult.
— Samuel Johnson

Old-timers with a Future

Japan, seen by so many as exclusively a land of high technology, may have to have its image revised. Engineers and farmers are trying to revive Japan's traditional, but much neglected, watermill industry.

In the first decades of the Meiji era, shortly after Japan was opened up to foreigners in 1853, watermills provided the main source of industrial power. Although steam power and hydroelectric power overtook the watermill by the beginning of the First World War, there were still some 40,000 waterwheels running in the 1930s. Yet by the beginning of the 1980s the number had shrunk to fewer than 800.

The attempts to revive this traditional rural technology owe much to a combination of nostalgia, small-scale energy economics in the wake of the 'oil shock', and the political concerns of a group of academics trying to promote pollution-free technology at the expense of the utility companies.

Evidence of the seriousness of the revivalist movement was a conference held in August at Kuroki, Kyushu, when more than 100 academics, farmers, and engineers met for two days to discuss the future contribution watermills could make to western Japan's energy economy. Delegates said that half the water-wheels in operation are used for pounding and polishing rice, while a third pump water for irrigation. The remainder have a variety of uses, such as milling and providing power for small factories.

Enthusiasts claim that recent technical advances will make water-wheels economic as a source of small-scale power. The Energy Saving Technology Design company, with the aid of a subsidy from Japan's powerful Ministry of International Trade and Industry, has produced a 2m diameter, 1m wide waterwheel weighing 1.2 tonnes. It can generate 3kW of power. The cost, with capital depreciation over twenty years, will be 3 pence(UK) per kWh. Utilities in rural areas charge 6p per kWh.

New Scientist, 1983.

World Bank Cools Sun Pump Rush

Whether sun pumps are an economic option now for farmers in developing countries, whether they are cheaper than diesel-powered pumps, whether reliability is a problem, are all questions guaranteed to cause debate and controversy.

To get some answers to these and a host of related questions, the World Bank commissioned a team of consultants to carry out a wide-ranging survey. In an effort to identify equipment suitable for use on the millions of family farms and smallholdings in developing countries, the consultants began their task with a state of the art review and a trial of different pumping systems.

Nine systems were selected for pumping trials in Mali, the Philippines and the Sudan, which are being conducted in close co-operation with the leading energy and agricultural agencies in those countries. The trials are the first independently conducted practical demonstrations and comparative testing programmes on the performance and reliability of small-scale pumps carried out under realistic field conditions in the developing world. They got underway in April and have about another six months to run.

Eight PV (photovoltaic) systems and one thermal have been subjected to rigorous testing and it seems that none of them have stood up to it too well. The team consider that it will take another ten years of development to evolve an economic, robust and reliable unit — a unit, moreover, that developing countries could manufacture themselves.

In economic appraisals it is common for sun pumps and diesel sets to be compared to see which is the cheapest. But it's difficult to compare like with like — there isn't, for instance, a diesel pump available with a power rating as low as 500W — and it is possible to load the equation to get the answer you want. The American NASA Lewis Research Centre reckons that PV systems are already competitive with alternative power sources for applications requiring 500kWh per year or less. On the other hand it is said that solar systems are going to be a poor buy until the cost of a unit goes down to around the US$600 mark. In the remoter regions of a good many countries the choice between solar or diesel is purely academic — oil supplies being unobtainable. It also matters not a jot if solar is cheaper than diesel, or vice versa, if the cost of either system is more than the small farmer can afford.

The study included working out how much small farmers could afford to pay for their water and then judged pumps against this performance criteria. The cost came out to $0.05 per m^3 of water delivered (at 1979 prices) — a target figure that none of the pumps tested came anywhere near achieving.

International Agricultural Development, 1980.

New Solar Infrastructure

Our future is a solar future, of that there can be no doubt. The question is whether we will continue in our old habits of thinking and futilely attempt to generate a high-technology, resource-intensive solar energy base that will hasten the degradation of the planet, or whether we will generate an energy base that, at every step of its formation and use, seeks to keep the flow of energy and resources at a mimimum.

Not surprisingly, the high-tech, resource-intensive mode is favoured by big business. Of the nine largest photovoltaics firms eight are now owned by large corporations, five of them major oil companies. According to the Solar Lobby, Exxon and ARCO will soon control more than half the industry between them. Other facets of solar technology are also being gobbled up by big business. For example, twelve of the top twenty-five solar companies are now controlled by corporate giants with annual US sales of $1 billion or more. Among them: General Electric, General Motors, Alcoa, and Grumman. Obviously, it is the goal of these companies to ensure that solar power is developed in as high-technology and centralized a manner as possible.

The 'big is beautiful' solar strategy is already leaving the drawing board and moving into actual production. Aerospace firms, for instance, are lobbying heavily to induce the government to fund 'Sunsat', a solar satellite that will be bigger than the island of Manhattan. And in Barstow, California, McDonnell-Douglas, backed by hefty federal funding, is completing work on its 'power tower'. The $130 million project consists of 2,200 giant mirrors that will focus sunlight on a boiler atop a 500-foot concrete tower. These schemes are obviously solar technologies developed by a fossil-fuel mentality; that is, they attempt to concentrate a diffuse solar flow as much as possible in the hopes of turning it into a centralized energy stock, much like coal and oil. The attempt, however, will only cause greater disorder than any possible value gained. The amount of nonrenewable energy resources that go into constructing the parts of a giant solar satellite and launching it into space where it must be assembled is far greater than the amount of energy the Sunsat could produce for many years. Concentrating solar rays in such high density and beaming them back to collectors on earth will cause microwave radiation pollution that will endanger the health of anyone living or working near the collector. Some areas of the country might be deemed uninhabitable because of the microwave danger. Once the power has been collected at a central location, it must then be shipped as electricity through power lines. This will require the use of more quantities of nonrenewable resources to construct this part of the infrastructure. The 'power tower' suffers similar problems. The more concentrated the collection of solar rays, the less net energy will remain.

Even on a smaller scale, there are important choices to be made. For example, in home units, solar power

'Perhaps we did make a mistake when we calculated the number of solar cells we needed . . .'
(*Hans Seim, GATE 1983*)

257

can be provided through either low or high technology. The higher the technology used, the less net energy will be provided, because the more nonrenewable resources must be used to build and maintain the collecting infrastructure. For instance, in the high-tech — or active — home system, sunlight is first concentrated in a collector made of nonrenewable resources; then the solar energy is stored in either air or water housed in containers manufacturered of nonrenewables; finally it is moved by fans or pumps to perform the work required. Another high-tech system is one in which photovoltaics concentrate energy and store it in batteries. Once again, nonrenewable materials form the base for the technology. While these systems clearly use a less intensive form of technology than do solar satellites and power towers, small-scale home units of an active nature must still depend ultimately on the supply of copper, platinum and other diminishing ores of which the solar utilization equipment is manufactured.

Passive home solar systems tend to be less ecologically damaging and provide the most net energy yield because they are based less on nonrenewable technology and more in the life experiences of the first Solar Age that preceded the fossil fuels era. In a passive system, homes are actually designed and constructed in such a way that they naturally remain cool in summer and warm in winter. Although many workable prototypes of passive solar homes have recently been developed by architects, anthropologists can point to workable systems that were developed hundreds and even thousands of years ago by peoples who had no other way to maintain their homes.

Jeremy Rifkin, *Entropy: A New World View, 1981.*

Before the windmill came people had to walk many miles to get water

VI

Technology for Development

Housing, Construction and Transport

ALONG WITH food, water and energy, shelter and the means of transportation
are among the most basic of all human needs. According to a recent UNIDO
report:

> If the poorer sections of the community are to be helped to build structur-
> ally durable and functionally adequate houses, the focus of attention will
> have to be on the provision of suitable building materials to them at a cost
> they will be able to bear. The materials should be such as are available
> locally and that do not require specialized skill in their use. The basic pur-
> pose of such a strategy should be to enable these people to build dwellings
> that would serve as a store of value and an appreciating asset, so that the
> time and money spent in continual maintenance and eventual replacement
> of their non-durable dwellings would be freed for useful alternative pur-
> poses and the dwellings themselves would increase in value owing to their
> durability and become capital assets that could be sold or mortgaged.[1]

It goes on to point out that it is unrealistic to suggest better ways of building
that require significant expenditure of cash because this is one resource that is
always in short supply in a subsistence economy. There are not enough
resources in the world to build housing for everybody according to conventional standards. Nor is it necessary to do so. There are many traditional and
improved techniques and materials which can provide the rural and urban-
fringe communities with decent accommodation at a fraction of the cost of
conventional methods.

It is on the use of local materials, skills and techniques that the housing section of this chapter concentrates. The extract by Miles and Edmonds sets the
scene by pointing out that the rate of housing construction in most developing
countries is totally inadequate, of little benefit to the masses of the poor, and
heavily dependent on the importation of materials and equipment. It recom-
mends therefore a move to reduce the overall investment per unit and to
make more use of local materials and resources. This is followed by examples
of low-cost alternatives to conventional housing strategies — extracts from
Hassan Fathy's *Architecture for the Poor* which talk of the dignity and beauty

---

[1] UNIDO, *Monographs in Appropriate Industrial Technology No.12: Appropriate Industrial
Technology for Construction and Building Materials*, 1979.

of the mud brick and the simplicity and naturalness of artisanal masonry techniques; an extract on lime and lime-pozzolana mixtures which are alternatives to Portland cement in some applications and can be manufactured by very simple processes at the village level; and the fabrication of clay or soil building blocks made by small enterprises or by poor community groups on a self-help basis.

Another area of the construction industry in which great potential exists for substituting labour-based techniques for expensive imported equipment is that of road building. In Section 2, Geoff Edmond's extract on labour-based road construction gives comprehensive coverage of this topic. It includes an indication of the objections that have been raised against labour-intensive techniques (too costly, too slow, low quality, etc.) and the steps taken to overcome them; an account of the difficulties faced during the implementation of major labour-based road construction programmes such as the Kenyan Rural Access Roads Programme, and the achievements realized; and the important role of training to enable efficient management of labour forces. Also in this section is an extract on using modern materials to improve traditional bridges in Papua New Guinea — an example of how the means of communication can be improved without great expense.

The final section of this chapter looks at the issue of low-cost vehicles. When the transportation needs of the poor are talked about at all, it is usually in terms of building more roads and bridges to 'open up' rural areas and link up rural communities with each other and with market towns. Little, if any, attention is given to the problems of movement between farms and homesteads or between water collection points and homesteads — yet an enormous amount of movement takes place at this level — usually undertaken by women — often along rough and dangerous paths. Similarly, although provision of feeder roads may ease the journey to market once the road has been reached, the poor may be unable to derive much benefit from these if no means of transportation is available to them.

Thus, most of the extracts in the transportation section deal with the neglected sector of low-cost transportation aids and vehicles for off-road and on-road movement of loads. It starts with an extract by Barwell and Howe which describes the transportation problems of the poor and defines the six categories of basic transportation aids available. This is followed by extracts on a range of appropriate technologies such as the Korean Chee-geh (a backpack); Chinese wheelbarrows, and one-wheel cycle trailers. Apart from their obvious benefits to users, these technologies also have the advantage of creating off-farm employment opportunities in small local enterprises.

The need for low-cost means of the transportation of people and goods is also an important issue for the urban poor — especially in the vast, sprawling cities of Asia. Here a variety of methods have been devised to meet the need — cycle rickshaws, auto-rickshaws, becaks, jeepneys and many other types of small, wheeled vehicles. An account is given of the thriving rickshaw business in an Indian city, and a cautionary tale by Ivan Illich on the choice of transportation technologies, completes the section.

1 HOUSING AND BUILDING MATERIALS

Shelter for All

Given the basic need for shelter, the output of housing is worth noting. Table 1 shows that in Africa less than 1 per cent of publicly financed construction work is in housing. The United Nations Economic Commission for Africa has suggested that, to meet the present and future needs for housing, developing countries should aim at an annual production rate of ten units per 1,000 population. This figure is somewhat optimistic in that few of the industrialized countries actually attain that level. However, the actual output of countries for which figures are available provides a gloomy picture. Most of the investment in housing comes from the private sector and may not necessarily benefit the mass of the population. Moreover, Table 2 shows that the actual output per annum in developing countries is derisory. Only when countries attain a level of GNP/capita greater than 1,000 does the level of housing output begin to represent a substantial figure. It should of course be recognized that the figures reflect only the formal, generally urban, sector. Dwellings constructed in the rural areas and/or by owner-occupiers and

Table 1
Division of Construction Output in Percentage in Developing Africa, 1970

Economic sector	All works	New work			Repair and maintenance		
		Total	Public	Private	Total	Public	Private
Agriculture, mining, quarrying	11	10.2	2.2	8.0	0.8	0.1	0.7
Manufacturing, construction	12	11.4	2.4	9.0	0.6	0.1	0.5
Gas, electricity, water	7	6.6	6.6	–	0.4	0.4	–
Communications	24	16.0	16.0	–	8.0	8.0	–
Commerce	5	4.5	0.5	4.0	0.5	–	0.5
Dwellings	35	31.8	0.3	31.5	3.2	–	3.2
Education	3	2.8	2.1	0.7	0.2	0.2	–
Health, welfare	3	2.8	2.3	0.5	0.2	0.2	–
Total	100	85.1	32.4	53.7	13.9	0.0	4.9

Source: UNECA, Economic Conditions in Africa (1971).

Table 2
Housing Construction, circa 1979

GNP per capita	Dwellings completed per 1,000 population
Less than 500	0.16
500-999	0.7
1,000-2,000	3.5
More than 2,000	6.5

self-help do not usually figure in the statistics. Nevertheless, it is the urban areas which have the highest growth rates of population. Unless the housing output keeps up with the population growth rate there will be no possibility of housing the urban population. As with many other aspects of the construction sector, the solution may not be more money but reorganization of the use of available resources.

Naturally, housing output is particularly dependent upon government policies. Thus, whilst the regression analysis shows that housing output variations are explained to a large degree by changes in GNP per capita, half of the variation has to be explained by other factors.

In most developing countries the construction industry is heavily dependent upon the importation of materials and equipment. This is a serious matter since materials alone account for between 50 and 60 per cent of the cost of construction output. In Africa, for example, as much

as 60 per cent of all materials used in the industry are imported. It would be over-optimistic to expect an even higher investment in construction to reduce this reliance. We have already noted that investment in construction is as high as it could reasonably be expected to be. More reasonably one can suggest that there should be a move to reduce the overall investment per workplace and make more use of local materials and resources. Certainly there has been no lack of research into the effective use of indigenous building materials. Examples abound of the possibility of substituting lime for cement, of using reinforced soil for brick and block making and of alternative roofing materials. Unfortunately, few of these alternatives have become commercially viable.

Construction equipment accounts for some 10 per cent of all imported equipment in developing countries, and the amount imported is growing each year.

G.A. Edmonds & D.W.J. Miles *Foundations for Change*, 1984.

 # The use of Mud Brick an Economic Necessity

We are fortunate in being compelled to use mud brick for large-scale rural housing; poverty forces us to use mud brick and to adopt the vault and dome for roofing, while the natural weakness of mud limits the size of vault and dome. All our buildings must consist of the same elements, slightly varied in shape and size, arranged in different combinations, but all to the human scale, all recognizably of a kind making a harmony with one another. The situation imposes its own solution, which is — perhaps fortunately, perhaps inevitably — a beautiful one.

Whatever the peasant may want to do, whatever rich

men's villas he may wish to copy, he won't be able to excape the severe retraint imposed upon him by his material. Whether, when he has lived in a truly beautiful and dignified village, he will still hanker after imported modernity, we shall have to wait and see. Perhaps, when he has no reason to envy the rich man anything at all — his wealth, his culture, and his consequence — then too he will cease to envy him his house.

Hassan Fathy, *Architecture for the Poor*, 1973.

Mud Bricks, an Adze and bare Hands

In each room there were two side walls, 3m apart, and an end wall somewhat higher against which the vault was to be built. The masons laid a couple of planks across the side walls, close to the end wall, got up on them, took up handfuls of mud, and roughly outlined an arch by plastering the mud onto the end wall. They used no measure or instrument, but by eye alone traced a perfect parabola, with its ends up on the side walls. Then, with the adze, they trimmed the mud plaster to give it a sharper outline.

Next, one at each side, they began to lay the bricks. The first brick was stood on its end on the side wall, the grooved face flat against the mud plaster of the end wall, and hammered well into this plaster. Then the mason took some mud and against the foot of this brick made a little wedge-shaped packing, so that the next course would lean slightly towards the end wall instead of standing up straight. In order to break the line of the joints between the bricks the second course started with a half-brick, on the top end of which stood a whole brick. If the joints are in a straight line, the strength of the vault is reduced and it

may collapse. The mason now put in more mud packing against this second course, so that the third course would incline even more acutely from the vertical. In this way the two masons gradually built the inclined courses out, each one rising a little higher round the outline of the arch, till the two curved lines of brick met at the top. As they built each complete course, the masons were careful to insert in the gaps between

Mud bricks . . . an appropriate building material in rural areas

the bricks composing the course (in the extrados of the voussoirs) dry packing such as stones or broken pottery. It is most important that no mud mortar be put between the ends of the bricks in each course, for mud can shrink by up to 37 per cent in volume, and such shrinkage will seriously distort the parabola, so that the vault may collapse. The ends of the bricks must touch one another dry, with no mortar. At this stage the nascent vault was six brick-thicknesses long at the bottom and only one brick-thickness long at the top so that it appeared to be leaning at a considerable angle against the end wall. Thus it presented an inclined face to lay the succeeding courses upon, so that the bricks would have plenty of support; this inclination, even without the two grooves, stopped the brick from dropping off, as might a smooth brick on a vertical face.

Thus the whole vault could be built straight out in the air, with no support or centring, with no instrument, with no drawn plan; there were just two masons standing on a plank and a boy underneath tossing up the bricks, which the masons caught dexterously in the air, then casually placed on the mud and tapped home with their adzes. It was so unbelievably simple. They worked rapidly and unconcernedly, with never a thought that what they were doing was quite a remarkable work of engineering, for these masons were working according to the laws of statics and the science of the resistance of materials with extraordinary intuitive understanding. Earth bricks cannot take bending and sheering; so the vault is made in the shape of a parabola conforming with the shape of the bending moment diagrams, thus eliminating all bending and allowing the material to work only under compression. In this way it became possible to construct the roof with the same earth bricks as for the walls. Indeed, to span 3m in mud brick is as great a technical feat, and produces the same sense of achievement, as spanning 30m in concrete.

The simplicity and naturalness of the method quite entranced me. Engineers and architects concerned with cheap ways of building for the masses had devised all sorts of complicated methods for constructing vaults and domes. Their problem was to keep the components in place until the structure was completed, and their solutions had ranged from odd-shaped bricks like bits of three-dimensional jigsaw puzzles, through every variety of scaffolding, to the extreme expedient of blowing up a large balloon in the shape of the required dome and spraying concrete onto that. But my builders needed nothing but an adze and a pair of hands.

Within a few days all the houses were roofed. Rooms, corridors, loggias, were all covered with vaults and domes; the masons had solved every problem that had exercized me (even to building stairs). It only remained to go out and apply their methods throughout Egypt.

Hassan Fathy, *Architecture for the Poor*, 1973.

264

Lime or Cement?

Before the development of modern cements, lime was used in many of the situations where cement is used today. For several centuries, virtually all permanent buildings were of masonry construction, and the masonry units, whether brick or stone, were usually laid up in lime mortar. Plastering, both internally and externally, was also done with lime plaster (usually with some fibrous binding material such as horsehair to hold the plaster together and to prevent the formation of cracks). Such concrete was used, for example in foundations, was also made with lime.

The replacement of lime by cement resulted in much more quick-setting mortars and plasters, and the consistency of the material gave the builder more confidence. At the same time, the messy lime-slaking operations which were needed to produce the best lime mortars were eliminated. Consequently, wherever Portland cement was available at a reasonable price, it tended to replace lime altogether.

But mortars made with Portland cement and sand alone, without lime, tend to be harsh and difficult to work with, and when they have reached their full strength, tend to be much stronger than the bricks that they bond. This is undesirable as it can cause cracking in the masonry. A much better mortar is one based on lime, to give the workability required, but with just enough cement added to provide adequate strength and a fast enough set to allow the work to proceed quickly. This is the standard mortar used in industrialized countries today. The availability of bagged hydrated (slaked) lime enables such mortars to be made on the site just as readily as cement mortars.

An alternative method of achieving the strength necessary for lime mortars and plasters is by the addition of pozzolanas. Pozzolanas are materials which are not cementitious in themselves, but which when mixed with lime will cause the mixture to set and harden in the presence of water-like cement. These materials are of ancient origin, and were used by the Romans to make the concrete for many of their most magnificent durable structures. The massive 42m dome of the Pantheon in Rome, for example, which has survived nearly 2,000 years, is made from lime-pozzolana concrete.

The significance of lime and lime-pozzolana mixture as alternatives to Portland cement in developing countries is that both can be manufactured by very simple processes suitable for village-scale technology. Lime is made by age-old technologically unsophisticated processes in most countries where limestone is available, in lime kilns of an enormous variety of shapes and sizes. However, most of these processes are highly inefficient, being based on intermittent or batch production of lime. Because of the inefficiency of the process, the price of lime is often as high or even higher, per unit weight, as cement and it is cheaper and more reliable to use cement alone in mortars and plasters. A 1:6 cement/sand mortar is common in many developing countries. Locally produced lime is frequently used only for lime washes. But with improved manufacturing techniques, it is possible to produce lime of a consistent and satisfactory quality at a price which will usually be competitive with Portland cement.

Pozzolanas are currently used in many developing countries. In Indonesia, huge deposits of volcanic ash or tuff provide one source of pozzolana which is used for mortars and for block-making. A pozzolana commonly used in mortars in India and Indonesia is pulverized fired clay, sometimes made by grinding up reject bricks and tiles from the brick and tile kilns; another is the ash from agricultural wastes such as rice husks. But a lack of scientific understanding of pozzolanas, and a lack of quality control in their manufacture has often meant that these potentially valuable but inexpensive, locally produced building materials are rejected in favour of modern factory-pro-duced materials, although recently developed manufacturing tehcniques and know-how make it possible for local pozzolanas to be used much more widely than at present.

The capital cost of the equipment to manufacture these alternative materials is very much less than that needed to set up the equivalent cement-making capacity, even when using small-scale plants. The equipment could be built locally or manufactured in small or medium sized workshops.

A further way of reducing the need for Portland cement and replacing it with more appropriate materials would be to design buildings in such a way that reinforced concrete is not needed, or the amount of cement needed in the reinforced concrete is reduced. Many techniques of this sort are available.

By adopting such strategies developing countries could reserve much of their supplies of Portland cement for the construction of dams, highway bridges and other reinforced concrete work for which no alternative material is available.

Robin Spence in UNIDO, *Monographs on Appropriate Industrial Technology No 12; Appropriate Industrial Technology for Construction and Building Materials,* 1980.

 # Self-help Housing in Brazil

In Campina Grande, a city 100 miles inland from Recife, the Programa de Aplicacao de Tecnica Adaptada nas Comunidades (PATAC) has been attempting to use appropriate technology to alleviate some of the poverty and appalling living conditions.

Its work is based on the use of appropriate technology devices as instruments of social and community education and it brings together many of the poorest community groups to use appropriate technology as a means of improving their lives without depending on outside financial help. One of the more successful technologies introduced is a block-making machine.

Behind the block-making machine is the very simple idea of families making the blocks for building their own houses, and paying for the materials used by making more blocks than they need and selling the extra blocks.

A family can produce about 1,000 blocks per day, and the average number needed to build a house is 2,000. In five days a family might make 5,000 blocks, of which 3,000 can be sold to pay for the materials used and for the cost of operating the machine. If they need money to pay for rafters, tiles, cement, etc., they continue to make extra blocks for sale so that they can buy these materials too.

This is not just a process for making blocks. It is a way in which people can join together to help each other, as well as themselves, and is an ideal educational device. It brings ownership of a solid brick house within the reach of people who have never even dreamed of the possibility. It demands group participation; the ideal number of operators is between four and eight. It is simple to use, and can call on the vast fund of unemployed and underemployed women and children. For the first time these small community groups are able to take the initiative in improving substantially their housing conditions.

Since the first machine was produced several problems have arisen because of the huge increases in the prices of transport, cement and sand. Furthermore, the need for the relatively expensive small wooden frames has exceeded the original calculations. For these reasons the block-machine has been redesigned.

The new machine will produce a block of the same outside dimensions, but with larger holes — this reduces the volume of material by 30 per cent and gives corresponding benefits in sand and cement savings. Cement content is raised slightly to give a similar maximum loading capacity, and this also improves the appearance of the block which in turn makes it more saleable. The new machine is also designed to function without the small wooden block frame, giving a further considerable saving. The gradual introduction of the new machine through substitution (35 per cent of the old machine's components are utilized) is significantly improving the flow of blocks and speeding up the process of house building.

Robert Mister, *The Courier*, 1980.

The human being, defined by Thomas Aquinas as a being with brains and hands, enjoys nothing more than to be creatively, usefully, productively engaged with his hands and his brains. Today, a person has to be wealthy to be able to enjoy this simple thing . . . He really has to be rich enough not to need a job. — E. F. Schumacher

Soli, The Block Maker

Soli learned how to mould building blocks at an early age; his mother died when he was nineteen and a year later his father died leaving the responsibility for the entire household on the young man's shoulders. He started work as an apprentice in a small building block factory near his home. They moulded building blocks made from a mixture of clay and water with a simple locally-made wooden moulding tool. Three semi-skilled workers and seven labourers were employed. Soli worked hard and was soon promoted to a semi-skilled labourer and was thus involved in actually making the bricks for a period of two years.

At this stage Soli left the factory in order to set up his own unit in a nearby village some ten miles away from his own home. He leased out his family's property and bought a smaller house in the village together with a metal-moulding machine for his own enterprise. He started to produce hollow building blocks in the backyard of the new house and employed two labourers to help him. The business flourished from the beginning since the demand was high, and the factory produced at its full capacity of around 100 blocks per day.

During the first year of operation the factory produced 36,000 hollow building bricks. Soli sold these for 20 Kobo per block and sales thus amounted to ₦7,400 during the year. Soli estimated that he used ten donkey loads of clay each day at ₦1 per load while he paid the labourers ₦2 each per day and withdrew ₦3 for his own use. Soli calculated that he had used ₦3,600 worth of raw material during the year and paid a total of ₦2,520 as wages to his labourers and himself. This the net profit at the end of the first year was ₦1,280. Soli reinvested this profit in his business and acquired a new manual block-moulding machine which was capable of producing 200 blocks per day. This machine cost ₦600 and he also built a small shed in which to store his finished stocks.

The business prospered during the first six months of the second year; towards the end of this year Soli faced certain serious problems. First, cement blocks started to offer serious competition. These quickly captured

Very effective building blocks can be made using simple machines

a large proportion of Soli's market and this had disastrous results on the demand for his products. Soli soon found himself with about 40,000 blocks in his shed ready for sale and there was no more space in which to store more blocks. Soli was nevertheless determined to survive; he laid off his two labourers and devoted all his time to finding a new market for his blocks. He found, however, that the cement blocks had penetrated into every market and were selling very well. The cement blocks equivalent to Soli's clay blocks were sold for 40 Kobo, which was twice the price of Soli's product but they were undoubtedly stronger and more durable.

A second problem arose from the fact that the rainy season was fast approaching. The engine room and the storage shed occupied more than three-quarters of the plot and the remaining piece of spare land was piled high with the finished blocks for which there was no room in the shed. These blocks would not survive the torrential downpours and strong winds which could be expected at the onset of the rainy season. In addition, the nearby source of clay was almost exhausted and the cost of transport made it quite uneconomic to bring clay from the next source which was about ten miles away, near Soli's original home.

Soli decided that his first priority was to clear his stock and he cut his selling prices from 20 Kobo to 15 Kobo. This price cut achieved very little and he only cleared his stock by making a further reduction down to 10 Kobo per block.

Soli then approached the Small Scale Industry's Credit Scheme Funds for advice on management and for financial assistance. This scheme had recently been introduced to assist in the establishment, expansion and modernization of small-scale enterprises which were considered technically feasible and commercially viable. The officials operating the scheme reviewed Soli's case and conducted a market survey in order to determine the feasibility of establishing a concrete block factory in the area. They were satisfied that there was a sufficiently large potential market and they approved a loan of ₦6,000 for Soli to purchase machinery and equipment. It was understood that Soli himself would contribute the cash he had realized from his cut price sale in order to provide working capital and to pay for an office and store as well as a new shed. In addition, the officials helped Soli to acquire a new plot and arranged for him to undergo a two-week training programme on the principles of management and basic book-keeping.

Malcolm Harper & T.Thiam Soon, *Small Enterprises in Developing Countries*, 1979.

If the people cannot adapt themselves to the methods, then the methods must be adapted to the people. — E. F. Schumacher

Labour-based Road Construction Comes of Age

The idea of using infrastructure works as a means of providing employment, and thereby income, to the unemployed is not a new one. This solution has been, and is still being used in both developed and developing countries. Generally, however, the emphasis has been solely on employment creation. What was new about the labour-based concept, developed principally by the World Bank and the ILO in the early 1970s, was the idea that labour-based techniques could be substituted for equipment. It was this idea that many found difficult to accept: labour-based techniques were fine as part of a relief programme, but to suggest that they could achieve the same results as equipment at comparable costs was something else entirely! Recognizing that in the long term it would be the technical people who would have the responsibility for implementing labour-based techniques, it was clear that it was these people who had to be convinced.The main objections to the use of labour-based techniques were that they were more expensive, took longer and produced an inferior quality. The question of cost is one which depends to a large extent on the basic assumptions used regarding availability and utilization of equipment (usually grossly exaggerated) and the productivity of labour (generally underestimated). The initial economic studies carried out in the early 1970s in Iran, Nepal, Thailand and the Philippines suffered from this lack of precision. They set a framework for analysis and showed that labour-based techniques were competitive: but they were open to the criticism that the conclusions reached depended greatly on the assumptions made regarding productivity.

The next step was to obtain a clearer idea of the potential output of labour. This was done through detailed work-studies carried out by the World Bank in India and the ILO in the Philippines and Kenya. The studies showed that, in experimental projects at least, labour-based techniques could produce at a rate which would make them competitive with equipment for a large range of construction activities. In particular, very few of the activities involved in rural road construction were more effectively executed by equipment, with the exception of compaction and long haulage. What was needed next was a full-scale demonstration of what seemed, on paper, to be a viable proposition. About this time, the Kenyan economy was suffering the effects of the 1973 oil crisis. Moreover, a general slackening of world prices for tea and coffee meant that unemployment was a serious problem. One of the measures the government decided on was the launching of a labour-based road-building programme. The Rural Access Roads Programme (RARP) was a major breakthrough, both in its concept and its actual implementation. The concept was that a network of over 10,000km of rural roads could be constructed and provided with a gravel pavement utilizing labour-based techniques. The Programme was the first of its type in Africa and many donors felt that the initiative taken by the government should be supported. In addition to assistance with the execution of the Programme, an advisory team was provided to develop the managerial, administrative and training aspects....

The major problems stemmed from the introduction of a 'new' technology. Supervisors were not trained to organize labour, payment systems could not accommodate large groups of casual labour, procurement systems were not adapted to the purchase of good-quality hand tools and no productivity data existed on the output of labour. It was on these issues that the management of RARP had to concentrate their efforts.

The RARP is now well-established, employing 10,000 people and producing some 1,200km of access roads each year. The average cost per kilometre is US $7,000 which would be very difficult to match using equipment. The success of the Programme should not only be measured by its effect in Kenya, however. On the strength of the physical evidence provided by the RARP, many other countries have decided to implement similar programmes. The Programme has also provided the basis for the development of the documentation and training material required for the setting-up of such programmes. Perhaps more than anything else, however, it showed that labour-based techniques could be used on a large scale for the building of roads, that they could achieve a good standard, that they were inexpensive and that overall output could be maintained at a reasonably high level. Parallel research, based on several programmes including the RARP, has shown that labour-based techniques can be viable for rural road construction even where wage rates are as high as US $5 to $6 per day. This, of course, would make them feasible in the majority of the developing countries. However, there are geographic differences as regards the application of these techniques. In the first place, there is a group of countries where the use of labour-based technologies is traditional and well-understood: India, Pakistan, Bangladesh and Sri Lanka. Massive unemployment in these countries is more or less endemic and the idea of using labour-based technologies is accepted as normal. In these cases, improvements could be made in the quality of the final product and, to a lesser extent, in productivity.

In South-east Asia, no comparable tradition exists. The model for the development of countries such as the Philippines and Thailand has been Japan rather than India. There is a natural resistance to the use of labour-based techniques. Pilot projects in the Philippines, however, have shown there is scope in these countries, particularly in the area of road maintenance which has become a major problem in so many developing countries.

Sporadic attempts at the introduction of efficient labour-based techniques have also taken place in Central and South America. Mexico, for instance, initiated a large-scale programme of rural road construction in the 1970s which succeeded in providing access to 60 per cent of the 12,000 villages previously having no access whatsoever. Smaller projects have taken place in Guatemala, Peru, Columbia, Honduras and the Dominican Republic. The problem is that these countries tend to look north for their technology. In addition, and this a general problem, the curriculae of most highway engineering courses are imported from the developed countries and take little cognizance of the environment in which the student will eventually design and construct roads.

In sub-Saharan Africa, the concept of labour-based road construction and maintenance was accepted more easily. Several factors explain this, not least the severe effect of oil prices and the very high growth rate of the population. In addition, many of the countries are very large, and construction and maintenance of an effective road network presents major logistic problems. There is no doubt that, at this stage of economic development in most developing countries, large fleets of equipment are a luxury which consume a large measure of scarce foreign exchange. Consequently, labour-based road construction and/or maintenance programmes have recently been set up in Mozambique, Lesotho, Botswana, Malawi, Benin and Ethiopia.... Given that labour-based techniques are new to most countries, it is clear that a certain measure of technical assistance has been necessary. However, documentation has been developed by the ILO and other organizations, in order to limit the external inputs to a minimum. Thus, basic supervisory training courses emphasizing the efficient management of large labour forces are available; also a guide for the manufacture of simple tools and equipment for labour-based programmes exists. In addition, some of the fundamental activities such as labour supply evaluation, basic site organization and management, payment systems and recruitment have been described in various

More roads, more jobs, more income

272

publications. In recent years, road maintenance has become a more important issue than road construction. In many countries, the road network is falling apart at a faster rate than it is being constructed. The reasons for this are many and varied. However, it must be stressed that it is generally not because of lack of concern by the governments involved. Most developing countries spend a higher percentage of their national income on road maintenance than most industrialized countries. The issue here is that it can no longer be assumed that more money and a computerized management system will solve the problem. It is necessary to look at the whole organization and administration and suggest alternatives. The key to solving the problem is to be found in using the resources that exist in the country, not only labour but also the goodwill of the people. Alternative systems of road maintenance are now being tried out in several countries. One word of caution: self-help will solve some of the problems, but people must be motivated towards self-help. If they have no involvement in the planning and construction of a road, it is unlikely that their assistance can be relied upon to maintain it.

In regard to road maintenance, Kenya is again showing the way in that the RARP is to be transformed into a minor roads rehabilitation and maintenance programme. This means that the methods used to construct and maintain access roads will now be used to deal with the minor roads of the country which account for over 40 per cent of the total network.

Geoff Edmonds, *Appropriate Technology*, 1983.

Better Bridges

Traditional bridges in Papua New Guinea (PNG) are constructed with materials usually found in the bush. The main materials are wood, bamboo, bark, vines, cane and stones. Traditional bridge construction has been changing in three ways: complete replacement by imported materials and design, obtaining traditional materials in new ways, and replacing some traditional materials with imported materials in basically traditional constructions....

Where motor vehicle roads have replaced walking tracks, traditionally constructed bridges have of course been replaced. But also on many walking tracks, traditional bridges have been replaced by footbridges using imported materials and design. An example is a bridge constructed with wood and vines being superceded by a larger bridge using steel cables and concrete anchorages....

Such 'imported' bridges, called 'wire bridges' in PNG, are more costly and require different knowledge to construct than traditional bridges, but they are supposed to be more durable. However, when Papua New Guinea's mighty rivers flood, they can take wire as well as traditional bridges. Wire bridges are also supposed to be easier to maintain, but the remnants of the bridge over the Sepik illustrate that the wire

273

The only way to market.
(*R. Witlin: United Nations*)

bridge may not have been appropriate because it could not be maintained properly.

With money and new forms of transportation, people have new ways of getting hold of traditional materials they may want to use for bridge construction. Rather than go far into the bush to find cane, the builders of the Sau River Bridge purchased it from another village. And the builders of the Kopeme Bridge brought their cane up from the Markham Valley by truck.

If imported materials such as wire, steel cables, or nylon rope are available, they are often used to repair or reinforce traditional bridges. The Kapolame Bridge uses wire for hold-back cables and the Ambui Bridge over the Lai River near Pompabus Mission has replaced vines with red nylon rope for suspended cables.

For the purist, these foreign intrusions into traditional technology are an eyesore, but for the people using the technology, they are an improvement. Perhaps this compromise is the future of traditional technology in Papua New Guinea.

J. Siegel, *Traditional Bridges of Papua New Guinea*, 1982

3 LOW-COST TRANSPORT DEVICES

The Transport Needs of the Poor

The most significant transport needs of the poor are those which relate to agricultural activities, since it is through the generation of marketable surpluses (and thus income) that other goods and services become affordable. However, transport needs at the farm level have very rarely been studied. Roadside surveys of the commodities carried by motor vehicles are a poor substitute. They are too far along the marketing chain to be able to isolate individual consignments and the distance over which they are being moved. None of the studies have been sufficiently extensive to give any adequate measure of seasonal fluctuations in travel; and they give no indication of on-farm transport needs.

One farm-level study was carried out in Kenya by the World Bank in 1976. This suggested that most transport needs could be characterized as the movement of small loads (10-150kg units) over relatively short distances (1-25km). On-farm the range of loads was likely to be the same, but the typical distances were shorter (1-13km). The amounts of water and wood required for household use were noteworthy (50 and 30kg respectively), since it was estimated that their collection occupied 3-6 hours per day. Since the farmer must follow a fairly rigid schedule to obtain optimum yields, it is important that on-farm transport for crop production and household needs should be so time-consuming as to delay operations. For example, if some crops are not sprayed on time

275

the results may be disastrous. Yet the spraying of cotton with insecticide required about 200-300 litres of water/hectare. For a four hectare plot this is 800-1200 litres and between 7-10 sprayings are normally recommended, i.e. between 6-12 tonnes of water, a formidable amount if headloading is the only available means of transport. In the studies in Kenya it was concluded that on-farm transport was already a burden if not an outright constraint on small farm activities.

Off-farm transport comprises two elements: between farm and roadside, and between roadside and collection point/market. An example of how large these elements can be is given by the definition in a study of Nepalese peasant agriculture of the terms 'on road' to mean at the roadside or within a few hundred metres, 'near road' to mean up to half a day's walk from the road, and 'off-road' to mean more than half a day's walk.

There is a dearth of information about the magnitude, frequency and duration of the small farmer's movement needs. What is clear, however, is that the transport requirements can be substantial, even when only small areas are planted.

Recent rural transport strategy has pursued development from the top downwards. There has been a progression from major primary, to secondary and only latterly to tertiary highways all built on the basis of design philosophies and technologies imported from the developed countries, a reliance on developed country motor-vehicles with only very recently a small step in the direction of lower-cost, but still motorized, vehicles and the complete neglect of traditional forms of transport. The result is skeletal road networks that in the poorer countries plainly do not serve effectively the majority of the population, and vehicles so expensive that they are beyond the means of all but the affluent. Moreover, past transport planning has failed to recognize that many people live remote from the (motor-vehicle) road system, and have movement needs that could never be satisfied by conventional vehicles.

Thus for the rural poor it would be difficult to conceive of a more inappropriate technology: often unrelated to the basic movement needs, inaccessible, scarce, expensive, difficult to use and maintain, and frequently wholly dependent on foreign resources in terms of manufacture, energy, spare parts and operating skills.

To define a more appropriate transport technology it is necessary to ask the question 'What are the appropriate vehicles for rural areas of developing countries'?

Given the variations in incomes, in topographical, road, farming and social systems, and in local resources and capabilities, there cannot be 'a universal vehicle' appropriate to all the rural transport needs of developing countries. Rather the need is for a graduated choice of vehicles whose performance matches need and whose cost is in sensible relation to income.

Consideration of the characteristics of the rural poor, their transport needs and the criteria of an appropriate technology leads to vehicles radically different in concept from conventional motor vehicles. The consequences of variations in operating environment, loads, cost, technical simplicity and the use of local resources lead to a progression of human, animal and, at the extreme, simple motorized means of movement. We term these collectively as basic vehicles, which may be defined as the range of devices from aids to goods movement by man himself up to but excluding, conventional cars, vans, buses and trucks.

Six categories of basic vehicles can be defined:
1 Aids to head, shoulder and backloading
2 Handcarts and wheelbarrows
3 Pedal-driven vehicles
4 Animal transport
5 Motor cycles
6 Basic motorized vehicles

Many such basic vehicles already exist in different parts of the developing world, though often their use is localized. Some are primitive, being traditional devices which have remained unchanged for many years. Almost all are capable of improvement, using contemporary technical knowledge, so as to increase significantly their efficiency and usefulness.

The present status of basic vehicles is that much good technology already exists which could be widely applied, but whose use is at present very localized. Where information on such technologies exists it is obscure, uncollated and certainly unknown to those who could make use of it.

While devices which meet the transport needs of the rural poor must be simple and low cost, this does not imply that their development is an easy task. Rather, experience suggests that the development of effective basic vehicles requires the application of contemporary technical knowledge and the very best technological skills.

Ian Barwell and John Howe, *Appropriate Technology and Low-cost Transport,* 1979.

In many parts of the world animals are the most appropriate form of transport.
(*UNICEF*)

Efficient Localized Low-cost Vehicles

The *chee-geh*, a traditional load-carrying frame worn on the back, is unique to Korea. Its main advantage over other human means of carriage is that the carrier can load and lift the device unaided.

The *Chinese Wheelbarrow* is of quite different design from the Western wheelbarrow found in most parts of the world. It has been proved the more effective device, but is not found outside China.

An alternative to the traditional but inefficient Asian *bullock cart* is manufactured in India. It uses a pneumatic tyred wheel running on ball bearings, the whole assembly being fitted to a specially fabricated steel axle.

Although the bicycle is very common in developing countries it is rarely used with a trailer. But the *bicycle and trailer* is widely used in Europe. For example, for the delivery of letters and parcels the Swiss Post Office uses bicycles and trailers since they are a cost-effective means of transport.

The *motor-cycle and sidecar* combination, which is one of the most popular means of moving goods and passengers in rural and urban areas of the Philippines, is not used in this way anywhere else in the world.

A range of simple locally-manufactured *motorized three-wheeled vehicles* has been evolved on the island of Crete. Most use an 8 to 12hp rope-started diesel engine, have a 1000kg payload and a maximum speed of 40-45km/hr.

Ian Barwell et al., *UNIDO Guide to Low-cost Vehicles for Rural Communities in Developing Countries*, 1982.

The Korean Chee-geh

Only some twenty per cent of Korea's land is farmable, and much of it is located where there is a mixture of flat and hilly terrain. These areas are divided up into many small farms of 0.5 to 3.0ha. As in all other parts of the world, farm materials, equipment and crops must be moved between field, home and market, but in Korea the need to move these goods is accentuated by the farm system, in which an individual's land is often split into several separate plots.

A Korean farm road's function is not simply to be a surface upon which goods can be moved, but also to be a property boundary or irrigation dyke. Thus, even in flat areas, routes are often narrow and winding with frequent ditch crossings. In hilly areas, steep inclines and rocks pose additional problems. Wheeled vehicles, such as handcarts, animal-drawn carts and single-axle tractors, are useful where roads are sufficiently flat and wide, but elsewhere loads must be carried. In other parts of the world, where similar conditions exist, pack animals or various methods of head loading, are employed. In Korea, the 'chee-geh', a simple pack frame carried by one man on his shoulders and back, has been playing a critical role in farm transportation for several hundred years.

The most significant advantage of the chee-geh is that loads can be picked up and set down without assistance. A pole is used to prop up the chee-geh while the user gets into or out of the shoulder harness. On the move, the pole is used as a walking stick and balancing pole. This enables exceptionally bulky and heavy loads to be carried — 60-80kg is common. Surprisingly, there is little difference in load-carrying ability between young and old users. Skill and experience are just as important as physical fitness. The chee-geh is used to carry all manner of farm goods, including barley, rice, firewood, manure and grass, but it is also used to move building materials and earth on construction sites. In urban areas it is used for deliveries in narrow streets and crowded markets.

Traditional chee-gehs are made by the user himself, or in small rural workshops, from pine branches, woven sticks and straw. The load container may be lined with cloth or plastic to enable loose loads such as sand and manure to be carried. The exact form of the chee-geh is different in different parts of the country, there being variations in the number of crossbars, the attachment of the backpad, the relative lengths above and below the fork, the overall length and curvature of the frame, the overall width at top and bottom, and the position of the shoulder straps. Some of these variations are simply the custom, others are functional. In mountainous areas the upper part of the frame is shorter because lighter loads are carried and overhanging vegetation must be avoided. The legs are also shorter to avoid touching the ground when descending steep slopes. Longer legs are required on flat ground to make picking up and setting down easier, and the upper part of the frame may also be longer to enable very bulky loads to be carried. Although the chee-geh seems simple, it has been very carefully modified to suit the physical character of the land and the body structure of the user, so that it will be of maximum utility.

A survey carried out by Han Nam University amongst Korean farmers indicated that over 90 per cent of them felt that the chee-geh was a primitive and troublesome device which should be discarded, yet almost 100 per cent felt that its continued use was inevitable because there was no alternative means of moving goods which could be used in the local conditions. Young farmers particularly resented using the chee-geh because of its 'backward' image. This ambivalent attitude among farmers was the main reason for concluding that modern technology ought to be applied to improve both the image and the performance of the traditional chee-geh.

A research team was therefore set up at Han Nam University in Taejon, in conjunction with the Georgia Institute of Technology. The method of developing an improved chee-geh was unusual in that it consisted of a series of some eight 'development conferences' spread over a sixteen month period. Participants in the conferences comprised chee-geh users (farmers), manufacturers and academics drawn from the technical and economics disciplines. The aim was to develop gradually an improved chee-geh based on the synthesis of ideas derived from the respective knowledge and experience of the conference participants. Six farmers (two each from lowland, hilly and mountainous regions), three metalworkers with long experience in chee-geh and farm equipment manufacture, two medium-industry machine manufacturers, four professors (of mechanical engineering, economics and management) and two technical high-school teachers (altogether seventeen members) took part in the conferences which involved seven model development meetings and two field surveys with their respective review and analysis meetings.

In order to investigate a wide range of opinion, especially from the farmers, a different group of farmers was invited to each conference. Also, local community leaders and many influential people of all ages were asked to participate in the field tests of the progressively improved models. The conferences aimed at unearthing as many of the participants' ideas as possible by means of 'brainstorming' — sessions in which everyone contributes ideas and they are discussed. The ideas presented, together with the opinions and matters of discussion, were recorded and were then compiled by the conference chairman for report and re-examination at the next conference. On this basis the participating manufacturers eventually made test models which were field tested. The results of each field test were reported in the following conference, beginning with a close analysis of items pointed out by farmers. A new model would then be designed for the next stage and in this step-by-step fashion the conference progressed.

In all, six models were produced and tested before a satisfactory improved chee-geh resulted. The overall arrangement of the new model was very similar to the traditional chee-geh so that it would function in the same way, but the use of modern materials (such as thin-walled steel tube) immediately increased its appeal. In addition, two wheels were added which can be folded down to convert the chee-geh into a simple handcart when route conditions allow. Before finalizing the design, the improved chee-geh was evaluated by one hundred farmers, to whom models were loaned for a brief period. This showed a positive reaction to the improved chee-geh and yielded criticism which was used in finalizing the design. Although the new model was necessarily somewhat heavier than a traditional chee-geh, the trials indicated that the increase in weight did not make the design less convenient since, whenever possible, it was wheeled rather than carried. A separate comparative trial between the new and traditional models — involving the transport of a 60kg bag of grain over distances of up to 3.5km — indicated that the new model gave rise to significant time savings and productivity increases. As part of the evaluation, farmers were asked at what price they would buy the improved chee-geh. Sixty-eight per cent said they would buy it at about half the price of a handcart, which is about three times the price of a traditional chee-geh. Since development work was completed a number of potential manufacturers have expressed interest in making the new model, but none as yet have been prepared to accept the risks of production without outside assistance.

Gordon Hathway and Dr. Seyeul Kim, *Appropriate Technology*, 1984.

Chinese Wheelbarrows

A wide variety of wheelbarrow types exist or have existed in China. The various types exist to suit different purposes and betray their origins in carrying methods which preceded the use of wheels. One type is derived from the packhorse with the load disposed around the wheel, and the wheel carrying most of the load but presumably is slightly biased towards the handles to maintain control. Another is a specialized version for earthmoving, enabling the load to be tipped out easily. Others are derived from the two-man stretcher, with the wheels substituting for one of the carriers, and would appear to be used mostly by street vendors.

Construction of the barrow is traditionally in wood, used also for the wheel, but modern versions use a steel tube frame. A characteristic of the majority of types of Chinese wheelbarrow is the use of a large (up to 700mm) diameter wheel for reduced rolling resistance. Another common feature of Chinese wheelbarrows is the use of straps of different forms, intended to lessen the strain on the operator's arms by transferring some load to the shoulders. The third major characteristic of Chinese wheelbarrows is the load disposition. Being closer to the axle, whether around or on top of the wheel, the load requires less effort for lifting than with the European type. For this reason, there has been increased interest in the Chinese barrow in recent years, concentrated mainly on the earthmoving type and its application in labour-intensive construction work.

Intermediate Technology Transport, 1980.

The Appropriate Transport Module!

A self-contained transport module has been designed especially to meet the needs of the small-scale agricultural business and road programmes.

Silent operation, except when the built-in audible alarm is triggered by some external hazard, it can be fuelled by a wide variety of locally available materials.

Far from polluting the environment its exhaust materials provide trail markers.

When this purpose has been fulfilled these markers can be collected up and used as a starter mechanism in the production of further supplies of fuel.

The module is largely self-steering so that the owner can safely concentrate on business planning or even take a well-deserved nap while being carried from place to place.

Two types of module are available. Both are entirely suitable for transport but in addition, by means of an ingenious process too technical to be described here, a matched pair can produce replacement modules which keep the owner's fleet up to strength without the need for further investment.

The maintenance schedule is simple. Apart from periodic refuelling it consists merely of an occasional rubbing down of the external coat which incidentally is almost completely corrosion proof. Some care must be taken during the process since unfortunately the maker has not provided any facility for switching off the donkey during idle periods.

The One-Wheel Cycle Trailer

The bicycle trailer has considerable potential for meeting many local transport needs in rural areas of developing countries because

— it allows heavy and voluminous loads to be transported safely by bicycle;

— it can be manufactured locally, made available cheaply, and can be purchased by existing bicycle owners to increase the utility of their vehicles;

— it offers flexibility because it is easily detachable from the bi-

cycle, and can be used separately as a handcart.

Two-wheeled cycle trailers are common in certain parts of Europe but are rarely used in the developing world. The evidence indicates that this is not because the technology is inappropriate, but because it is unknown. An innovative approach, which has particular potential in developing countries, is the use of a single-wheeled trailer. One of the reasons for the popularity of bicycles in developing countries is that, hav-

A one-wheel cycle trailer — where there are no roads.

ing no significant width, they can be used on narrow footpaths which form the largest part of the communication network of rural areas. Thus the use of a bicycle is not dependent on the availability of motorable roads. The use of a single-wheeled trailer has the advantage of retaining the 'two-dimensional' characteristics of the bicycle.

Intermediate Technology Transport, 1983.

Cycle Rickshaws for the Poor

The West German Protestant Central Agency for Development Aid had earmarked funds for cycle-rickshaw pullers (rickshawalas) in Nagpur to buy their own autorickshaws.

The 'Jinrikisha' as it was originally called, has come a long way from the slow, two-wheeled vehicle pulled by a single runner, which was invented in Japan in 1880. Gradually, it became upgraded to the trishaw or cycle-rickshaw: half a bike with a two-wheeled passenger carriage tacked on behind, which is still one of the most inexpensive and widespread means of transportation in India's towns and villages. The next stage ws the autorickshaw, an ingenious adaption of a cheap piece of Western technology to Eastern needs. It is a scooter attached to an often gaily decorated cab with a canvas cover.

Admittedly, passenger comfort leaves something to be desired, but comfort is the least of a poor man's worries, especially those accustomed

283

to using leg power to earn their daily bread. The stumbling block was the price of an autorickshaw at around Rs.19,000 — a horrendous sum for the average Indian who can count himself lucky if he earns as much as Rs.200 a month.

It was decided that Nagpur's 18,000 rickshawalas could best be helped by enabling them to own their cycle-rickshaws, so that if they should die their families would no longer become destitute overnight. At that time, most of the rickshaw pullers were organized in eighteen small unions, which were ostensibly there to help them, but, in reality, most of them were mainly concerned with reaping benefits for themselves alone. The only answer was to persuade the unions to join forces in a United Front — an undertaking that took many years to succeed.

The first problems that beset the United Front were daunting and seemingly insoluble. Almost all Nagpur's cycle-rickshaws were concentrated in the hands of 100 people, companies, groups or garages, popularly known as the monopolists.

Clearly, if the rickshaw pullers could be helped to own their vehicles, then they could escape from this ruthless form of exploitation. But how was the problem to be tackled? The leaders of the United Front put their heads together and agreed that their best plan of action was to take the matter directly to the Maharashtran Government in Bombay to persuade it to change the law.

With this in mind a number of strategies were adopted, ranging from bargaining with the government to agitation and hunger strikes. One particularly effective tactic was the so-called 10 paise strategy. Each cycle-rickshaw puller was asked to buy a postcard costing 10 paise, write on it 'Answer our demand, abolish

Rickshaws are still a much-used form of transport in Asian cities. (*Hoddy*)

284

monopoly, make rickshaw man the owner — long live puller-owner' and send it off to the government in Bombay. Confronted with such a deluge of postcards, the politicians bowed to the inevitable and acceded to their demands.

But the United Front had to face setbacks too. Success was finally theirs only after six and a half years of hard struggle. On 4 April 1979, the government amended the 1920 Public Vehicles Act, abolishing all monopolists in Nagpur and in the rest of Maharashtra. In future, anyone who applied for a licence to own a cycle-rickshaw would have to be the puller, too.

The immediate result was a sudden rise in the demand for cycle-rickshaws and another problem surfaced. Nagpur's two cycle-rickshaw manufacturers spotted the chance to earn a quick buck and promptly increased their prices from Rs.2,600 to as much as Rs.3,000 per rickshaw — a sum far beyond the means of a poor man.

But was this price justified, especially as a bicycle in India costs no more than Rs.4,000? This was the question the United Front began asking itself and it was not long before they resolved to find out for themselves. They borrowed some funds from the Industrial Service Institute, bought all the necessary parts and assembled their own cycle-rickshaw. It cost exactly Rs.1,125.

This did not get them very far, though, for they were still forced to purchase their cycle-rickshaws from the only two manufacturers in Nagpur. Their only hope was to set up a manufacturing centre of their own and thus push down the price and convince the banks that they need release loans of Rs.1,200 at the most for prospective cycle-rickshaw owners.

A development agency came to the help of the Industrial Service Institute by providing a grant of Rs.37,000. This money was used to build a temporary shed, buy parts from the local market and engage eighteen rickshaw pullers to work shifts all round the clock, producing twenty cycle-rickshaws every twenty-four hours.

Elizabeth Hoddy, *Development and Co-operation*, 1982.

Illich on Transport

Under President Cardenas in the early thirties, Mexico developed a modern system of transportation. Within a few years about 80 per cent of the population had gained access to the advantages of the automobile. Most important, villages had been connected by dirt roads or tracks. Heavy, simple, and tough trucks travelled over them every now and then, moving at speeds far below 20 mph. People were crowded together on rows of wooden benches nailed to the floor to make place for merchandise loaded in the back and on the roof. Over short distances the vehicle could not compete with people, who had been used to walking and to carrying their merchandise, but long-distance travel had become possible for all. Instead of a man driving his pig to market, man and pig could go together in a truck. Any Mexican could now reach any point in his country in a few days.

Since 1945 the money spent on roads has increased every year. It has been used to build highways between a few major centres. Fragile cars now move at high speeds over smooth roads. Large, specialized trucks connect factories. The old, all-purpose tramp truck has been pushed back into the mountains or swamps. In most areas either the peasant must take a bus or go to the market to buy industrially packaged commodities, or he sells his pig to the trucker in the employ of the meat merchant. He can no longer go to town with his pig. He pays taxes for the roads which serve the owners of various specialized monopolies and does so under the illusion that the benefits will ultimately spread to him.

In exchange for an occasional ride on an upholstered seat in an air-conditioned bus, the common man has lost much of the mobility the old system gave him, without gaining any new freedom. Research done in two typical large states of Mexico — one dominated by deserts, the other by mountains and lush growth — confirms this conclusion. Less than one per cent of the population in either state travelled a distance of over 15m in any one hour during 1970. More appropriate pushcarts and bicycles, both motorized when needed, would have presented a technologically much more efficient solution for 99 per cent of the population than the vaunted highway development. Such pushcarts could have been built and maintained by people trained on the job, and operated on roadbeds built to Inca standards, yet covered to diminish drag. The usual rationale given for the investment in standard roads and cars is that it is a condition for development and that without it a region cannot be integrated into the world market. Both claims are true, but can be considered as desirable only if monetary integration is the goal of development.

Ivan Illich, *Tools for Conviviality*, 1975.

The bicycle was a cheaper alternative, but many of those who could afford to purchase a bicycle preferred not to use it because of the loss of status involved, whereas most of those who would have been willing to use a bicycle could not afford it.
Tapan Kumar Das Gupta, 'The Role of the Rickshaw in the Economy of Dhaka City', 1981.

VII

Technology for Development

Manufacturing, Mining and Recycling

MOST DEVELOPING countries are characterized by a rapidly growing population and labour force. While much of this labour force will be absorbed in traditional agriculture, an increasing number of people will seek employment in non-farm occupations. A recent World Bank report estimated that during the last quarter of the twentieth century, this would account for two out of every three job seekers.[1]

Productive employment opportunities are needed in the urban areas to help absorb the slum dwellers who have been driven to the cities by increased landlessness and diminishing employment prospects in the rural areas. They are even more desperately needed in the rural areas themselves to help stem the flood of rural-urban migrants which is a major contributing factor to the urban problem[2]

While the housing, construction and transportation industries will absorb part of the growing labour force, there is a need for still more opportunities in as wide a range of productive activities as possible. The extracts in this chapter have been chosen to illustrate the various opportunities which exist along with the experiences involved in creating non-agricultural employment in urban and rural areas.

Markets for non-food products arise in satisfying the needs of the masses of the population for basic goods such as clothing; household durables (furniture); household commodities (paper, soap, candles, pots); farm equipment and transport devices; and, to a lesser extent, in satisfying the needs of large firms for intermediate and producer goods (mineral ores, engines), and the demands of local urban elites and foreign buyers for specialist items (crafts, gems, minerals). The challenge is to meet this demand in the most economically efficient way in terms of employment, output and cost per unit.

There are those who argue that production of many of these commodities is not possible in small-scale establishments in the urban informal sector or in rural areas, or that profitable operation is feasible only if such units are given

---

[1] World Bank, *Employment and Development of Small Enterprises* (Washington, 1978) p.11
[2] In Africa, urban populations have mushroomed overall by 6 per cent a year, and 8.5 per cent annually for 35 major capitals — a rate at which they will double in size every nine years. Much of this is the result of rural migrants searching for employment. (World Bank, Accelerated Development in Sub-Saharan Africa, 1981)

substantial subsidies and other assistance. Several of the extracts in the small-scale manufacturing section offer proof against this argument and show that, providing the appropriate technology is available to them, small-scale units can make better economic sense than larger establishments. Of particular interest in this respect are the extracts on soap making in Ghana and low-cost pharmaceuticals in Bangladesh.

Three factors are worthy of note in the section on manufacturing. First is the role of technology institutions in developing or adapting processes which become the basis for profitable small-scale industry. This is particularly clear in the cases relating to the Technology Consultancy Centre in Ghana. Second is the ability of technologists to provide the means of reviving traditional crafts and industries. Such industries have faced competition from three major sources — large-scale modern industrial plants producing similar products; firms of all scales and levels of technology producing competitive products: and imported substitute products. The introduction of an improved technology can prevent total decline by raising productivity in traditional activities or by enabling traditional skills to be used in the manufacture of non-traditional, more marketable products. The extracts on Manipuri weavers in Bangladesh and lost wax casting in Ghana are good examples of this. And third, although technology is important, many of the extracts show that, on its own, it is often insufficient to ensure successful industrial activity in the small-scale sector. Other factors, such as direct government regulations and assistance measures, and the overall economic environment, can and do have a significant effect. This can be seen in the case of soap making in Ghana and engine manufacture in India.[3]

As pointed out in the extract by Tom Wels, small-scale mining is a non-agricultural activity which is much more widespread than normally assumed. The tendency is to think of mining (much more so than even manufacturing) as a large-scale, highly capital-intensive venture, with small-scale activities providing, at best, income earning opportunities to a very few prospectors and artisans. In fact, as the extracts in the small-scale mining section show, a very substantial amount of the mineral output (about 10 per cent worldwide) comes from the small-scale sector. Besides providing useful employment opportunities in remote areas, this sector has other useful characteristics such as enabling the working of very small, scattered mineral deposits which are of no interest to large-scale mining companies.

Another under-acknowledged sector of the economy is that of garbage recycling. Many industries, both in developed and developing countries, are based on a variety of waste products including paper, rubber, metals and glass. According to the extract by Kirkpatrick Sale, some dumps in the US have more copper in them than do some mines. 'Garbage' tends to be wasted in the Western world: recycling is more commonplace in the Third World where almost every material that can be recycled is brought into use — fre-

---

[3] These issues are also covered in some depth in Chapter IX which deals with the diffusion of appropriate technologies.

quently providing much needed income and employment for the rural and urban poor. The extract by Vogler on 'trash technology' and that by Sale on 'garbage recovery' make the interesting point that, because of the scale of technology applied, the recycling of waste products in the West has been far from successful. Even the process of garbage recovery makes sense only if done on a small, human scale.

1 SMALL-SCALE MANUFACTURING

Wool Spinning in Pakistan

The Department of Industries of the Government of Baluchistan started a wool-spinning centre in Mastung in 1969 in order to create employment for the local population. The factory uses locally-produced wool to make yarn for the various carpet factories which are also located in the province, and forty seven people are employed.

The factory has 500 spindles for spinning yarn and the capacity is about 500,000 pounds of woollen yarn per year, working on a double shift basis. In 1977 the factory was operating at less than 10 per cent of this capacity, and the following figures show that ever since it was started in 1969 the business has been running at a substantial loss.

There is a large quantity of unsold completed woollen yarn, which is likely to deteriorate in storage, and this clearly implies a potential for further losses. Supplies of raw wool have always been erratic, and the wool which is supplied by the contractors, who are responsible for collecting it from the farmers, is generally adulterated with sand and dirt. There is also a serious scarcity of water in the area: the only tube-well available to the factory produces 6,000 gallons of water over the 16 hour daily operating period whereas the machinery actually requires 11,200 gallons during this period. In addition to these problems the machinery in the factory is not balanced. The initial washing machine is of insufficient capacity

Year	Production in lb.	Cost of Production in Rs.	Loss in Rs.
1969-70	10,783	68,743.00	19,788.00
1970-71	20,551	132,235.00	20,935.00
1971-72	19,368	133,858.01	45,927.30
1972-73	87,482	815,903.70	83,714.71
1973-74	20,814	417,829.79	124,560.52
1974-75	63,597	1,355,873.50	377,584.30
1975-76	38,850	559,440.00	79,642.50

to match the throughput of the machines which follow it and there are other similar problems. The manager of the business is an employee of the Government and has little knowledge of the technology of wool-spinning. Furthermore, labour relations in the factory have never been satisfactory.

The Government of Baluchistan is currently reviewing the history of losses in order to decide whether it should continue to make further investments in this business or refuse any further commitments to it.

Malcolm Harper and T. Thiam Soon, *Small Enterprises in Developing Countries,* 1979.

 # Weaving Technology in Bangladesh

Most Manipuri girls marry later than Bengali girls, and are regarded as valuable members of their society, as they spend many useful years at home involved in weaving, household chores and work in the fields. Normally a girl would weave on a traditional backstrap loom on which complex designs can be picked out by means of a series of leashes. It is a slow process even for skilful hands, and the distinctive cloth produced is expensive by local standards. This imposes a severe restraint on the amount of weaving that can be sold locally.

For the last five years one hundred women from four villages have helped their family income by selling their cloth through HEED Handicrafts. HEED (Health Education and Economic Development) is a development organization working in Bangladesh with programmes including health, agriculture, fisheries, leprosy and community work. The handicrafts section seeks to create employment for several groups and to market their products in Dhaka and abroad.... Each village has a field worker, chosen by the villages, who collects the finished work, distributes payment and delivers cotton for the new order. He is paid collectively by the weavers.

In this way each weaver earns on average 180 Taka per month (£4.80, US $7.20), not a large sum but a welcome supplement. To achieve this HEED Handicrafts has had to develop a system of quality control, and designs, colours and products to suit the wider market. These quality products are not cheap. It takes a weaver a week to weave a 54 x 108in bedspread made up of two strips 27in wide and for this she earns about £2. Smaller items are more popular but less remunerative. Even in the broader markets the final product is comparatively expensive. Many weavers have small bamboo counterbalance looms as well but these can only produce very lightweight fabrics. A few have larger wooden fly shuttle countermarch looms, but though weaving is faster on these, they are not suitable to produce most of the traditional designs. If weavers want a more dependable income they need to look elsewhere and that is why they expressed interest in different looms.... The dobby loom is ideal for the quick production of complex

290

designs without a difficult system of pedalling. Two pedals rotate a series of pegged wooden bars (lags) which determine the lifting of shafts. An American firm Ahrens and Viollette Looms (AVL) have developed the dobby for the handloom weaver and their design would seem ideal for the Manipuris' needs. If effectively introduced it would increase productivity, thus lowering prices while maintaining or increasing earnings. Manipuri cloth is attractive to the Bengalis but is usually too expensive. It is of little use competing in already saturated local markets such as saree manufacture. Many looms suitable for saree production lie idle in Bangladesh. But it may well be that good quality fancy cloths can be competitive with imports. The export of traditional cloth has been limited in the past because of its cost rather than design, and a lower price should boost sales.

At the end of 1982 HEED Handicrafts imported AVL's 12-shaft, 36in folding dobby loom. Though limited, it has been a useful and not too awesome stepping stone to the two 16-shaft fly shuttle dobbies to follow. As a skilled weaver from one of the poorer families a girl called Rambha was chosen to have a 36in loom. After the initial training Rambha with two others from her village came into HEED Handicrafts office to learn the new warping and threading-up techniques, how to follow a pegging plan and how to operate a dobby loom. Now Rambha has a loom in her own home where it fits on their raised mud verandah. Training and supervision continued there for some months. She and her sister are now familiar enough with the process to need minimal supervision. The loom has aroused great interest and delight as the Manipuris see some of their traditional designs produced so quickly, yet they realize the small loom's limited production capacity and so a desire for the fly

Improved looms mean more money for village weavers

291

shuttle loom is growing. Rambha herself is particularly keen to use one. Though further training will be required in sectional warping and use of the fly shuttle, the women are already proving that this will not deter them.

Of course looms cannot be imported regularly from America to Bangladesh! But another traditional skill of the Manipuris is their carpentry. With a minimum of equipment they make excellent furniture and the existing wooden looms. If the VAL 60in dobbies prove efficient after a trial period and if weavers are keen to continue with them, then work must begin on a local prototype. This will involve weavers, carpenters and the AVL loom-makers with any other technical advisers necessary as modifications are worked through, bearing in mind local availability of

materials, technical skills and the final cost. The prototype will then be tested and further modified if required. Only then can it be decided whether the project can continue. It will be a lengthy process and in the meantime other decisions must be taken. For production on a large loom several women will have to work together. In the past weavers have chosen to work as individuals in their own homes but now some sort of co-operative will need to be developed. Group ownership of looms, whether within an extended family or of wider membership, must be worked out. Some will earn by production while all will reap the benefit from profits from sales. Because many weavers are unmarried the group structure must be flexible enough to replace any who marry and move away.

Eva Pettigrew, *Appropriate Technology*, 1984.

Handmade Paper

In 1935 the All-India Village Industries Association, founded by Mahatma Gandhi, made some efforts to revive the handmade paper industry. It was, however, only after the Khadi and Village Industries Commission was established in 1975 that serious steps were taken to develop it. The handmade paper industry at present accounts for 0.5 per cent of the total production of paper in the country and is not expected to rise above 1 per cent unless something dramatic and revolutionary is introduced in the programme.

Production of handmade paper has

been mainly concentrated on writing and printing papers from recycled waste and bond and drawing papers from rags and blotting paper. These products have been marketed mainly to protected markets such as Central and State Government offices, Khadi and Village Industries Commission offices and publications, State Khadi and Village Industries boards and other nationally minded institutions. Because the manufacture of writing and printing papers is not economically viable, the handmade paper industry has now shifted its attention to the manufacture of non-competi-

Varieties of and Raw Materials for Handmade Paper

Variety	Raw material
Art and engineering drawing paper	White rags, tailor cuttings
Permanent document papers	White rags, tailor cuttings
Water-marked certificate paper for universities and commercial firms	White rags, tailor cuttings
High-grade stationery, card sheets, greetings of fancy colours	Coloured and white rags, cotton and jute
Album papers	Gunny waste, rags and paper cuttings
Filter papers used in commercial firms, mercantile companies and distilleries	Old rags, paper waste, cotton
Blotting paper (white and pink)	Old rags, paper cuttings
Packing boards (plain and laminated)	Grasses, bagasse, paper waste, rags, citronella grass, bengal grass, gunny waste, jute waste, etc.
Electrical insulation boards and paper	Bagasse, rags, citronella grass, gunny waste
Cover papers	Rags, paper waste, gunny waste, etc.
File boards	Mixed colour rags, paper cuttings
Grey boards	Road sweepings and any cheap fibrous material locally available
Straw boards	Locally available straws with no fodder value

tive decorative, commercial and industry-grade papers. In addition, there is a large demand for drawing papers, cloth-lined paper, special paper for preserving records, packing boards, electric insulation papers, filter papers, invitation cards, etc. The raw materials and the varieties of the paper that can be manufactured from them in the handmade paper industry are shown in the table.

The major handicap in the economic viability of handmade paper is the lifting operation. The pulp, after being prepared mechanically, is suspended in water in a tank from which a work lifts individual sheets of paper onto a special screen frame. The pro-

ductivity is very low. The cost of lifting is high; only high-grammage paper is possible, and uniformity in quality depends on highly individual skills. Diversification to common-usage paper is technologically not possible.

The other handicap concerns the raw material for pulp-making. Waste paper and tailor cuttings and rags are now much more in demand by the medium-scale paper industry. Their availability is uncertain and their cost is inflated to a level where even speciality paper manufactured by the handmade paper industry is gradually becoming uneconomic. This means that if the development effort made over the last three decades is to be sustained, further development and improvement of the technology will have to be carried out.

The Appropriate Technology Development Association (ATDA) in Lucknow has investigated this problem in depth. Discussions with various technologists and experts in this field lead to the following conclusions:

(a) When all the other processes for producing handmade paper have been mechanized, why should lifting still be done by hand? If this process can also be mechanized, the manufacture of low-grammage paper, especially for rural school exercise books, text books, typing and other general purposes can be undertaken. This requires the development of a 1-tonne/day paper-lifting machine. Already 2 to 3 t/d cylinder machines are being manufactured for making duplex and triplex boards as well as packing boards. Earlier attempts to make low-grammage paper from this type of machine did not prove successful. A prototype of a Fourdrinier paper-lifting machine with a 2 t/d capacity has been developed. If it can be incorporated in the handmade paper industry, low-grammage paper for rural school exercise books and textbooks could be manufactured. The introduction of this machine would revolutionize the present rural handmade paper industry. It becomes a technically feasible and economically viable proposition to produce a large variety of daily-usage paper competitively with large-scale paper machines;

(b) The other problem is raw material. Any meaningful programme of paper manufacture on a cottage scale cannot depend solely on recycled waste. Moreover, long-fibre pulp is vital for making low-grammage paper. Furthermore, in order to keep the cost of competitve agricultural waste pulp down, the pulp volume will have to be increased. The technology of both kinds of pulp-making, their capacity and required capital investment is beyond the capacity of the rural, small or cottage units. As stated earlier, the viable capacity of the agricultural waste pulp unit starts at l0t/d and goes up to 30t/d. If rural paper units are standardized on 1-t/d, then one agricultural waste pulp unit will serve 30-100 rural paper units. This assumes that only about 30 per cent of the pulp will be used along with recycled and long-fibre pulp.

The problem with a long-fibred chemical pulping plant is still more difficult. The capacity generally recommended for such a plant nears 100t/d which requires an investment of more than 100 million rupees. ATDA has been investigating the possibility of scaling down the size of the plant and has had some support for designing a long-fibred chemical pulping plant with a capacity of 30 to 50t/d if bleaching is not adopted.

ATDA has therefore worked out

the following programme for upgrading present handmade paper units;

(a) Set up a 30-50t/d capacity chemical long-fibred plant by development organizations and government institutions and arrange distribution of the pulp to rural handmade paper units. Such a plant can be at the State level. The pulp will be supplied to the small units;

(b) Set up a regional pulping plant from agricultural wastes covering two to three districts to be owned and operated by local development agencies, both governmental and non-governmental. The pulp to be supplied to small units;

(c) Set up a lt/d rural paper unit complete with a recycling pulping plant on the pattern of the Khadi Commission but with the 2t/d prototype paper-lifting unit mentioned above. The paper will be manufactured with a suitable mixture of long-fibre pulp obtained from the State pulping plant, agricultural waste, short-fibre pulp and recycled pulp;

(d) The handmade paper units will also benefit by obtaining part of their pulp requirements from these State-level and regional-level pulping plants. It is not the intention completely to convert the handmade paper units to the mechanized lifting process. Handmade paper sections will be a useful adjunct to the mechanized-lifting units. The special-quality paper, which is best manufactured by the hand-lifting process, could continue at a better economic level because of the supply of good-quality, lower-cost pulp.

Such a programme will increase small-scale paper production to a meaningful level and will meet the local paper requirements in rural areas, generate employment in them and improve their capital formation and economic condition.

M.K. Garg and M.M. Hoda, *Appropriate Industrial Technology for Paper Products and Small Pulp Mills*, 1979.

Local Soap Manufacture in Ghana

The Technology Consultancy Centre in Kumasi originally became involved in the problems of soap manufacture during 1972 in response to enquiries from local traditional manufacturers who wanted a chemical analysis of their soap, and advice on how to improve its quality which was somewhat unsatisfactory. Experimentation started in the TCC's workshop, with technologists from various faculties of the University of Science and Technology working with one of the local soap-makers. A satisfactory formula was eventually evolved and the prototype soap plant installed at the TCC's workshop started production for sale in September 1973. The prototype boiling tank had a capacity of 500 lkg bars per day, and was electrically heated, with an outer tank for clarifying the

295

oil and an inner tank for boiling the soap. Average daily production during the first year was only 160 bars (much less than capacity), because of a combination of problems arising from the inexperience of the soap-makers, technical difficulties needing continuing experimentation and fluctuating demand. Monthly output varied from 420 to 6,870 bars during this period. The product (Anchor Soap) was a basic pale soap, suitable for both washing and cleaning. Sales fluctuated according to the availability of other brands of soap. When Lever products were available they were bought in preference to Anchor soap — mainly because they had the advantage of familiarity and were also being sold at a slightly lower price. In April 1974 a second plant was installed and started production. This also had a capacity of 500 bars a day.

Plans had already been made to construct a soap pilot plant near Kumasi, the purpose being to produce soap for sale, to serve as a demonstration project showing the technical and commercial viability of the intermediate technology, and to continue research and development. Construction of the plant started at Kwamo in January 1974.

As caustic soda had become increasingly scarce and expensive it was decided to begin caustic soda manufacture so that, as far as possible, the supply to the proposed soap pilot plant could be guaranteed. A caustic soda plant was designed by a member of the Department of Chemistry and Chemical Technology and production started in early 1974.

Caustic soda was produced at a cost of ¢35 per 50kg, which was only a third of the prevailing market price. The plant proved to be adaptable for other purposes: one was produced at the TCC workshop for the manufacture of insecticide and two more for domestic bleaching fluid.

The prototype soap and caustic soda plants attracted a good deal of favourable attention from manufacturers and entrepreneurs and, during 1974/75, orders were placed with the Centre by entrepreneurs located in Ho, Sekondi, Akim-Oda, Sunyani, and Tarkwa. During this year, plant operators were trained at the prototype plant for transfer to the sites of these entrepreneurs and also for the soap pilot plant, where production began in June 1975. Later in the year, the prototype on the campus was renovated and sold.

During the following years, the soap pilot plant and the entrepreneurs who had bought soap plants faced difficulties relating to shortages of palm oil and rising costs of this and other raw materials. In 1976, production averaged only 30 per cent capacity at the pilot plant and fell even further during 1977/78 because of raw materials shortages. Some of the entrepreneurs who bought soap plants have closed down operation because of supply and cost problems, but some keep going, and orders for new plants have continued to arrive. Twenty soap-boiling tanks and associated equipment were supplied to six entrepreneurs in 1976. Most also bought caustic soda plants to ease their importation problems.

Sally Holtermann, *Intermediate Technology in Ghana*, 1979.

Low-cost Pharmaceuticals

Gonoshasthaya Pharmaceuticals Limited (GPL) is unique in Bangladesh and probably in the Third World in being a private company with its production entirely geared to meeting the needs of the people. All the shares in the company are owned by the charitable trust and under its charter profits are limited to 10-15 per cent (after payment of duties and bank charges). Half of the profits will be ploughed back into the company to expand production, reduce prices further and fund research. The remainder will be used to fund health and development projects.

Over the first three years of production, the factory will build up to formulating a range of about 30 drugs all included in the WHO Selection of Essential Drugs. Every effort will be made to keep prices down by purchasing raw materials by competitive tender on the world market. Despite its initial high overheads, the retail prices of GPL's drugs are considerably lower than those of equivalent products on the local market. Comparative prices are ilustrated in the table below, which also shows GPL's deliberate policy of setting different profit margins — lowest on the drugs considered most useful

Prices will not be kept low at the expense of quality. GPL is a modern factory, with quality control facilities comparable to the local big name producers. GPL aims to be competitive by going for larger scale production, taking advantage of modern

Comparative Retail Prices Between Gonoshasthaya Pharmaceuticals Limited (GPL) and Other Manufacturers in Bangladesh, April 1982

Drugs	Unit cost to GPL	GPL Profit (per cent)	Maximum retail prices per capsule/tablet in paisa (100 paisa = 1 taka) GPL	OTHERS
Ampicillin (250 mg)	76.2	6.57	100	Hoechst 186 Square 175 Beecham 169
Tetracycline (250 mg)	38.4	5.26	50	Squibb 110 Pfizer 106 Albert David 77
Metronidazole (250 mg)	25.6	22.7	40	BPI 79 (200 mg tab) Square 65
Paracetamol (500 mg)	11.7	3.41	15	Fisons 24 Square 25
Diazepam (5 mg)	7.1	36.6	12.5	Roche 55 Square 30
Frusemide	26.0	85.6	60	Hoechst 125

machinery, production and management techniques. The factory has 42,000 ft^2 of floor space, making it one of the largest in Bangladesh. It was built with capital provided mainly by the Dutch charity NOVIB, a loan from the Bangladesh Shilpa (Industrial) Bank and further contributions from Oxfam and Christian Aid. Although GPL has had to rely on foreign donor agencies to provide the initial capital and the International Dispensary Association in Holland for technical assistance, the underlying objective of the project is self-reliance. Designs for the factory building, air-conditioning and machinery layout were all planned and executed by Bangladeshis. Similarly, the production, quality control and marketing managers are all Bangladeshis who have gained valuable experience in the past working for big foreign manufacturers.

There are plans to carry out research into using locally available raw materials as excipients (the non-medical ingredients like starch that are mixed with the active ingredients to make up a medicine). Attempts are also planned to develop better dosage forms to suit local conditions — for instance, to take account of the nutritional status of the poor, and humidity during the long months of the monsoon. This sort of research tailored to meet local needs is largely neglected by the bigger foreign-owned companies. Patents on both processes and products are protected by law in Bangladesh. But it is stressed that GPL would not want to patent any new process discovered because they do not believe in 'the monopoly of knowledge'.

To achieve greater self-reliance and leave their operations less vulnerable to external pressures, the team at GPL are keen to expand into the more complex production of raw materials. These would have to be produced on a large scale to make the operation cost-effective. But the main difficulty to be confronted would be in obtaining the necessary technology. This is seen as an important wider objective behind setting up the factory.

The unskilled labour force is drawn from the villages of Savar and skilled labour from the capital. Most of the unskilled and semi-skilled production work is being carried out by local women who received a year's special training before the factory started production.

GPL expects to sell about 60-70 per cent of its production to the government health services and the voluntary health sector. This distribution through organized health services is seen as 'the safest and quickest way to channel the benefits of cheap drugs to the people most in need'. The remainder will be sold on the open market. This raises a major problem for GPL in trying to prevent middlemen from stealing the advantages intended for the poor by jacking up the prices. Cases of retailers charging more than the maximum retail price for GPL products have already been reported. One doctor comments: 'A lack of confidence in anything that comes from Bangladesh itself is part of our sad colonial heritage, and pharmacists, having heard something of Dutch financing, charge excessive prices claiming that this is a new 'Bilan' (European) medicine'.

The team at GPL are well aware

298

that past attempts to promote the use of cheap generics have failed because doubts about their quality are easily whipped up in the minds of doctors and patients. GPL has its own sales representatives to promote its products to doctors and pharmacists, but priority is also given to the need to popularize wider health issues. This is done through the health project's monthly magazine, *Mashik Gono-shasthaya* printed on its own presses. The magazine is written in simple Bengali and aims to be lively and informative. It covers many health issues including appropriate non-drug treatments and warns against the socially damaging effects of existing drug sales and over-medication. Fifteen thousand copies of the magazine are distributed each month.

Dianna Melrose, *Bitter Pills: Medicine and the Third World Poor*, 1982.

Vast Needs and Small Firms

The craft sector and small firms are very important in terms of employment in both towns and country areas, and they are equally important when it comes to the total production of manufactures and added value. The data now available suggests that this sector represents 70-96 per cent of the labour force and more than a third of the added value in the manufacturing industry. In Jamaica for example, 27 per cent of jobs in manufacturing are accounted for by individual craftsmen and small firms, which create 25 per cent of the value added. In Somalia, the figures are 50 per cent and 40 per cent respectively, in Ghana 87 and 39 per cent, in Sierra Leone 96 and 44 per cent.

The needs of the craftsmen are considerable, ranging from the raw materials needed for production to transport facilities to shift the finished products. If any given product is missing — and this often happens — a whole branch of activity is condemned to a long period of stagnation. If a craftsman fails to get the money he needs to update his equipment then there are a couple more people unemployed. But as craftsmen are self-employed or only work in small firms they find it difficult to make themselves heard.

But very little is required for people who have obtained know-how and a high degree of technical ability on their own, particularly in the services sector (e.g. car and motor-cycle maintenance, radio repairs, refrigeration techniques, photography) and the functional craft sector (carpentry, brick-making, dyeing, weaving and so on). The priority should be an organization to meet their needs. In Conakry in Guinea, for example, there is a mill and hulling machine workshop with only one welding set and one home-made sheet metal bender, which employs three craftsmen and eight apprentices. It currently turns out 15-20 machines a month (adjustable hullers for maize, rice and coffee, peanut shellers and mills). These machines are entirely made from scrap and can be driven by an electric motor or fitted with a manual system of operation. The driving belts are made from strips of old truck tyres. This firm is unable to meet the demands of the local market and of neighbouring Mali because it is short of input and equipment and there are constant power cuts. If these problems were solved then 100 machines could be turned out every month.

This is a particularly good illustration of the link between craft and industrialization. If this organization could get properly equipped and organize its management, then tomorrow it could be an industrial unit. And what goes for mechanical construction in Guinea also goes for the production of farm implements in Upper Volta and for footwear and textiles in other countries. Craft can and must lead to industrialization, and this means winning new markets outside the immediate neighbourhood and opening the doors wider to the poorer sections of the population who already make up the bulk of the clientele. However, these people must themselves have greater purchasing power, in particular via an increase in the price at which agricultural produce is bought, for incomes in rural areas are so low at the moment that they hold back any significant development of the craft sector.

Amadou Traoré, *The Courier*, 1981.

The Nikart

In 1981 I spent two months in India advising a small engineering company on the manufacture of an animal-drawn tool-carrier, the Nikart. The Nikart was designed by the Overseas Division of the National Institute of Agricultural Engineering (NIAE) in England. It is a wheeled tool-carrier drawn by a pair of oxen. The design philosophy behind the Nikart was that it should provide a one-man ride-on implement, have fair versatility without incorporating costly features not needed by most users and be capable of manufacture at reasonable cost.

By mid-1980 the Nikart was ready for small-scale manufacture in India, and a small engineering company in Hyderabad, Mekins, that was enthusiastic about producing the machine was identified.

The primary objective of my involvement was to design and construct a set of jigs and fixtures for the manufacture of the Nikart. During my seven-week stay in Hyderabad, five welding jigs, three drilling jigs and a series of marking-out templates were built. We applied conventional engineering principles in designing the jigs, but were careful to ensure that:

— they were simple to use, and that components could only be located in the jigs in the correct position;
— they were of sufficient accuracy to allow good quality Nikarts to be produced off them;

— they accommodated the dimensional variations of locally available steel sections, while locating the critical points on the Nikart accurately;

— they were simple to make.

Except for the grinding of the drilling jig bushes, all the jigs were made in the Mekins factory. This was felt to be important in transferring skills in the design and use of jigs. It also minimized the investment cost of the tooling.

The use of these jigs reduced the time that was needed to mark and cut the components and to fabricate the various subassemblies. A much greater degree of precision was also achieved, making fitting at the final assembly stage easier. These jigs minimized the manufacturing time and the number of reject components, thus also reducing the cost of producing the Nikart. They ensured that the products were of a satisfactory level of quality and consistency.

Consistency is important in facilitating the supply of spare parts once the machines have been sold. If the spare parts are produced on the same jigs as the original machine, there should be no fitting problems.

To complement the introduction of jigs, we defined simple planning procedures for the production of batches of Nikarts. This involved defining the manufacturing operations for each of the individual components and the various stages of assembly. Based on this, the sequence of operations on the different items of manufacturing equipment was specified so that work began on those components with the longest lead time and so that machine set-up time was minimized.

To some extent these two criteria were in conflict, but the net result was that both labour and equipment were used more efficiently, and the time taken to manufacture a batch of Nikarts was minimized.

The engineers involved in the

Nikarts have many uses — and can be manufactured locally.
(*David Kemp, NIAE*)

design and development of the Nikart put considerable effort into evolving a technology suitable for local manufacture; particularly in terms of materials, components and production processes used. Nevertheless, in the course of designing the jigs, a series of modifications to the Nikart were identified to simplify manufacture. These were all detail design changes that had no effect on the performance of the Nikart and that were imperceptible to the casual observer. However, taken in total, they had a significant influence on the ease and cost of manufacture. For example, several components were specified as being made from 40 x 12mm steel strip. According to the steel suppliers catalogues this was a standard, available section, but in practice it was very difficult to obtain. By analyzing the dimensions of all components and making minor changes where appropriate, it was possible to utilize those material sections that were easiest to process and to minimize the range of different material sections required to produce the Nikart. The latter helped to reduce the management problems of purchasing and stock control and the working capital tied up in raw materials.

Similarly, another feature of the original Nikart design, the adjustable-height wheel mounting, required a high-quality channel section of specified tolerance that, although manufactured in India, is allocated on a quota basis, and it is very difficult for a small manufacturer to guarantee regular supplies. Mekins was therefore forced to use a lower quality specification. This necessitated a series of detail design changes to allow the wheel leg assembly to be produced to a satisfactory standard with the lower quality material that was easily available.

Two other examples of design modifications that simplified manufacture are of interest. By making a simple design change, it was possible to make the foot-rest in one piece instead of two and, by altering the configuration of the joint between three components, it was possible to allow a wider tolerance on the dimensions of the parts and to simplify the design of one of the welding jigs.

In the last year Mekins has produced and sold more than 70 tool-carriers, including some for export. The Nikarts have proved to be of good quality, have performed satisfactorily and are very competitive in cost with other products available in India. To follow up this initial success the marketing company with whom Mekins collaborates is about to embark on a major sales drive in one Indian state.

Ian Barwell, *Ceres*, 1983.

Time to the land is a bringer of gifts, but time to a machine is the seed of destruction. This is why industrialized time is short — the machine must be paid for before it dies. — Wendell Berry

Problems of Expansion

Two entrepreneurs with a background in both mechanical and agricultural engineering started a small-scale diesel engine manufacturing plant in 1962; they rented a ready-built factory on the industrial estate at Hyderabad in Andhra Pradesh, and their initial investment in the company, which they called Kisan Engines, was Rs300,000. The factory could produce up to 300 engines per month ranging between five and ten horsepower. A few of the key components were manufactured in the factory and the rest were obtained from sub-contracting firms in Kolhapur in Maharashtra. As a result of Kisan's initiative a number of sub-contracting firms also started in Hyderabad and these provided part of the firm's needs.

Kisan established a reputation for good quality workmanship, and in 1967 the entrepreneurs were able to expand the capacity to 500 engines a month through an investment of a further Rs200,000. Turnover rose from Rs400,000 in 1964 to 1,200,000 in 1968 and the technical and managerial ability withn the firm, combined with the advice the partners received from development institutions, seemed to promise a secure future.

However, the demand for diesel engines is cyclical and depends on the prosperity of farmers and their ability to obtain subsidized loans for installation of irrigation. In 1969 there was a severe slump in demand which lasted three years and a number of diesel engine manufacturers, both large and small, were in serious difficulties. Kisan managed to survive, partly because the Government of Andhra Pradesh had included its engines on their approved list for subsidised loans in order to encourage local industry.

In spite of the cyclical demand the overall prospects were good particularly in areas which had not yet been reached by rural electrification. The profit margins were also generous, and in the late 1960s a number of enterprises started to manufacture diesel engines in various states including Andhra Pradesh. In 1970 a large-scale firm called Shri Ram was established in Hyderabad with a capacity of 3,000 engines per month. Kisan, Shri Ram and the other manufacturers in Andhra Pradesh still obtained the bulk of their components from Kolhapur in the north where there were over 500 small-scale firms sub-contracting to the diesel manufacturing industry.

In spite of difficulties and competition Kisan expanded its turnover to Rs2,000,000 during 1972. Because of the increasing local competition and the cyclical nature of the business the partners thought it might be advisable to diversify their product lines. They decided to remain in the agricultural area and to manufacture power tillers in collaboration with a Japanese company.

With the help of this Japanese collaborator Kisan put forward a proposal for manufacturing power tillers. This involved an additional investment in machinery of

303

Rs500,000 which would raise the capital invested in the business to Rs1,000,000.

It would also be necessary to increase the factory space and there were two ways of doing this: one alternative was to rent another shed on the factory estate and the other possibility was to build a power tiller factory in the industrial development area where land was available. In the latter case the building would have to be built at Kisan's expense whereas if they remained on the industrial estate they would be able to rent the necessary space.

The partners preferred to invest money in manufacturing rather than in buildings and they applied to the Industrial Estate Authorities in 1972 for another shed on the estate.

This request posed the Estate management with a problem: if Kisan made the envisaged investment the company would go beyond the official limit for a small industry and would therefore not be eligible for factory accommodation on the estate. Any deviation from this rule would mean a basic change in government policy and the Estate management therefore preferred the company to locate its new factory in the industrial development area. They suggested that Kisan should build this factory with a loan to be obtained from the State Financial Corporation and the problem was then referred to the Directorate of Industries for a decision.

Malcolm Harper & T.Thiam Soon, *Small Enterprises in Developing Countries*, 1979.

 # Brass Casting and Metalwork in Ghana

In several villages centred on Kurofofurom, eight miles south of Kumasi, most of the adult male population is engaged in a craft industry — making brass figures by the lost wax process. The ornamental figures vary in size from single figures no more than three inches high to large urns and elaborate tableaux portraying chief and attendants. Each figure has some symbolic significance.

Prices vary from ¢3 for the small figures to ¢100 for the large ones, which may contain several pounds weight of brass. The figures are sold in the villages to passing traders, and in craft stalls in the towns. The raw material is recycled scrap brass, obtained from traders in the towns.

The lost wax process has been in use in some parts of the world for 3,000 years. It has several stages:

— The figure is first modelled in beeswax, exactly as it should be when finished in brass.
— The beeswax image is covered with a wet mixture of powdered charcoal and local clay, built up in layers until a nearly convex object is formed. A final thick layer of clay reinforced with palm fibre is added to strengthen the block.

- The third stage is to heat the block gently over a small wooden fire so that the wax runs out.
- Pieces of scrap brass are packed on top of the mould and enclosed in another layer of clay and palm fibre, making one large block with the mould for the figure at one end and the brass at the other, connected by a passage through which the molten brass is to run.
- The block is then put into the kiln with the brass at the bottom and heated until the brass is molten. The charcoal-fired kiln is open at the top so that the craftsman can judge whether the brass is molten by the changed colour of the rising smoke. It is open at the side to allow the mould to be tightly packed among the charcoal. Up to six hand-operated bellows blow air through holes in the base of the kiln. As a result of this design, consumption of fuel is very high, and the temperature needed to melt the brass is reached only with difficulty.
- The block is removed and upturned so that the molten brass runs into the mould.
- When somewhat cooled the mould is broken and the brass figure is taken out. About 10 per cent of the figures are faulty and the brass is reused.
- Finally, some simple hand polishing is done.

This skill has been handed down from generation to generation, but the market for the artifacts is insufficient to provide a good living for all of the population, so there has been the usual migration of young people in search of greater opportunities.

The Technology Consultancy Centre in Kumasi has attempted to expand the range of activities open to the craftmen by showing them how to turn their

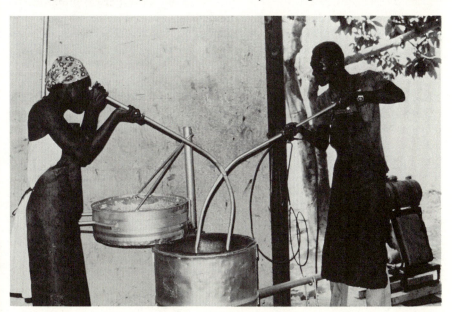

Lost-wax casting: an old skill with new uses

skills to the manufacture of different objects, such as valve parts for water pumps.

TCC was originally approached in 1975 by the chief of the main brass-making village, Kurofofurom. He asked for help in securing cheaper and more reliable supplies of scrap brass. The Centre was able to provide some temporary help with this but there is only a limited amount of scrap brass in the country and it has become increasingly scarce and expensive. It occurred to the TCC staff that the craftsmen's skill in the lost wax process could be applied to new products, thus enlarging the employment opportunities in the villages and giving the artisans a new source of income. An experiment was undertaken by the chief brass-maker of Kurofofurom to see whether he could turn his skills to making brass parts for engineering use. The first products made were valve parts for water pumps being made by the Faculty of Engineering, and bushes for rice threshers made in the TCC workshop.

The Centre provided the patterns and the brass-makers copied them in wax and made brass castings using the usual lost wax process described above. Because of inexperience they had some difficulty in adapting to the geometric shapes but the parts were successfuly made. Turning on a centre lathe to achieve precise dimensions, turning of the thread and the polishing were done at the University's Mechanical Engineering Workshop.

A problem with the traditional method is that for each item produced the wax image has to be made individually, and, of course, this is time-consuming and limits the scope for applying the technique to the production of a large number of identical objects. The TCC are now working on the development of a method of making large quantites of identical wax images which can then be cast by the lost wax process. For example, a plaster of Paris split mould for making door handles has been developed. The wax used in the villages is too soft for this and shrinks during the casting, but successful results have been obtained by mixing some candle wax with the beeswax and lining the mould with light machine oil to prevent sticking.

Now that some of the technical problems have been solved, it is hoped that a range of products can be developed for manufacture by the lost wax process from wax images made in plaster of Paris moulds. Apart from the door handles and similar domestic fittings, there could be considerable scope for the development of engineering components, such as the pump valves, for use in plumbing and vehicle repair work. Virtually all parts of this kind are currently imported into Ghana, and consequently are in very short supply. Many of them are suitable for the lost wax process which is still used in developed countries for highly intricate objects. Many would be made in a developed country by an automated sand-casting process, but little of this is at present undertaken in Ghana. It is therefore appropriate in the Ghanaian context to make them by the lost wax process, as the skills already exist. It would be possible to achieve import saving and provide extra local employment.

Sally Holtermann, *Intermediate Technology in Ghana*, 1979.

The Forgotten Partner

Small-scale mining is much more widespread than is normally assumed. It has been estimated that it produces as much as 10 per cent of the global mineral output, and much more in developing countries. In Chile, for instance, 20 per cent of the copper used to come from small mines, and in Sri Lanka 100 per cent of the gems. It is not unusual to find that, in a developing country, all the industrial minerals are mined on a small scale. It is less prevalent in industrialized countries because of the high labour costs and environmental restrictions, but there are exceptions, such as Finland, New Zealand and some states in the US.

It is very difficult to find quantitative information. By nature, small-scale miners do not take kindly to government regulations and frequently escape the fiscal net. Government employees are occupied with bigger matters and leave them to their own devices. Unless they get together and form pressure groups, they do not appear in statistics. Yet one cannot doubt their social importance. They provide employment, full-time or seasonal, in depressed agricultural areas, and reduce the drift to towns; they import new skills to the local population; they create a demand for transport and service industries, such as small-scale engineering; they provide the raw materials for many mineral-based domestic industries; and they do not interfere with the social fabric of the area. On a national level they often provide goods or materials that are either exported or save importation. In short, they can substantially improve the self-sufficiency of a country or area.

The importance of the mining-farming relationship has been recognized in Mexico, where the Agrarian Reform Secretary has a Non-renewable Resources Director, whose main objective is to promote mining projects within the Ejidos (State agricultural small land grants). The idea is in keeping with the policy of creating part-time and permanent employment for the Ejidatorio groups in their own localities. On the other hand, mining and agriculture can have conflicting claims on resources and it is not always possible to arrive at a mutually beneficial arrangement — the pollution of rivers is an example.

Like all enterprises, small-scale mining can thrive only if conditions are right, and one or more of the requirements listed below must be fulfilled.

Exploration. It must be possible to find the orebody without sophisticated equipment, or the Geological Survey must be willing to assist.

Mining. The mineral should be near the surface to obviate the use of expensive equipment or an advanced knowledge of rock or soil mechanics.

Grade. Must be sufficiently high to counteract the other increased costs.

Beneficiation. The mineral should require no treatment other than hand-sorting or be amenable to simple metallurgical processes before it is shipped to the buyer. Unconsolidated placer deposits are particularly suitable.

Transport. The weight to be moved must be small or the buyer must be near the mine.

Infrastructure. The cost of housing, roads, power and water supply must be minimal.

Marketing. There must be a stable demand at reasonable prices.

Government. The authorities should be sympathetic and active.

Tom Wels, *Transactions of the Institute of Mining and Metallurgy,* 1983.

Barefoot Geologists

While international organizations must continue their role in supporting the professional geoscientific communities in developing countries, they must also take the initiative and spread their support to lower technical and subtechnical levels of activity, especially to the uneducated, rural villager. It is this group that may play a very important role as barefoot geologists or prospectors in locating potential resources, as the legendary, and still active, prospectors have done in Canada, the US, Australia, Finland and elsewhere.

Village-level participation in development can be accomplished by using local human resources as barefoot geologists or prospectors, who have intimate knowledge of the land and the traditional methods of prospecting. This group can be encouraged, supported and trained at a very low cost and with potentially very high returns. Training courses for prospectors could run at a fraction of the cost of training professionals. In fact, professionals could assist at various training centres in the outback, bush or hinterland, and this would also serve as an important, relevant field link for the urban geoscientific group.

Michael Katz,
*Appropriate
Technology,* 1983.

Small is Expensive

Imports of agricultural machinery to Africa in 1978 cost US $2 billion. Well, heavy goods don't come cheap. But a closer look shows that $300 million was spent on hand-tools alone, highlighting a real need for locally produced implements, better adapted to local conditions.

*Development
Forum,* 1982.

Old Methods to Meet New Needs

Kaduna, with an estimated population of over 500,000, is probably Nigeria's fourth largest city and is the capital of Kaduna State. It is a modern city founded by the colonial administrators after 1910 because the railway lines from Port Harcourt and Lagos meet on the south bank of the Kaduna River before crossing and going north to Kano. Today Kaduna is a thriving, bustling, growing city with much industry including a new petroleum refinery capable of supplying half of Nigeria's refined products' needs. In the heart of this busy industrial city can be found a flourishing small-scale mining project; one that has apparently been active for many years, but even more surprisingly it seems there is no foreseeable end, either to the reserves of the 'ore', or to the demand for the product. It is true that each year's production tonnage is limited but this is counter-balanced by the assurance that reserves will always be available next year. The secret of this phenomenon is the Kaduna River. Each year its flow varies dramatically from a high at the end of the rainy season in August and September, to a mere trickle at the end of the dry season in March and April. This leads to a drop of 6m or more in the water level flowing under the railway bridge in the virtual centre of the city. As the water level falls, in about late November, the sand bar just downstream from the railway bridge first appears and by 1 January it is usually over 1m above water level and the first workers start digging the 'ore'.

The ore comprises generally clean, coarse-medium grained, sub-angular sand grains, ideal for construction purposes. It is derived from the coarse grained granites and basement rocks outcropping upstream from Kaduna.

The sand is first dug by hand labour with shovels from a single face 70-100cm high that is moved systematically along the length of the sand bar (which measures about 100 by 45m). After this first cut, or slice, and as the river level continues to fall, the diggers rework the same area by scraping together the sand in conical heaps about one metre high which allows it to dry. The sand is loaded into wicker-work baskets, about 60cm at the top and tapering to 30cm with a depth of about 20cm. These baskets are carried on the heads of porters for about 200m from the sand bar up a specially constructed path on the river bank, under the railway bridge approaches and dumped in piles in a flat area to which 7-10 tonne trucks have access down a short road off Kaduna's main street. The trucks are hand loaded and the sand is transported to sites of consumption. As many as ten labourers may be working on the sand bar shovelling, while another thirty carry head loads up the river bank. The average daily earnings are about US $5.50 but a hard worker can earn as much as US $12 which represents over 200 basket loads at the present rate. No tally is kept of individual workers' efforts but at the end of each day the porters state the number of baskets they have carried 'in the name of Allah' and they are paid on this.

The sand bar is estimated to contain a minimum of 4,000m³ of exploitable sand *in situ* and this will all be dug during the dry season. During the period of the high water the sand bar reforms so that next year the same mining scene will be re-enacted. The fortuitous hydrological conditions that lead to this redeposition are clearly related to the pattern of rock outcrops and the curve of the river. The need for clean sand in Kaduna is never ending due to the constant building activity and the present gross value of this innocuous-appearing sand bar in the middle of the city is more than US $100,000 p.a. The present production is controlled by two businessmen/contractors who meet their own requirements and sell the excess to other users. Sand is dug from other parts of the Kaduna River and its tri-butaries but this example is of particular interest because of its location in the heart of the city. Small-scale mining in Kaduna does contribute to the industrial development, however incongruous the head-baskets of sand may look against a background of modern factories. Perhaps the purist will argue that since sand is not classified as a mineral under the Minerals Act of Nigeria, then the exploitation of sand is not truly mining. But it does represent a practical and effective combination of old methods to meet new needs and is surely, in a country with a high level of under-employment for the unskilled and a woeful lack of mechanical maintenance facilities, good utilization of available resources. One might however propose changes in the organization of ownership and distribution of benefits, perhaps through a workers co-operative.

M.E. Woakes, *Earth Sciences Programme Newsletter*, 1984.

Mica Mining in Sri Lanka

In a small mine in the Rattota district, north of Kandy, a 2ft-wide vein of vermiculite was being mined with hand steels down to a depth of approximately 30ft. The vein continued below this into unweathered gneisses but to go further down would have required explosives and involved higher costs. The strike extension outlined was about 60ft before the vein wedged out and the reserves were an estimated 200t — enough for about three years at the present rate of mining. This mine is being worked by a rural co-operative under the general supervision of the Government District Office. The mining is done by three or four local men under the direction of an older man with many years' mining experience in other parts of Sri Lanka and with a strong commitment to the utilization of mineral resources on a village level. After mining the mica is sun-dried and trimmed by about 15 village girls who work on a

part-time basis. It is then bought by the Graphite Corporation, whose geologists provide technical advice and control the overall quality.

One can, of course, criticize this kind of mining as wasteful because it does not extract deeper ore levels, or unsafe because the miners work under fairly primitive conditions. But it does provide a living, however modest, for some people; it generates cash revenue for the village (profits are now being invested in a truck to transport the mica); and it clearly gives a sense of self-sufficiency which can be extended to other fields of endeavour — all this in a district which is by no means a wealthy one.

The role of the State Graphite Corporation is to assist in the location and opening up of new deposits, to purchase mica from these mines and quarries, to sort, break and classify this in central mica-curing warehouses (one near Kandy employs 50 girls at present, at piecework rates considerably higher than their counterparts in the private sector), and to handle marketing and sales arrangements. Indeed, the young geologist who developed the first Cor-

Mica mining: a source of cash revenue for Sri Lankan women

poration mine, did so almost on a one-man basis from exploration to mine supervision, through quality control, processing by hand, and even to the actual marketing!

At present the total work force employed in mica mining approaches 150 in both the private and public sector. Much of the current production — a little over 300t per year, mostly by the Corporation plus a smaller amount from private mines — is exported to Japan where it is processed into micanite sheets and mouldings. Some mica is also used as an additive to paints and to wallpaper, giving a sparkle which customers in the Middle East like. The Corporation geologists are confident that production can be expanded and that further processing and even mica-based manufacturing can be done on an economic basis in Sri Lanka.

A.R. Berger,
*Appropriate
Technology*, 1979.

3 RECYCLING

 New Jobs from Old Rubbish

Waste is one of the world's largest industries, though this could never be discovered from any book or statistics.

All over the world, especially the Third World, the collecting and recycling of used tyres, lumber, glass, metals, cloth and plastics provides jobs for millions. And many of these millions are women, children, the handicapped, and former prisoners — people who are unlikely to find work elsewhere.

The re-use of materials saves governments valuable foreign exchange and saves clean-up bills, as well as providing employment and basic industrial training. Despite all of this, few governments do anything to give workers in waste the minimum training they need.

A recent eight-nation survey, in which the British Government sponsored me to look at waste materials and how they are used, proved that in the case of almost every material someone has come up with a way to recycle it and make money from it. There is a need to spread the word of these small-scale technological breakthroughs.

Take Oliver Moxon, a northern Jamaica restaurant owner, and all the dead coconut trees rotting in his country due to yellow leaf disease. Moxon was apparently the only man in Jamaica to wonder why this wood could not be used for anything. Now a village workshop turns out some of the world's toughest wooden parquet floors, as well as ceiling fan blades for the North American market.

Take the scrap steel business. The expanding steel industry, which feeds

312

and is fed by Mexico's oil boom, finances an army of the unemployed poor who dredge scrap metal from demolition sites, ditches and roadsides. In Cairo, youths flatten oil drums for use as roofing sheets by the dangerous technique of hauling them into the middle of the road for passing trucks to run over. Out in the desert, hoards of men with oxyacetylene torches cut through the valuable, high alloy steel or rusting tanks and armoured cars left by the many wars to sell to the city's steel mill.

Take rubber tyres, a nightmare to European waste disposal engineers who find them too tough to chop up, too smoky to burn, too elastic to stay buried long and too buoyant to dump into the ocean. Yet in the Third World there is no such problem. In many countries a large truck tyre is the nearest thing there is to hard currency. Shoes, sandals, stool seats and bedsprings are all made from strands of rubber skilfully cut from tyre carcasses.

In India, village cobblers absorb so many tyres that the vigorous large-scale reclamation industry whose sixteen different factories produce every conceivable type of rubber product from this scrap has toured Europe in the hope of importing more old tyres.

Entire cars are recycled in many Third World nations. Throughout Latin America one rarely finds a derelict car because scavengers dismantle any abandoned vehicle and re-sell or re-use the parts. A London-based intermediate technology development group has devised methods, based on those used by the scavengers of Latin America, whereby two men with simple hand tools can cut a car into transportable bits in a few hours, and sell enough scrap steel to run a lorry and earn modest wages.

Similar projects involve the recycling of plastics in Kingston, Jamaica, and retrieving valuable tin from can manufacturers' scrap in Kenya. Sixty mission hospitals have asked for help in recovering silver from X-ray wastes, and a simple recovery kit suitable for use in a bush hospital is being planned.

It would be wrong for the wealthy world to preach to the people of developing countries to work in rubbish. But as this already happens on a grand scale to the benefit of the poor and of governments, then workers in waste should be helped to do their work more safely, more efficiently and more profitably.

Jon Vogler, *Ap-tech*, 1982.

Man does not live by bread alone, but by faith, by admiration, by sympathy. — R. W. Emerson

---

With the exception of the instinct of self-preservation, the propensity for emulation is probably the strongest and most alert and persistent of the economic motives proper. — Thorsten Veblen

Trash Technology

In Baltimore, US, the 'Langard' plant, built by the huge Monsanto company to produce energy from the garbage, emitted such severe pollution that millions of dollars were spent on modifications. Finally, the company ignominiously withdrew from the contract, paying a hugh penalty for so doing.

Out of fifty-five US plants built with the intention of recovering materials or energy from garbage, only seventeen actually operate — what can only be described as a gigantic failure of high technology. In an age when money is needed to invest in job-producing industry and agriculture, or for health and education programmes, such plants cost too much, create too few jobs and draw money and municipal attention away from the main priority of collecting garbage from homes and litter from the streets.

Jon Vogler, *New Internationalist*, 1982.

Social Structure and Solid Waste

Cairo has no elaborate waste disposal system. In poor areas of the city, residents and goats dispose of what trash there is. In middle and upper income areas, which produce potentially valuable trash, scavenging has been developed into a large, elaborate system. It is motivated by self-interest, rooted in powerful social structures, and is impressively effective in keeping important parts of the city clean.

The traditional Cairo system depends on two groups: Muslim 'Wahiya' and Coptic Christian 'Zabaline'. The Wahiya act as brokers. They purchase waste collection rights from building owners, paying commissions that vary with the size of a building and the value of its waste. The Wahiya brokers collect fees from the residents. They also rent to the Zabaline the actual right to pick up the trash. The Zabaline Christians are socially-marginal people in a Muslim country. About 40,000 Zabaline families live in eleven shantytown satellites of Cairo. They derive their income from collecting and sorting waste into marketable components, and by breeding and selling pigs (pigs are not unclean to Christians).

Nothing is wasted. About 2,000 tons of paper are recycled every month. Cotton and wool rags are reprocessed for upholstery and blankets. Tin is pressed and soldered into vessels, toys, and spare parts for machinery. Glass and plastic are reused. Organic matter is fed to pigs or used for compost. Bones are used to make glue, paints, and high-grade carbon for sugar refining. Hundreds of workshops and factories within the city depend on the Zabaline for raw materials.

Some forms of recycling can be profitable — but not necessarily healthy

In terms of sanitation, the Cairo system is efficient. Solid waste is collected and reused. The Christian Zabaline also sell about ten tons of pork each day. The Zabaline make money out of trash. Some retire in financial security.

But the system is not without its costs. These include a high infant mortality rate — estimated at 60 per cent — among the Zabalines who live amid wastes in the desolate fringes of the city, and are often uprooted as Cairo's borders expand. Health centres, schools, water, electricity and other municipal services do not exist.

The Governorate of Cairo has considered importing incinerators, waste collection vehicles and other sophisticated waste disposal technologies. These could satisfy immediate waste-disposal needs. At least one development agency has considered the Cairo trash situation as an opportunity for a technology-based waste-disposal project. Trucks with trash compactors would collect solid waste. But the Zabaline and Wahiya would be deprived of income, and Cairo would lose a source of raw materials.

Tom d'Avanzo in R. Mitchell, *Experiences in Appropriate Technology*, 1980.

Clearly the attitudes and values which make production the central achievement of our society have some exceptionally twisted roots. — J. K. Galbraith

Garbage Recovery on a Human Scale

This process of recycling waste has dividends beyond the agricultural one of restoring the food cycle, and they are supremely important for a properly balanced ecosystem. Most of what this country (US) throws away — or, rather, tries to — is easily reusable, either as fuel or compost or raw material. Municipal garbage, on average, contains approximately 40 per cent of recyclable paper, 10 per cent glass, 10 per cent ferrous metals, 3 per cent rubber and leather, and 4 per cent wood — a marvellous, and essentially untapped storehouse. And whatever it would cost to put these materials back into productivity is many times less than trying to create them from scratch, and the energy savings are considerable as well.

Aluminium, for example — it makes up 2 per cent of the nation's total garbage stream and at present is mostly thrown away. But it can be separated and recycled quite simply, at one-twentieth of the environmental (energy and pollution) costs of mining and smelting virgin ores. Or copper — fully a quarter of the nation's copper consumption, about 375,000 tons a year, could be obtained by extraction from municipal garbage. What's more — incredible as it may seem — some municipal dumps have higher percentages of copper than some mines that are operated profitably right now in the Rockies. It is obviously very close to criminal to discard resources such as these in a world of growing scarcity.

The logic of recycling — not to mention the current profits therefrom — is so overpowering that virtually no-one now disputes the virtues. However, the general governmental response has been to apply technofix methods, particularly the huge 'resource recovery' plants that have been built in a number of big cities in the last few years. They certainly have the right idea, but so far they have proven to be dismal failures — they are very capital intensive, they are wasteful of resources, they turn out to be inefficient and susceptible to recurrent breakdowns, and they even end up adding to pollution by creating dust during their shredding processes and air pollution during their incineration phase. Moreover, since they all need a tremendous volume of garbage to be able to produce sufficient amounts of recycled materials for resale, they place a premium not on conservation but on waste, and the more the better; the city of New Orleans, for example, has had to *pay* its recovery plant for every ton of garbage below the contracted level that it is unable to deliver. Not one of these plants has so far operated as expected, and several of them — in Baltimore, St. Louis, New Orleans, and Seattle — have gone sharply into debt or even been abandoned as hopelessly uneconomic. It will come as no surprise that the Federal government has spent about $100 million to perfect these high-techology plants, but virtually nothing for other recycling systems.

A small-system, community-based approach, however, has worked out very well in a number of towns and cities across the country. According to the solid-waste expert at the Institute of Local Self-Reliance in Washington, there were between 150 and 200 cities with source-separation and recycling programmes as of 1978, up from only twenty in 1975. (Most of them in smaller places; experience has shown that easier communications and more established patterns of social cohesion, make smaller towns easier to work with in setting up recycling operations.) The ORE plan in Portland is among the most successful of them. There the residents of four neighbourhoods pre-sort their garbage into four types — paper, glass, metal and organic — and place it on the curb once a week. It is picked up by separate pick-up trucks, one for organic, another for inorganic, and hauled off to be composted or sold to bottlers and metal processors. The system has been found to be the most energy-efficient municipal garbage system in the country, and it even offers the household subscribers a lower collection rate than they would have to pay for ordinary private haulers.

Small-scale recycling operations in the future can be made even more efficient, by the application of both communitywide support and alternative technologies. Studies have shown that the optimum efficiency for a recycling programme is reached when every household in a community of 5,000 participates — at that level, a once-a-week collection can produce a large enough quantity of various materials to make a 15-20 per cent profit on resale, and it will still not cover so much territory that it loses money in collection and transportation costs, which account for 70-80 per cent of the overhead of an ordinary system. Collection could be even more economically done with small electric carts or trailers — the ORE system has used golf carts profitably — or ultimately with a simple network of underground, solar-powered conveyor belts that, in small areas, could be operated as efficiently as such belts are in any large factory today. That would be the most efficient and equitable separation and recycling operation of all.

A small-scale system would be most economic, too, if the community were to reuse everything within its own borders as raw material for its own products. Small-scale aluminium and de-tinning facilities can now be constructed for less than $5,000 and, using ordinary cans, could turn out metal for local production of bicycles, wheelbarrows, machine and auto parts, and the like. . . . Paper-processing plants could easily convert wastes into newsprint or reusable paper or even into cellulose-fibre insulation. Bottling centres could either recap and reuse the bottles as they come in or crush them to use in road or building construction or, with currently available machinery, remould them into canning jars for neighbourhood use. A somewhat more complex, but still small-scale, process can be used to convert plastic wastes into building materials, furniture, auto-body parts, or fish tanks. Human and food wastes could be fed to community livestock — pigs as everyone knows, are marvellous waste-eaters and process even human excrement to a valuable

317

manure — or composted collectively for community gardens; 'greywater' (household water used for washing or bathing) could be piped from each house and, after a simple filtering, used for irrigation.

There seems to be a scale at which even such a process as garbage recovery can make economic and ecological sense. It is small, and human.

Kirkpatrick Sale, *Human Scale*, 1980.

In Peru thousands of people are employed in making shoes from old tyres.

(*Jon Vogler*)

VIII
Generation and Transfer of Technology

THE PROCESS of technology generation, transfer and diffusion is complex, involving a wide range of agencies and individuals including farmers, entrepreneurs, governments, research institutes, commercial firms, transnational corporations, commercial banks, bilateral and multilateral donor agencies, and national and international non-governmental organizations. This chapter attempts to cover some of the more important issues involved in the generation of technologies and the transfer of technologies between countries. The process of dissemination of technologies within countries is dealt with in Chapter IX.

By the end of the 1970s 'Research and Development' had become a 150 billion US dollar global enterprise employing some three million scientists and engineers. There is however an enormous gulf between rich and poor countries. It has been estimated that developing countries do only 2 per cent of all R & D conducted in the non-communist world[1] and that despite the recent expansion of university education in some countries, they have only a tiny fraction of the world's pool of researchers.[2] According to one report:

> While these disparities simply mirror many others between rich and poor countries, they nevertheless have important implications. As long as the world's R & D capacity remains highly concentrated in the industrial world, the focus will contine to be largely on the problems of the rich countries, and the developing world will remain dependent on imported — and often inappropriate — technology for its economic development.[3]

These statistics and statements, however, refer only to the formal R & D system, and this is indeed very small in most developing countries. The formal system also tends to mirror the kind of activity carried out in the industrialized world: it caters to the better-off elite who can back up their needs with purchasing power, and ignores the basic minimum needs of underprivileged people. But this is not the only, or even the main, source of innovation in the Third World. Technological developments and adaptations continuously arise from outside the formal sector — in commercial firms, small workshops and

---

[1] Lawrence J. White, 'Appropriate Factor Proportions for Manufacturing in Less Developed Countries' in Austin Robinson (ed) *Appropriate Technologies for Third World Development* (Macmillan, 1979).
[2] Colin Norman *Knowledge and Power: The Global Research and Development Budget,* Worldwatch Paper 31, 1979.
[3] *Ibid*

rural development organizations. In addition, there are millions of individual, innovative entrepreneurs and farmers contributing to the process of technical change in the Third World.

It is to the indigenous sources of technology in developing countries that the first section of this chapter addresses itself. It begins with an extract by A.K.N. Reddy on the formal R & D sector which gives a useful explanation of how this sector is biased towards the needs of the better-off elite in such countries. This is followed by an interesting description by Nicholas Jequier of the 'invisible' actors in the 'appropriate' technology-generation game. These fall into two groups. First, the organized but peripheral sector — the large industrial corporations and public and private engineering and consulting institutions which, although not concerned with appropriate technology, have in fact developed such technologies. Second, the 'informal' or 'unorganized' sector which represents a vast pool of largely unsurveyed appropriate technologies. Jequier gives examples of an indigenous starch separation plant in the Philippines, a locally developed razor blade plant in Algeria, and small-scale pump manufacturers in Pakistan.

Subsequent extracts look further at the nature and scope of the 'informal' indigenous R & D sector. These include accounts of rural innovation in Indonesia and Thailand, the innovators movement in Nicaragua; and agricultural innovations by farmers in Bangladesh. Also included are two more general extracts on the issue of indigenous technical change. Martin Bell's extract explains how technical change can take place in a variety of ways, many of which do not involve knowledge-creation by R & D. These include adaptations of existing production systems, adoption by producers of new methods developed by others, and incorporation of existing technical knowledge into an item of technical equipment.

Such technical change does not normally require the sophisticated facilities of formal sector R & D laboratories. Robert Chambers' extract re-emphasizes the value of rural people's knowledge drawn from experience and suggests that while professonal scientists and technologists can contribute to rural development, they should also be prepared to learn from the 'unglamorous poor'.

The section ends with one of the best current examples of a Third World success story of appropriate indigenous technological development. This is Raphael Kaplinsky's account of the development of open pan sulphitation (OPS) sugar technology in India. Developed by M.K. Garg — a remarkable Indian engineer — first at the Planning Research and Action Institute and then at the non-governmental Appropriate Technology Development Association (ATDA) in Lucknow, the story of this 'mini' technology demonstrates how, with determination, it can be shown that 'small is possible'. Most developing countries now have agencies similar to ATDA which are devoted to the development and dissemination of technologies appropriate to the needs of the poor. Many of these were mentioned in Chapters III to VII, along with some of their technological accomplishments. The work done by such agencies is receiving increased recognition by governments and donor agencies. Their resources for R & D are, however, miniscule by comparison with those of the formal R & D sector.

Of great relevance when assessing a country's technological capability is the effect of external agencies such as transnational corporations and bilateral and multilateral donors on the extent and nature of indigenous technological development and adaptation. Until recently, the development literature has tended to assume that transnational corporations are a hindrance to development because they exploit resources and labour in host countries, transfer inappropriate products and processes, and destroy local small-scale industries and pre-empt local technological development. Similarly, donor governments are frequently identified as a medium for transfer of inappropriate technology, with aid doing more to assist industry in the donor country than to contribute to self-sustaining development in the Third World.

The second section of this chapter looks at the process of technology transfer — particularly by transnationals and donor agencies — to the Third World. The first extract by Carl Dahlman and Larry Westphal gives a good introduction to some of the issues at the heart of the international debate on technology transfer. It makes the important point that there is often an implicit notion that technology transfer gives the recipient three broad types of capability: the capability to operate a technology; the capability to expand existing productive capacity or establish new capacity; and the capability to develop new methods of doing things. In practice, however, this is rarely the case. The capability to operate a technology is different from the ability to develop the means of implementing it. Similarly, having the capability to implement a technology is different from having the capability to create a new one.

This general extract is followed by a more specific one on the disadvantages of imported technology and by an account of the problems experienced with the transfer of technology in the Caribbean. These problems include balance of payments crises stemming from excessive payments for imported technology, reduction in export potential as a result of clauses limiting exports, maldistribution of income between modern and traditional sectors, lack of employment-creation possibilities, and the use of restrictive business practices limiting the use of indigenous technology as well as technological transformation possibilities.

Lists of problems such as these are commonly used to point out the harmful effects of transnational corporations (TNCs). But are TNCs really as bad as has been made out? The extracts by Sanjay Lall and Paul Streeten suggest that there may be some redeeming virtues. For example, Lall looks at the extent to which TNCs are actually strengthening the R & D capability of developing countries. Also, many more TNCs are now established by the developing countries themselves and evidence suggests that they tend to score better than European or American TNCs in terms of job-creation possibilities and the use of local resources.

The section ends with a look at the impact of aid on developing countries. Again, the problems are mentioned first. There is an interesting account by

Julius Nyerere on why Tanzania is now worse off than a decade ago, despite billions of dollars of foreign aid; and an interesting extract, too, by Tarzie Vittachi on some of the myths of aid, including inappropriate, expensive consultants, inappropriate projects and inappropriate technologies.

Finally comes the issue of the attempts by the aid agencies to come to grips with the effect of technology choice on the impact of their programmes on Third World Poverty. The extract by Fluitman and White gives a good overview of the approaches and policies of bilateral and multilateral donor agencies towards 'appropriate' technology and technology choice. This covers the 'narrow' approach of the bilaterals — funding AT Centres in their own countries and funding specific AT projects — and the 'broader' approach of the multilaterals — demonstrating a willingness to assess the appropriateness of technology across the whole range of their programmes and projects.

1 SOURCES OF TECHNOLOGY : FORMAL AND INFORMAL

Generation of Appropriate Technologies

All social wants are not necessarily responded to by the institutions responsible for the generation of technology, viz., the educational, scientific and technological institutions. There is a process of filtering these wants, so that only some of them are transmitted as demands upon technological capability and the rest are bypassed by these institutions. In other words, there are ignored wants which institutions do not seek to satisfy by research and development.

This filtering process is usually operated by decision makers, firstly, in the bodies which control the research and development institutions, and secondly, within the institutions themselves. These decision makers are either conscious agents of social and economic forces, or are unconsiously influenced by these forces.

In untempered market economies, only wants which can be backed up by purchasing power become articulated as demands upon the research and development institutions and the remaining wants are bypassed, however much they may correspond to the basic minimum needs of underprivileged people. Thus, like all commodities in these economies, technology too is a commodity catering to the demands of those who can purchase it, and ignoring those who cannot afford it.

The generation of technology involves the so-called 'innovation chain' which is the sequence of steps by which an idea or concept is converted into a product or process. This sequence of steps varies with the circumstances, but can often be schematically represented thus:

Formulation of research and development objective → idea → Research and development → Pilot-plant trial → market survey → Scale up → Production/product engineering → Plan fabrication → Product or process.

It is essential to note that socioeconomic constraints, and environmental considerations, if any, enter the process in an incipient form even at the stage of formulation of the research objective, and then loom over the chain at several stages. These constraints in the form of guidelines or preferences or paradigms, for example 'Seek economies of scale!'; 'Facilitate centralized, mass production!'; 'Save labour!'; 'Automate as much as possible!';

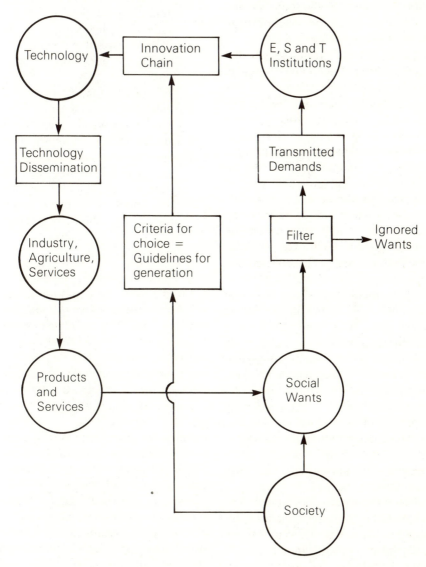

Generalized scheme for the development of technology

'Don't worry as much about capital and energy (in the days before the energy crisis) as about productivity and growth!'; 'Treat polluting effluents or emissions as externalities!', etc. (These guidelines for generating technologies are only another representation of the criteria for the choice of technologies — guidelines stand in the same relation to the generation process as criteria to the selection process).

Thus, every technology that emerges from the innovation chain already has congealed into it the socio-economic objectives and environmental considerations which decision makers and actors in the innovation chain introduced into the process of generating that technology. It is in this sense that technology can be considered to resemble genetic material for it carries the code of the society which conceived and nurtured it, and, given a favourable milieu, attempts to replicate that society.

The technology that emerges from the innovation chain will become an input, along with land, labour and capital, to establish an industry or agriculture or a service, if and only if the aforesaid socio-economic and environmental constraints are satisfied. Thus, it is not only the technical efficiency of the technology, but also its consistency with the socio-economic values of the society, which determine whether a technology will be utilized.

Social wants are not static. The products and services that are produced create new social wants, and in this process, the manipulation of wants through advertising, for example, plays a major role, and thus the spiral:

Social wants → Products/Services → New Social Wants →...

The widespread generation of appropriate technologies depends, therefore, upon the fulfilment of three important conditions:

— A filter which transmits basic human needs, particularly the needs of the neediest, (viz., the urban and rural poor), to the technology-generating institutions;
— The introduction of a new set of guidelines into the innovation chain, a set which is consistent with the criteria of appropriateness; and
— The existence of the requisite technological capability (trained and competent personnel, laboratories, workshops, test facilities, etc.) to complete the innovation chain.

The crucial question, therefore, is to what extent these conditions are satisfied in the developing countries. An exploration of this question can begin by noting that a causal spiral of the type represented in the figure is too simplistic in many ways, but particularly with respect to the social homogeneity that it implies. In point of fact, almost every developing country is polarized into a dual society; an elite consisting of the richest 10-20 per cent of the population, which usually includes industrialists, businessmen and feudal landlords, politicians, bureaucrats, rich peasants, professionals such as doctors, engineers and scientists, and the bulk of organized white-collar labour; and the poorest 80-90 per

324

cent most of whom live in the rural areas, and the remainder in urban slums. In other words, dual societies are characterized by islands of affluence amidst vast oceans of poverty. Thus, in effect, a developing country consists of two 'societies' which may not be spatially isolated from each other, but are separated by a wide chasm of incomes, consumption patterns, attitudes and life styles.

At the same time, the elite of developing countries practise a philosophy best described thus: 'all that is rural is bad, all that is urban is better and all that is foreign is best', which means that there is a strong influence of the life styles of the developed countries upon the life styles of the elite in the developing countries.

The characteristics of these dual societies are such that the filters do not emphasize the transmission of basic human needs, particlarly the needs of the neediest (the urban and rural poor), as demands upon the technology-generating institutions. The magnitude of the R & D funding for problems related to basic needs is usually a clear indicator of this bias, for it is very often significantly less than that for problems related to defence, to glamorous technologies and to those aspects of the industrial, agricultural and services sectors devoted to the demands of affluent elites. Even if this funding bias did not exist, and even if these institutions made deliberate efforts to respond to the basic needs of the urban and rural poor, there is a serious problem in the identification of these needs. This problem arises because the areas in which the urban and rural poor live, i.e., the slums

and villages, are not virgin territories uncontaminated with the demonstration effect of urban life styles. So, it is not simply a question of asking slum-dwellers and villagers what their needs are — such a 'questionnaire' approach only results in their demanding needs similar to the urban elite.

Further, the intellectual domination of the developed countries over the educational, scientific and technological institutions in the developing countries leads, in the latter, to a virtually unexamined and unquestioned introduction of alien and inappropriate guidelines, preferences and paradigms into the innovation chain, for example, the implicit faith in 'economies of scale'. Unfortunately, these guidelines are largely unexpressed and unstated. In fact, the participants in technological innovation are rarely conscious that they cannot avoid using preferences. The net result of not revealing, exposing, and evaluating the guidelines used in the process of technological innovation in (or for) developing countries is that the participants in innovation fall back on the preferences of the industrialized countries. But the factor endowments of developing countries may be fundamentally different from those of developed countries. Under these circumstances, the transfer of all those preferences related to factor endowments is incompatible with development. Besides, developed countries have largely satisfied the elementary minimum needs for most of their populations, hence their technology has been increasingly oriented towards other objectives (mainly towards non-essential luxuries and military applications). In developing countries, however, the main preoccupation has to be with elementary minimum needs from which large segments of their population are disenfranchised. Thus, guidelines and preferences related to products and services must necessarily be different in developing countries.

Finally, there is the problem of the thrust of technological capability. The task of generating appropriate technology appears to be impeded by the type of technological capability that developing countries have and are currently growing; and by the nature of linkages that their educational, scientific and technological institutions have and are forging with domestic and foreign societies.

Thus, most developing countries have followed a standard approach of establishing universities, institutes of science and/or technology, technical institutes and industrial laboratories modelled on the corresponding institutions in the developed worlds, with even their staff emulating counterparts in the industrialized world. As for the institutional linkages, the strongest links are with the demands of the elite, with counterpart institutitions in the developed world, and with western technology. Furthermore, because of the inevitable financial stringencies, these institutions — like naturalized technologies — become, at best, cheaper and cruder versions of the corresponding western institutions, and at worst, complete parodies of the latter.

This predicament is an inevitable consequence as long as institutional linkages with the needs of the urban and rural poor and with traditional technologies are virtually non-existent, and are very strong with elite demands, with institutions in the developed world and with western technology. The situation is worsened by the fact that most teachers, scientists, and engineers are drawn from, and/or become part of, an elite which, in the dual societies of developing countries, is virtually cut off from its countryside and its rural poor, as well as from its slums and urban poor.

A.K.N. Reddy in A.S. Bhalla (ed.), *Towards Global Action for Appropriate Technology*, 1979.

Research institutes often have little contact with poor people

Innovation Effort in Appropriate Technology

The total amount of money spent on developing and diffusing appropriate technologies on a world-wide basis by the organizations which view themselves as 'appropriate technology' institutions is currently (1975) of the order of $10 million a year.

Of this expenditure, less than half (i.e.,under $5 million) is spent on research and development. Compare this R & D expenditure with the $60 billion or so spent on developing new modern technologies. Given the time needed for an R & D programme to be translated into a viable innovation, and the difficulty of reorienting both men and resources to other priorities, it is obvious that the appropriate technology movement as such cannot make a significant impact in the very immediate future.

Alongside the information-diffusion and research activities of the formally organized appropriate technology groups, there are a number of organizations involved, in fact if not in name, in this area. These can be divided into three groups. First are the large modern industrial corporations which have developed new products or new technologies which in one way or another can be considered as particularly appropriate to the developing countries. One example here might be the simplified plant for assembling radios and television sets, which was developed by Philips, the Dutch electronics firm. Another is the 'basic vehicle' or the 'developing nations tractor' (DNT) of Ford Motor Company. At the same time, a number of

firms are developing soft technologies which are of particular interest to the industrialized countries: Boeing for instance is applying its aerospace technology to the development of new types of windmills.

The second group includes a large number of public and especially private institutions working in such fields as engineering, consulting, management assistance, information or the provision of services to small industry. Although they are not concerned primarily with appropriate technology, many of them have in fact developed appropriate or intermediate technologies. This is the case for instance of CENDES (the Centre for the Industrial Development of Ecuador), which has contributed to the development of a small-scale inexpensive production technology for polyurethane flexible foam. In Colombia, the Fundacion para el Fomento de la Investigacion Cientifica y Tecnologica (FICITEC) has developed an electricity generator based on a modified bicycle as well as several small machines for rural and cottage industries. In France, the Société d'Aide Technique et de Coopération (SATEC) has developed among other things a small rotary tiller which is now manufactured by a wide number of village artisans in Madagascar. In Senegal, the Institut de Recherche et de Formation (IRFED) has experimented with the use of plastic-covered roofs to collect rainwater in the villages of semi-arid zones.

Hundreds, if not thousands, of

similar examples could be mentioned here. The obvious conclusion is that the current innovation effort in appropriate technology is in fact considerably wider than generally suspected. The nature of the organizations involved is equally varied. It ranges from large multinational corporations to charitable organizations like Oxfam or the Christian Relief and Development Association, from small companies making agricultural implements and machines to public development corporations like the National Industrial Development Corporation of Switzerland. It also includes, and this often tends to be overlooked, the public and private aid agencies of the industrialized countries which have tried, and often been quite successful in developing and introducing technologies which are particularly appropriate to local conditions. For the moment, there is no overall inventory of the appropriate technologies developed by these organizations, let alone any global or even piecemeal evaluations of the successes, failures and problems of innovation in this field.

The two groups mentioned here — large industrial corporations and organizations such as SATEC, CENDES or FICITEC — represent what might be called the 'organized but peripheral' sector in appropriate technology in the sense that their primary activities do not lie in the field of technological innovation. When trying to draw an overall — and necessarily very sketchy — picture of appropriate technology, one must take into account a third group which might be described as 'informal' or 'unorganized'. It includes individual innovators — small industrialists, inventors and tinkerers — and an immensely large number of peasants, artisans, teachers and tradesmen who have developed, transmitted, or who are currently using some form or another of appropriate technology.

This 'informal' technological potential is not of course specific to developing countries. In the industrialized countries, most of the big technical innovations of today stem from the organized R & D efforts undertaken in the laboratories of private industry and government and in the universities. The industrial laboratory however is a relatively recent institutional innovation which goes back to the end of the last century. The industrial revolution started long before laboratories were set up, and many of the important innovations which contributed to transforming Western society, from the automobile to the aircraft, from the steam engine to the railway, originated from the innovative effort of individual inventors and entrepreneurs working either alone or, when within an industrial firm, without the support of a formally organized laboratory.

The same is still partly true today. In the industrial firm, a relatively important proportion of the innovations still stems from outside the R & D laboratory, and the total research effort of a country as measured by its national expenditures on R & D tends to overshadow the importance of the 'informal' innovation system. The large number of innovations which stem from this informal sector, small or big, include incremental changes in production methods, new forms of organization,

328

rediscovery of old knowledge, transfers of technology from one sector to another or the better utilization of existing resources.

In developing countries, the 'formal' innovation system, as institutionalized in the research laboratory, is very small, and contributes proportionately much less to innovation than in the industrialized countries. However, apart from this 'formal' innovation system, which belongs to the modern sector, there is in all developing countries a large 'informal' innovation system represented by thousands of small industrial workshops, individual entrepreneurs, innovative farmers in the rural communities, and institutional entrepreneurs in the service sector (e.g. missionaries, charitable organizations, private associations, money lenders, etc.). This 'informal' sector represents a vast pool of technology, both in hardware and software, which is often relatively simple, but which plays an immensely important part in the economic system.

The technology of this informal sector is more sophisticated than the traditional technology inherited from the past, but less capital-intensive and generally simpler than the technology used in the modern sector. Its very survival and development, despite the strong competition from modern technology, testifies to its appropriateness and its vitality. Two or three examples can be given here as illustrations. One is the case of a Philippino entrepreneur who built up a starch separation plant using as only equipment $10,000 worth of second-hand washing machines. The efficiency of his plant was so great, and his production costs so competitive, that the 'efficient' plant set up in the same region had to be closed down. Another case is that of a small Algerian factory making first class razor blades with completely outdated equipment. A modern plant, imported from abroad, would undoubtedly look much better, but stands little if any chance of outcompeting the appropriate technology of the indigenous plant. In Pakistan, there are several small pump manufacturers who have not only survived but also managed to improve their technology without any help from the government planners, without any access to the cheap credit facilities accorded to modern industry and without any opportunity to purchase the inexpensive imported raw materials which are channelled in priority to the modern sector.

Thousands, if not millions of similar examples can be found throughout the developing world. This vast pool of appropriate technology has never really been surveyed, and few if any systematic attempts have been made to facilitate its diffusion, even at the local level, improve it where possible and integrate it into a national development effort. This technology, which represents an enormous potential resource, is to a large extent an unexploited, or dormant resource. It plays an immensely important role in the survival of thousands of small and medium-size enterprises and helps provide its customers with products and services, which would otherwise not be available. Yet it is almost totally neglected by national planners and policy-makers.

Nicolas Jéquier, *Appropriate Technology: Problems and Promises,* 1976.

Rural Innovation in Indonesia . . .

The bankruptcy of strategies which stress institution-based R & D as the prime mover in rural technology development is made all the more clear when we consider the normal level of inventors in a population. For the San Francisco Bay Area, for example, there are an estimated 100,000 inventors among the 5 million inhabitants (2 per cent). Considering the much higher daily involvement with tools of all kinds demanded by rural life in developing countries, it appears reasonable, despite a lower average level of formal education among the population, to assume a 2 per cent or higher level of innovators is present. Visual evidence of innovation supports this. It is this level of innovation activity that stands a chance of quickly generating solutions to the rural technology needs of the poor. In Java, for example, this 2 per cent level would mean 1.7 million inventors among the population of 85 million. A hand-ful of highly-trained researchers cannot possibly match the innovative capacity of such a population in generating relevant, low-cost, incremental village technology improvements.

Observation of rural communities and their technology suggests that there is a high level of innovation going on. In examining the low-cost, locally-built sail windmills of southern Thailand, for example, it is seen that crucial parts such as the hub, which is under the most stress, have been built in an astonishing variety of ways that work successfully. It is commonplace for rural communities to have three of four different technologies to solve any of the more common problems. Potters in an area of south Sulawesi, Indonesia, have four different ways of carrying pots 10-20km on foot to market — and these people are from the same village.

Ken Darrow, *UNESCO, Technologies for Rural Development*, 1981.

. . . and In America

A lesson from the America experience is that contrary to what happened in most European countries, a high proportion of the inventors and entrepreneurs came from the rural communities. Oliver Evans, the inventor of the automatic milling machine, was brought up in a Delaware farm; Eli Whitney, who was to play a crucial part in the development of the textile industry, and later the machine industry, grew up to manhood in his father's farm in Connecticut; Cyrus McCormick, whose name became the major trademark in agricultural machinery, was also a farmer's son, and Henry Ford himself came from a Michigan farm.

Clearly, the American farming community of the nineteenth century was very different from the peasant societies of many other countries: the farmers were free men, and they knew that the future would be what they wanted it to be. These few examples are given here to suggest that development is not necessarily an exclusively urban phenomenon and that inventiveness and entrepreneurship in the rural sector are extremely important. This point must be emphasized, since more than 70 per cent of the world population today still lives in rural communities.

Nicolas Jéquier, *Appropriate Technology: Problems and Promises,* 1976.

Innovators' Movement in Nicaragua

Nicaraguan workers, faced with the prospect of widespread unemployment, have responded by developing their own machines and industrial processes. Such activity, which expresses the hope of science by the people, has spawned the Movement of Innovators.

Innovators are workers who develop new tools, develop new production processes, or solve other technical problems and thereby allow industrial production to continue or cut costs. . . .

The most common type of achievement (of the innovators) is in the manufacture of machine parts and tools that were previously imported. At the Agrarian Reform Repairshops, for example, workers have reconstructed two lathes that were abandoned since 1933 and they have used them to make replacement parts for over 600 tractors as well as for other agroindustrial equipment. At the German Pomares Engineering Works, Jose del Carmen Filetes used pieces of scrap to make a sheet metal bender, which is a tool that was normally imported for a cost of 90,000 cordobas. (The government-set exchange rate is 10 cordobas per dollar, but the practical rate is approximately 40 cordobas per dollar.)

At the MACEN factory, workers were confronted with a need to find new products that could be made with the materials on hand. MACEN makes woven bags for storing such farm products as beans, rice and corn. Since the floods in late May destroyed much of those crops, the demand for storage bags decreased significantly and the factory, which is state-owned, began to operate at a loss. In an effort to boost sales, the workers had a contest to develop new products from the same materials that they use to make bags. Over fifty of the 380 workers participated in the contest, making purses, lamps, hammocks, clothing, curtains and upholstery fabric.

In the same factory technicians have also been looking for substitutes for the imported Spanish cord used in the bag-making process. (The factory normally uses over 200 tons of cord per month.) After testing cords made from different Nicaraguan fibres, technicians believe they have found a substitute called 'junco'. The replacement of the more expensive Spanish cord with 'junco' will increase the profitability of MACEN and create new jobs in the manufacturing of cord.

Even more impressive is the invention of a cacao stripping machine. Previously in Nicaragua, cacao nuts were stripped by hand. The new machine will greatly improve productivity and make possible the development of a chocolate industry. This will help save foreign exchange, since all chocolate currently consumed is imported; and it will also encourage the fledgling dairy and tool-making industries.

Another important invention is the development of a process for fabricating plastic airplane cabins. Three workers at the Plastic Record Company decided to apply their experience from making plastic lamps and table decorations to the problem of making aircraft cabins. They made moulds, prepared materials, and tested their prototypes until they developed a successful cabin. They say that with their process the cost for each cabin is 32,000 cordobas ($3,200 at the official Nicaraguan rate of exchange, $720 at the black market rate), compared to the import cost of $9,000 each. And they claim that they can modify their process to make cheap automobile windshields.

Workers who develop new products receive full support from the government. They receive access to machine shops in order that they may work on their ideas, they meet with other innovators to compare and co-ordinate research, and they get preference for admission to technical education courses offered by SINAFORP (Sistema Nacional para la Formacion de Profesionales). For a worker to be eligible for such government support, s/he need only be recognized as an innovator by her/ his co-workers who inform the central office of the CST of their decision.

Each aspect of the Innovators' Movement exhibits the same spirit of democracy, from the government's opening of resources to the Innovators, to the election of Innovators by their co-workers, to the expectation that Innovators will share their knowledge and skill with their co-workers.

That spirit of democracy has even invaded such traditionally conservative fields as medicine. A series of health-related technical seminars open to all health workers (lab techs, nurses, doctors, etc.) took place in Managua in September of 1982. The director of the Medical School, Dr. Oscar Flores, stated that such meetings used to be for doctors only. Alberto Sequeira, a delegate of the health workers' union FETSALUD declared that the opening of such seminars to include health workers other than doctors is important because it helps 'to break the doctors' monopolization of scientific knowledge'. Such efforts to broaden the accessibility of knowledge are common throughout Nicaragua.

Peter Downs, *Science for the People*, 1983.

Informal R & D in Rural Communities

In generating new agricultural technologies, two related processes are involved: natural selection processes (the survival over time of the fittest plants and animals in specific agro-climatic environments); and human experimentation and selection processes when people, whether farmers or scientists, intervene and help generate, select or promote certain technologies rather than others. In the early days of crop domestication natural selection dominated, although farmers did, and still do, influence the process by choosing in their fields those plants that have more desirable characteristics than others. As formal crop-improvement programmes developed, human activities became relatively more important although natural selection still plays a very important role when, for example, plant breeders place their new plants in disease-prone environments so as to 'screen out' those lines that may be susceptible to different types of rusts.

One of the problems for some plant-breeding programmes is that experimental stations are not located in representative agroclimatic conditions for the crops they are developing. As a result, the new plants are not being screened for their suitability for local representative soil types, pests, diseases etc.

In the development of non-genetic agricultural technologies, such as irrigation techniques, cropping patterns, pesticides, herbicides, etc., the human factor has always been dominant, although of course, natural selection processes have impinged because agricultural production involves biological organisms at one stage or another.

In recent years, considerable effort and money have gone into strengthening national and international agricultural research and development (R & D) systems. While these formal systems are important, there is also a need to give attention to the significant role of informal R & D systems found in rural communities, and discuss ways in which formal and informal systems can be more closely linked.

There are numerous examples of how such an informal R & D system operates and why it is important.

In Bangladesh in the late 1970s, a group of researchers found that some 'traditional' varieties of deep-water rice had yield potentials of four tons per hectare of rough rice without modern inputs. By the standards of experimental station yields, these were extremely good. It was also found that farmer-named varieties were specific to particular hydrological zones in water depth, and that the overall yield potential of local cropping systems without modern inputs was high. Thus, the local informal R & D system had developed varieties and cropping systems of considerable yield poential that were relevant to local conditions.

In recent triticale trials in the Himalayan hills, it was found that the 'local check', i.e., the crop that farmers would have grown if they had not been working with the researchers, yielded in some locations at least as well as the new triticale varieties and the high-yielding wheat check in the trial. Indeed, in some maize trials in Bulandshahr, Uttar Pradesh, India, when scientists were testing their varieties and recommended practices for target groups of farmers, they found that the local composite variety and farmers' practices gave a yield of 7.2 tons per ha compared with 6.1 tons for their own package. While recognizing in these cases that some of the genetic materials in the 'local' varieties may have come from early work at experimental stations, the point is that, in specific agroclimatic environments in Asia, there may already be far more yield potential in the existing crops and cropping systems than is realized. The genetic technology that exists today owes as much to the informal R & D system as it does to formal research.

As regards the specific effects of informal R & D on recent varieties of rice and wheat from experimental stations, we find that new developments have come about because of the informal system. For example, after 'dwarf' IR8 rice was introduced in Bangladesh, farmers were found to be actively selecting from the dwarf material those plants that had longer stems and were more suitable to local conditions. Recently a group of CIMMYT scientists noted that there were farmers with some plants of the improved variety of wheat, Sonalika, which were far less susceptible to leaf rust than others. To capitalize on these developments the CIMMYT group suggested that researchers collect rustfree heads of wheat and multiply the lines that are resistant.

The fact that farmers conduct agronomic research is borne out by the rapid spread of improved varieties of wheat in India and Bangladesh in the 1960s and 1970s. After official demonstrations were made in farmers' fields to show the potential of the new seeds, often under optimal or high-input conditions, it was frequently the farmers themselves who adapted those packages to their own conditions. Since the mid-1970s, about 70 per cent of the high-yielding wheat varieties in Bangladesh have been grown without irrigation. This rapid spread of wheat has surprised many scientists and extension agents who did not appreciate its potential on the residual moisture conditions in Bengal. Unfortunately, scientists have frequently overlooked the importance of farmer R & D to agronomy practices. Only too often they have seen the non-adoption of the full package as a sign of backwardness on the part of farmers or as a result of inadequacies in pricing policy, the supply of inputs, etc., instead of monitoring the creative way in which farmers have modified and adapted inappropriate packages of practices and then capitalizing on such new developments by passing the information on to extension agents.

Other inventions from Bangladeshi farmers include: the testing of wheat seed germination on bamboo leaves before planting to determine what quantity of seed to use; the application of fertilizer after occasional rains in winter; the use of tins, polythene bags and local drying practices to keep seeds in good condition during the monsoon; the adaptation of the rice paddle threshers for threshing wheat; the sowing of wheat on ridges (like potatoes) to facilitate irrigation on impermeable soils; the division of plots with ridges and channels to optimize the use of scarce irrigation water.

The case of the bamboo tube-well in Bihar is an example of informal engineering research. Farmers and rural artisans responded to the introduction of a 'lump package' of two steel tube-well and fixed pump set — inappropriate to local conditions — by developing a low-cost bamboo tube-well and serving several of these by a single diesel pump set mounted on a bullock cart. Another engineering example of research by rural people comes from Bangladesh where the standard hand-pump — originally introduced to provide drinking water — is now being used for irrigation purposes. It is interesting to note that the use of hand-pumps for irrigation is also spreading in India in areas north of Patna, Bihar.

These examples show that the rural communities in different parts of Asia are not mere passive recipients of technology that is transferred to them from Western countries or formal research and development programmes. In agricultural communities there continues to be a dynamic and productive informal research system in its own right, and this interacts with any new technologies or organizational structures introduced from outside.

In agricultural development programmes in the past, the view has been that technologies can be developed and widely adapted to a broad range of environments. This view was supported in part by the fortunate event of the dwarf wheats developed in Mexico and spreading through large areas of Asia. However, this may well have been an exception. Increasingly, it is being recognized that closer attention has to be given to specific local agroclimatic and socioeconomic conditions before and while new technologies are being generated. Some of this response has resulted from numerous studies investigating why farmers have or have not adopted various parts of technological packages, developed on research stations and the production and socioeconomic impact of the adopted technologies.

The studies have helped renew our interst in the importance of farm and village-level research programmes. Local research programmes are now regarded as critical because scientists working on rural poverty problems in developing countries are dependent on the reverse flow of information and knowledge to assess local agroclimatic and socioeconomic conditions adequately so as to establish research priorities and allocate scarce resources effectively.

Three interrelated types of information emanate from informal R & D systems.

— Technical and organizational innovations that make efficient use of resources scarce to farmers. The bamboo tube-wells in Bihar and hand-pumps for irrigation in Bangladesh are examples that could be directly extended to other similar agroclimatic and socioeconomic environments and quickly evaluating them against national development goals for productivity, employment, etc. In some circumstances, only minor adaptations may be needed before these new local technologies are promoted in other areas.
— Signposts for new research that scientists in the formal R &D system

might start to work on. Examples here are the methods of growing wheat on ridges and some of the new cropping systems that Bangladesh farmers are developing by trial and error. Farmers selecting rice for longer stems also emphasized for scientists the importance of this characteristic in rice varieties to be grown in Bangladesh. In addition, the relatively high yields of deep-water rice with no chemical fertlilizers pointed scientists in the direction of investigating the importance of blue-green algae as a source of nitrogen.

— Methods for conducting cost-effective research and classifying knowledge. For instance we know that a farmer often tries out a new variety, fertilizer or other agronomy practice on a small patch of land next to his normal crop to see which responds the better. Similar yes/no trials on farmers' fields by researchers can quickly identify whether a broad area of research is relatively important before fine adjustments are made for different types of complex treatments in the design of on-station or on-farm experiments. Far better use could be made by scientists of indigenous technical knowledge. In some situations farmers can often suggest proxies for different sets of circumstances that can greatly reduce the time and cost of analyzing situations by more conventional methods.

One of the most difficult rural development problems today is the effective communication of information from the poorest groups in local rural communities to influence the decisions taken by formal R & D programmes. This flow of information from the grass roots is important, whether it is of the type by which scientists learn and capitalize on R & D carried out by rural people or whether it is information that helps scientists to diagnose and solve local problems better. Strong, pragmatic, interdisciplinary, locally designed farm- and village-level surveys and experiments are a means of improving the transfer of information and technology from the grass roots up to higher levels of decision making.

Stephen D. Biggs, *Ceres*, 1980.

Indigenous Technical Knowledge

Discussion of the uses of ITK (Indigenous Technical Knowledge) tends to focus on its use in R & D. Technical change takes place in a large variety of ways, many of which do not involve knowledge-creation by R & D. Also, even when R & D is involved, a wide spectrum of other, complementary, technology-using activities is involved. To illustrate the wide range of possible uses for ITK, in addition to its use in R & D, the range of different kinds of technical change process is collapsed here into five categories.

— Technical change may take place as a result of adaptations or improvements of existing production systems which are car-

336

ried out autonomously by those operating those systems. Such forms of technical change involving learning to produce more efficiently are common in dynamic 'modern' production sectors. However, even so-called 'traditional' sectors are seldom as technically static as is often suggested — the adaptations and improvements to production methods which are affected by those operating such production systems can be significant. However, as production systems come to embody new types (and perhaps increasingly complex forms) of technical knowledge, the ability of the system operators to adapt and modify these depends on their accumulating relevant technical knowledge in parallel with the acquisition of new production techniques. Without this parallel indigenization of the technical knowledge which underlies techniques, technical change must depend increasingly on specialized agents of change rather than on the autonomous minor innovative activity of indigenous system-operators.

— Technical change may take place as a result of the adoption by producers of new methods developed by others. Very often adoption must be accompanied by adaptation and modification in order to fit the new method efficiently into the existing production system.

Even if technical change is based on methods developed outside a particular social group, an indigenous technological capability is needed to assess and evaluate the new method (to decide on whether and how to use it), to adapt the production activities within which the new method is to be used, and possibly to modify the new method itself. This modification may be carried out by the adopter of the new technique or may have to be carried out, at last partially, by specialists. Even with outside specialists, however, technical knowledge within the group may be needed to identify, specify and communicate the modifications that are necessary.

— Technical change involving the use of some new method of production may take place through incorporating existing technical knowledge into an item of capital equipment. Such technical change may involve the use of technical knowledge already available to capital goods producers. These specialists in using technical knowledge to embody it in capital goods may be large machinery and equipment producers, small machinery workshops, or village blacksmiths, carpenters and masons. Technical change can be effected in this way without involving anything resembling R & D.

If ITK is to be applied in effecting technical change, then its use in this form of change may be important. One would have to be concerned about the acquisition and use of relevant technical knowledge, and maybe about the prior development of the types of non-R & D capability needed to acquire, use,

337

and embody it in capital goods.
— Technical change may take place as a result of design activities. If a substantially new type of capital good is to be produced, some form of design activity (however rudimentary) is likely to be involved. Design activity may also be involved in effecting types of technical change which involve no new capital goods at all, such as the design of a new cropping pattern, the specification of (new) patterns of fertilizer application, the design of a new set of procedures for using water supplies, and so on.

Design is often closely linked to R & D in order to incorporate newly-created knowledge in a new system of production. However, the specifications of new production systems are often based on existing knowledge. Existing knowledge is selectively drawn upon and synthesized to design a new way of doing things. Such 'existing' knowledge may have been created by R & D at some earlier time, but can be used in designing new methods of production without any further R & D activity. One should also take note of the relative importance of new, R & D-created knowledge in the specifications for most innovations. Even for very advanced technical systems new knowledge immediately derived from R & D is often only a small part of the total knowledge which designers incorporate into the new system.

So the use of ITK to effect technical change may mean the indigenous use of technical knowledge to carry out this design-based type of technical change. This may imply the prior acquisition of relevant technical knowledge and the prior accumulation of indigenous capabilities to use it.

— Technical change may take place as a result of R & D. Where existing knowledge is inadequate for specifying efficient and appropriate new methods of production, new knowledge has to be created. Almost anybody can experiment to create new technical knowledge — peasant farmers as well as capital goods producers and design engineers. Knowledge-creation by R & D can then take place entirely inside or entirely outside the indigenous system. Alternatively R & D may take place partly within and partly outside the indigenous system, with different elements of the knowledge needed to develop a new system flowing inside out and outside in.

The variety of forms of interaction between the indigenous and non-indigenous systems is widened further when we take account of all the non-R & D activities which are needed to effect technical change — even R & D-based technical change. The knowledge created by R & D must be incorporated, with large amounts of pre-existing knowledge, into designs and specifications. Parts of these specifications may have to be transformed into capital goods, and the new technique may have to be modified, as may the broader pro-

The Million Dollar Man

Yes, I am the expert
As you can well see
Arriving by jet plane
*With my HYV ***

It's straight to the hotel
Decisions today!
I'm meeting those chaps
*From the M o A ***

Let's grow carnations!
—Now wouldn't that be
A nice way of raising
*Their low GNP ***

Then off to the village
(I hope it's not far)
First on with the shorts
*And my SLR ***

I can't speak the lingo
Phew, what a hot day!
—But I saw a nice farmer
And things seem OK

So I write my report
And I pocket my fee
Then it's time to depart
*For the next LDC ***

The Village

Now who was that man?
And why did he come?
Stayed just half an hour
Only met the chief's son

The Government

What's this he's left us?
Four volumes or more!
We'll need a new expert
To say what it's for

KEY TO THE EXPERT

HYV	High Yielding Variety
MoA	Ministry of Agriculture
GNP	Gross National Product
SLR	Single Lens Reflex (camera)
LDC	Less Developed Country

Drawing by Clive Offley

(New Internationalist, 1981)

duction system within which it is to be used. Technical knowledge must flow between these different activities, and if, as with R & D, some of these activities are carried out within the indigenous system and some outside, technical knowledge may flow both inside to out and outside to in.

In four of these five categories of technical change, then, R & D is not directly involved. In the other, it is not the only activity involved. This is not to state that R & D is unimportant, but rather to indicate that R & D is by no means the only, or necessarily the main, use for ITK. Other uses may be far more important in order to arrest, and hopefully reverse, the decay of ITK, and to build up greater involvement in and control over technical change on the part of rural people — both as an objective in its own right and as a means to effect more appropriate forms of technical change.

Martin Bell, *IDS Bulletin*, 1979.

The Unglamorous Poor

The links of modern scientific knowledge with wealth, power and prestige condition outsiders to despise and ignore rural people's own knowledge. Priorities in crop, livestock and forestry research reflect biases against what matters to the poorer rural people. They concentrate on what is exotic rather than indigenous (e.g. exotic cattle over local goats; rubber, coffee and cocoa over millet and sweet potatoes), mechanical rather than human, chemical rather than organic (e.g. chemical fertilizers over goat droppings), marketed rather than consumed (e.g. crops for export over subsistence crops), male-based rather than female based.

Whilst science has much to contribute, rural people's knowledge drawn from experience is often superior in their own context to that of outsiders. Indeed it has been described as the single largest knowledge resource not yet mobilized in the development enterprise.

Examples of their superior wisdom can be found in their preference for mixed cropping, their knowledge of the environment, their fine abilities to observe and discriminate (plants, climate, animal behaviour, for example) which can provide more accurate data on which to act, and their inventive agricultural experiments.

Modern scientific knowledge and rural people's knowledge are complementary in their strengths and weaknesses — rural knowledge is at its strongest with what is observable (although local beliefs and practices can be harmful, especially in health and nutrition), and scientific technology is superior in being able to measure precisely and examine microscopically. Combined they may achieve what neither would alone, but for this, outsider professionals must be prepared to stop, sit down, listen and learn.

In this move towards really helping

340

the rural poor help themselves, processes which deprive them and maintain their deprivation must be slowed down, halted and then reversed. Three reversals of this kind should in particular be considered: spatial reversals away from concentration of professionals in urban areas to rural areas, and a general decentralization of resources, power and funds; reversals in professional values and preferences, putting what is now regarded as 'last' first and as 'first' last (see box); and reversals in specialization.

This is not to say that Third World professionals should abandon the 'first' list, nor that poor countries should abstain from developing the expertise to manage technological hardware or from negotiating sales in this area, since so-called 'high' technology has had and will continue to have important applications in attacking rural deprivation. However, it is commonplace that the pursuit of 'first' values favour the richer and neglects things directly important to the poor. Reversal in specialization is also needed to enable the identification and exploitation of under-recognized resources and opportunities often lying between disciplines, professions and departments. New economic niches and livelihoods can be generated by exploiting slack resources and applying new technology to ways whch enable poor people to establish rights to these resources and flows of income from them.

These reversals require professionals who are explorers and multi-disciplinarians. Such new professionals who put the last first do already exist — the hard question is how to multiply their number.

Professional values and preferences

A. *For technology, research and projects*

FIRST	LAST
Urban	Rural
Industrial	Agricultural
High-cost	Low-cost
Capital-using	Labour-using
Mechanical	Animal or Human
Inorganic	Organic
Complex	Simple
Large	Small
Modern	Traditional
Exotic	Indigenous
Marketed	Subsistence
Quantified	Unquantified
Geometrical	Irregular
Visible and seen	Invisible or unseen
Tidy	Untidy
Predictable	Unpredictable
Hard	Soft
Clean	Dirty
Odourless	Smelly

B. *For contacts and clients*

FIRST	LAST
High status	Low status
Rich	Poor
Influential	Powerless
Educated	Illiterate
Male	Female
Adult	Child
Light-skinned	Dark-skinned

C. *For place and time*

FIRST	LAST
Urban	Rural
Accessible	Remote
Day	Night
Dry season	Wet season

Robert Chambers,
Development Forum, 1983.

The Process of Indigenous Technical Change

The development of modern sugar processing technology is a relatively recent phenomenon. Well into the nineteenth century crystal sugar production took place under conditions which are very similar to current open-pan sulphitation (OPS) technology. The first major improvements undertaken concerned the crushing process. Wooden vertical rollers had initially been used in the fifteenth century, with iron being substituted in the mid-seventeenth century. It was only in 1794 that the first isosceles-shaped roller-crushers were introduced.

At about the same time beet-sugar technologies were being developed. In 1799 Achard presented Frederick William III, King of Prussia with a sample of sugar and asked for a ten year monopoly on producing sugar from beet. By 1812 there were twelve beet factories in Prussia but these sank into oblivion and it was only in the late 1820s that beet sugar production was reinvigorated. Even then the survival of this technique was uncertain and in 1844 the eminent chemist Justas von Liebig concluded 'This fine manufacture cannot keep going long ... it has no future ... it is not worth the great sacrifice and offers no advantage ... it is like a costly hothouse plant'. but it was in France that beet-sugar technology prospered, largely due to an attempt by Napoleon to circumvent the English sea blockade. However, after the Treaty of Paris in 1814, French markets were reopened and the 334 factories which had been in operation by 1813 were almost all closed down; only in the 1830s did the French industry begin to expand again. Thereafter the technology thrived, with the sugar-beet recovery rate rising consistently.

Technological improvements in cane sugar arose partly as a response to the increasing competitiveness of beet sugar. Although largely based on technologies invented for other purposes, the development of the industry made an important contribution to the emergence of the science of chemical engineering. For example, the first school of chemical engineering in the USA, the Audoban Sugar School, was founded as an experimental station on a sugar mill in the 1890s. Indeed the chemistry of inversion was only understood after the 1850s.

The milestones in the industry's maturation can be classified into three groups, those which increased recovery, those which promoted energy efficiency and those which improved quality. In the first was the introduction of multiple-effect evaporation. The first possibility of the use of steam was patented in 1692, but only in 1813 was the vacuum pan invented, and the multiple effective system (that is, incorporating more than one vacuum pan, and optimizing the use of energy) was first installed in 1844. Despite its advantages the technology spread slowly. The vacuum pan was invented in 1813, yet by 1827 only six English plants used it; it was first used in France in 1824, and only in 1834 was it introduced into a sugar colony. Centrifuges were invented for textile drying in 1839 and first used in the sugar industry in 1843.

The introduction of seed grains into crystallization was patented in 1865, but was preceded by the introduction of agitation techniques to increase the rate of crystallization in 1860. With respect to quality, the first major improvements occurred with the introduction of sulphurous acid to neutralize lime in 1810, with bone-char decolourization invented in 1820.

We can see therefore that the emergence of modern beet and cane sugar processing technologies has taken almost two centuries. It was a process which had its ups and downs, depended heavily on protection, and spread slowly. Technological improvements reflected not only the application of scientific and technological principles to production, but also stimulated the emergence of chemical engineering as a specialized scientific endeavour. By the 1970s the technology was 'mature'. Sucrose extraction in best-practice plants exceeds 85 per cent of potential and the product is almost entirely free of impurities. Technological improvements are now orientated towards the introduction of process controls to reduce the labour input, minimize the operational losses of sucrose and maximize fuel efficiency.

By contrast, compare the development of OPS technology. Only in the 1920s was attention given to improvements, with a brief attempt to increase the efficiency of boiling. It was around 1925 that OP and khandsari technology first diverged, with the introduction of small centrifuges, modelled on those of the VP mills. Although sulphitation clarification techniques were discussed in the 1930s, it took almost twenty years for lime-sulphitation technology to be tried. but in 1952, the first demonstration OPS plant (built in 1948) was closed, declared uneconomical, and the research programme was ended. Then in 1955 the Planning Research and Action Institute (PRAI) was set up by the Uttar Pradesh Government, with Rockefeller Foundation funding. It built a demonstration plant at Ghosi in 1955-7 and organized the first Technical Seminar to discuss and disseminate OPS technology in 1957. By 1960, following the Second Technical Seminar in 1959 over 100 OPS plants were in operation and this was followed by the first introduction of the hydraulic crusher (again modelled on those of the VP mills) in 1961. Then, in quick succession, double-stage crystallization and four-pan boiling systems were developed, interspersed with two further Technical Seminars.

By the mid-1970s over 6,000 OPS plants were operating, representing a very rapid pace of diffusion. Yet, it was questionable whether this OPS technology was viable without state intervention. A number of studies were undertaken which made this very point and Forsyth went so far as to conclude:

Moreover, as innovation proceeds in sugar machinery, the advantages of sophisticated methods will almost certainly increase for nearly all R & D is being done at the capital-intensive end of the factor proportions spectrum and, in any case, the scope for technical improvements at the labour-intensive end of the spectrum is strictly limited; the very considerable improvements in labour-using methods required to overtake even existing capital-intensive methods would appear to be outside the bounds of technical possibility.

Faced with the unviability of OPS technology, the Indian engineers responsible for its development drew a dramatically different conclusion: namely, precisely because little R & D had been put into OPS technology, it would be feasible to improve it with relatively little effort. In the words of M.K. Garg, the key Indian development engineer:

> The large-scale technology took about 125 years to develop to the present efficiency. Most of the present day resources are on its side, *viz*, rich infrastructure, capital and finances, political influences and demarcation of market. The mini-sugar of OPS khandsari is only twenty-two years old. Even if a small part of these resources and development facilties are provided to mini-sugar technology, it can even compete with the large-scale sugar complexes having their own cane cultivation.

So a programme of R & D was instituted in 1964. This programme identified the need to improve the furnace and crushing technologies and after various obstacles were overcome, took seventeen years to mature. The result was the development of a screw expeller and a shell furnace. The consequence is that OPS is now a viable technology, able to operate profitably in India, Kenya and probably many other LDCs, without state protection.

As to the future, the Indian engineers accept that it is probably unlikely that they will be able to exhaust more sucrose from juice as long as the plants operate under atmospheric conditions. Therefore, together with the Ecole National Supérieure de Industries Agricoles and Alimentaires in France, they are seeking to exhaust sugar from molasses. Through a further lime-phosphitation process, supplemented by ion exchange, it is hoped to remove impuri-

Sugar processing technology has developed slowly. (*West Indies Committee*)

344

ties from molasses to provide liquid sugar which has industrial uses. By further concentration this can then be converted into a fine powdered sugar called 'bura' in Hindi. If these attempts are successful the sugar recovery rate of OPS technology looks set to match, or even possibly exceed, that of the VP technology.

What relevance has this story got for the wider question of Indigenous Technical Change? We believe that there are six major conclusions which can be drawn.

The improvement in OPS technology is a much more recent phenomenon than that for VP mills. It can also be seen that the recent development of the screw expeller and the shell furnace have had a major impact in closing the gap between the two technologies. Thus the error of judgement of observers such as Forsyth and Baron lies largely in their static viewpoint; without a wider time-perspective it is easy to misjudge the relative worth of the alternatives.

It is diffficult to know whether a similar picture emerges with respect to other 'Third World technologies'. Certainly the Indian engineers appear to have been correct in their assessment that precisely because OPS technology had been neglected over the years, the returns to marginal inputs of R & D would be substantial. Thus, insofar as a similar story can be told for other technologies emerging in LDCs, it is important to have a wider time-horizon than is often displayed in their evaluation.

The literature of development economics is replete with incantations against the costs of protection and the refusal of infant industries to 'grow up'. While many LDCs indeed suffer from static, inefficient industries, this need not always be so, as we have seen from the case of OPS technology. Indeed it is often forgotten that beet-sugar technology in Europe, and efficient VP mills in India, Kenya and elsewhere, benefited from, and grew to maturity during extended periods of protection. Similarly, without state protection, OPS technology would have been crushed in its infancy. Therefore in pursuing policies which enhance the potential of ITC, it will often be essential to protect or subsidize emergent technologies.

This conclusion has particular relevance for the industrial strategies of LDCs at this point in history. Following the rapid growth in Third World debt in the 1970s many LDCs are being forced to accept the intervention in policy-making of multilateral aid agencies such as the IMF and the World Bank. Amongst the conditions often specified in loan packages is a reduction in protection offered to new industries. Indeed one of the most influential advisers to the World Bank suggests that infant industries should only be given 'five to eight years' to grow up. Yet the experience of VP sugar-cane technology, beet-sugar technology and OPS technology all suggest that ITC takes a lot longer to mature. A five or eight year period may be adequate to reduce the 'X-inefficiency' with which imported technology is used, but the experience of the sugar industries suggests that it is wholly inadequate for the development of a new type of technology or, in the case of LDCs, for the maturation of ITC. Of course great care must be taken in this matter since . . . protection may be a necessary component for ITC to occur, but it is by no means a sufficient condition.

For many decades the Indian scientific infrastructure was criticized for being marginal to the needs of domestic industry and for pursuing research programmes which were designed to maximize international prestige rather than meet local needs. This was patently not the case in the development of the screw expeller and the shell furnace; moreover it was only because the technological improvements were directly linked to particular plants which served as guinea-pigs, that they were brought to fruition. Related to this endogenization of Indian science and technology has been the emergence of a domestic capital goods sector which has specialized in meeting the needs of local, small industry. In the most successful cases, these capital goods firms — which are increasingly science-based, employing engineers in design departments — are directly linked by ownership to OPS plants. This allows them not only to identify technological problems, but also provides a fertile ground in which they can test improvements.

Although the development of VP sugar cane technology was partly responsible for the emergence of the science of chemical engineering, it also progressed by taking up inventions from other industries. Thus the roller mills were first developed for other uses, and the centrifuge was originally invented for the textile industry.

OPS technology has, in a similar way, leant selectively on the 'technological shelf', especially that existing in developed countries. it was a conscious move by M.K. Garg to develop a plant that was rooted in the traditional labour-intensive practices of the old khandsari manufactueres, but using down-scaled and suitably adapted versions of modern, larger-scale plant — crushers, crystallizers and so on. In addition to this, several novel units needed to be developed — latterly the shell furnaces and expeller — where inspiration was sought from a variety of bodies of existing technological expertise. Thus the expeller, originally based on a variant of an Indian oil-seed press by I.B. Patel in the early 1970s, was progressively redisgned by M.K. Garg, drawing on the experience of the French Bagasse dewatering presses. At one stage a retired engineer, tracked down in the Canadian backwoods, described the engineering rules of thumb he had utilized over the years. Similarly, the shell furnace has gone through a lengthy period of evolution, and in its current version uses a development of the well-known 'Ward'-type hearth as a central feature.

Thus, in pursuing Third-World technological improvements, there is much to draw on from the experience of the rest of the world; the skill lies in being able to locate it, draw the relevant lessons, and adapt and integrate it into suitable designs for local use.

When comparing VP mills in DCs and LDCs there is generally little different in the technology, save perhaps in the degree of automation and hence the labour-intensity of the plants. However, there is a clear difference between VP and OPS technologies, since by virtue of the vacuum-boiling the former is inherently capital-intensive and operates at a 'higher' level of technology. It is also inherently open to economies of scale in production. As a consequence

OPS technology is much more likely to be appropriate for Third World operating conditions — with large supplies of labour, weak technological infrastructures and balance of payments constraints — than for DC environments. It represents an example of Third World technology, not just because of its origins but also because of its operating characteristics.

The question is whether this is a unique phenomenon, or one example of a kind. It is still too early to know given the embryonic state of ITC in much of the Third World. But there are increasing signs of a divergence between DC and LDC technologies, with the former being geared to systems-gains based upon the widespread diffusion of micro-electronics in process, and the latter concerned more with incremental technical change at the sub-process level. The evidence also shows that technology exports from LDCs tend to go to other LDCs, thus suggesting that their competitiveness arises as much from appropriateness as from price. Certainly this is the case for sugar, where technical change in VP technology is orientated to the introduction of electronic process controls to automate production further, by linking various sub-processes into a coherent system. In OPS, as we have seen, past and future technical progress is focused on circumventing bottlenecks in particular states. Automation, and the use of electronics, is a long way down in priority. But as a consequence of these varying foci, we find the LDC technology increasingly appropriate for LDC operating conditions.

In the development of OPS technology international agencies played a crucial role in identifying relevant technologies already used in other industries. In itself this is an important function, saving LDC technologists from the necessity of re-inventing the wheel. But there is an equally important role yet to be played in disseminating these and other technological advances to wider users. The need to do so arises largely from the underdeveloped nature of the LDC capital goods industries. Unlike their multinational DC counterparts, these Third World firms do not possess the institutional infrastructure required to generate markets and transfer the technology. Thus ITDG has been instrumental in nurturing OPS sugar production in East Africa, and a similar institutional role is still to be played in other parts of the Third World.

One of the obstacles which will have to be faced is the coalition of parties with a vested interest in pushing VP technology to the Third World. Not only does this include the machinery suppliers and foreign investors, but often also bilateral aid agencies (interested in assisting their country's machinery suppliers) and multinational agencies who have a strong, historic commitment to large-scale sugar production, using VP technology.

Raphael Kaplinsky, *Sugar Processing: The Development of a Third World Technology*, 1984.

The thing looks simple and easy. Do not deceive yourself. To make sugar is really one of the most difficult things in the world. And to make it right is next to impossible. — Mark Twain

The Acquisition of Technological Capability in Developing Countries

Market failures in the creation and the diffusion of technology are at the heart of the international debate about technology transfer. Technology has characteristics of a public good in that, once produced, it is not depleted through further use by others. It is usually presumed that the cost of transferring technology is zero and that additional uses of the technology do not detract from its value. On those grounds, achieving optimal welfare requires that technology be available to all potential users without charge. This argument is the basis for claims that developing countries should have free (or cheap) access to developed countries' technology. But free diffusion pre-empts markets that the creator might have served, and may thus remove the incentive to innovate. The patent system permits the diffusion of technology while attempting to protect the proprietary rights of the innovator. In exercising these rights, technology suppliers seek to restrict use of the technology so as to maximize their returns. Control over the supply, plus the buyer's ignorance regarding the true value of technology, can lead to excessively high prices.

High prices for technology and restrictions on its use are the basis for many developing countries' call for an international code of conduct on the transfer of technology and a revision of the international patent system. But no satisfactory agreement has been reached on either because of the inherent conflict of interests between the suppliers and the demanders, which mirrors society's fundamental dilemma between the need to stimulate the creation and the need to encourage the diffusion of technology.

Many governments in the developing world have adopted 'defensive' measures — that is, measures that control contractual technology transfers — in order to redress the bargaining asymmetry and protect the development of local technological capabilities. Although such regulations have helped reduce the price and improve the terms of the contractual inflow, they may also have affected the character of the foreign technology that can be imported. Foreign technology suppliers are unwilling to sell when they consider the returns too low. Moreover, direct foreign investment is often the only means of obtaining access to closely guarded technological assets. It is also not clear that regulating formal inflows has stimulated the development of local capabilities. Such development requires technological effort on the part of local firms, which is not ensured by regulation of or protection from technology imports.

Less developed countries typically obtain many elements of technology from more developed countries. But there are various combinations of foreign and local contributions. Information, means, and understanding can be provided by foreigners who retain ownership; purchased from foreigners; or acquired through indigenous efforts to translate foreign technological knowledge into specific methods. And technology can be transferred with varying degrees of human capital accumulation and institutional development.

At one extreme, a package consisting of all the elements is transferred, with indigenous involvement limited to an unskilled labour force — as with direct foreign investment, or to operating the technology — as with 'turnkey' projects. (In the latter case, a foreigner contracts to provide all the elements needed to design and establish a production facility in the local environment, but ownership is local.) At the other extreme the underlying knowledge is assimilated and then used to create the necessary elements. The knowledge can be acquired through education, experience, experimentation, research, or purchase.

The modes of technology transfer that are most often discussed are those where foreigners play an active role and provide information in an immediately operational form — direct foreign investment, turnkey projects, licensing, know-how agreements, and technical service contracts. But modes in which foreigners play a passive role, and where locals acquire the knowledge and later translate it into technology, are very important channels of technology transfer. These channels include sending nationals for foreign education, training, and work experience; consulting foreign technical literature; and copying foreign processes and products.

In discussing what is acquired through technology transfer, it is useful to distinguish three broad types of capability: production capability — that required to operate a technology; investment capability — that required to expand existing productive capacity or to establish new capacity; and innovation capability — that required to develop new methods of doing things. There is often an implicit notion that technology transfer gives the recipient the first two if not all three types of capability. That is rarely the case. The capability to operate a technology is different from the ability to develop the means of implementing it. Similarly, having the capability to implement a technology is different from having the capability to create a new one.

Production capability is not achieved by passively importing technology. It requires local participation and considerable indigenous effort to master the technology's use. Research shows that in most cases where the technological elements are imported as a 'black box', the recipients are not able to take full advantage of it because they do not understand how or why the black box operates as it does. This hampers their ability to improve productivity or to adapt to changing circumstances — such as shifts in input prices or demand patterns — that affect how it is best used.

The understanding that underlies production capability is also an important aspect of the capabilities to invest and to innovate. Thus the accumulation of local production experience can provide the understanding necessary to carry out some but not all, of the tasks involved in investment and innovation.

Part of the increase in local capabilities acquired through transfers spills into related activities. For example, the capabilities gained from establishing one industry can enable greater indigenous participation in subsequent transfers of related technologies, increasing their effective assimilation. The accumulation of such experiences can also lead to the creation of specialized firms which, in turn, permits greater local participation in future transfers. More generally, the increased capability contributes to an economy's capacity to undertake independent technological efforts, including replication or adaptation of foreign technologies as well as creation of new technologies.

But unless carried out with the explicit objective of doing so, some modes of technology transfer do not provide the experience that is critical to the development of indigenous technological capability. Tasks involving project design and the manufacture of capital goods, for example, which could be performed locally, may be carried out by foreigners. This precludes local learning through experience — experience that may be directly relevant to the industry's subsequent development. Moreover, project costs may be higher: cheaper local services may not be used; and intimate knowledge of local conditions, required to optimize project design and to take advantage of available raw materials, may be ignored.

Any project entails much iterative problem solving and experimentation as the original concept is refined and given practical expression. Important elements of the technology appropriate to the project are developed through applying existing technological knowledge and engineering principles to specific local circumstances. There may even be some minor innovations or adaptations in the technology being implemented. Whether the elements of technology should be obtained locally or from abroad ought to depend on the relative costs and benefits involved. Few would argue that foreign technical knowledge should be eschewed, so the issue ultimately concerns the division of labour between foreigners and locals in transposing technological knowledge into concrete form.

An economy's capacity to provide the necessary elements depends on the stage of development of the relevant sector and those closely related to it. Firms engaged in well-established activities may often acquire technology locally — either through their own efforts or through the diffusion of expertise from other domestic firms. Hiring personnel with expertise from previous work experience plays an extremely important part in the diffusion of knowledge among firms, as does the interchange of information among suppliers and users of individual products, especially for intermediate products and capital goods.

Firms in new or relatively new industries can rarely take advantage of previous local experience or the diffusion of expertise or information from other domestic firms. Such firms are likely to find it more cost-effective to rely initially on foreign technological 'packages' in the form of direct foreign investment and 'turnkey' contracts. As a country develops its technological capabilities, it can disaggregate these packages to import more cheaply or efficiently only those elements that it cannot obtain locally.

The relative merits of different ways of acquiring various elements of foreign technology depend on several factors. First, the costs and terms at which elements can be obtained from abroad may be affected by the competition among alternative sources of supply and the negotiating power of the recipient, including the degree of government support. The second factor is the technological capability of the recipient and stage of development of local technological infrastructure. The third is the size of the market for which the technology is to be applied.

There are trade-offs — involving risks, short- and long-term considerations, and private versus social costs and benefits — between attempting to supply some of the elements locally and importing them. A rational firm is unlikely to use inexperienced local engineering services or untested capital goods, for example, unless their use brings long-run developmental benefits that more than compensate for the greater short-run risks and higher costs of using such local inputs. The social benefits from increasing technological capability generally exceed the private gains that an individual firm can expect to capture. There are many avenues along which technological capability can move to other firms, and not all of these are controlled by the firm that finances the initial acquisition. This discrepancy between private and social value

often leads to underinvestment. Moreover, firms may value the private benefits that they do capture at less than their true social worth, or consider that the cost of securing them exceeds the true social cost.

Furthermore, firms may opt for more expensive monopolistic sources of foreign technology, such as those that confer a well-known brand name, because such sources confer monopoly power. There is than a convergence of interests between the domestic firm and the foreign supplier since the domestic firm can off-set the promise of domestic monopoly profits against the excessive price paid. Thus, the motives that give rise to technology imports can sometimes conflict with social objectives. In turn, where imports are consistent with social objectives, domestic firms may prefer importing technology without considering ways of increasing domestic technological capability. Even the simplest form of participation — intelligent observation of activities carried out by foreigners — entails a cost to firms.

The central issue of strategy is how to build upon what can be obtained from abroad to stimulate the development of local capability in selected areas. For many reasons, timing is of critical importance. Since all capabilities cannot be developed simultaneously, and since the accumulation of any one capability takes time and experience, the sequence in which various capabilities are developed is crucial. And the required capabilities change as a firm or country matures, because of changes in existing capabilities, and because of changes in market conditions.

If the market is small and growing

slowly, so that investments in new plants are infrequent, the best strategy may be to acquire only production capability. But if the market is large or growing rapidly, it may be economic to acquire some investment capability. Furthermore, if technology is changing rapidly, it may be desirable to insure the capability to assimilate new advances quickly or even to innovate new products or processes. Or a decision may be made to rely on direct foreign investment in dynamic areas where it would be too costly to keep up with rapid world technological developments.

There are various situations in which it may be cost-effective to develop basic product and process knowledge as an element of local innovation capability. These include instances when foreign technology is not appropriate or does not exist for the needs at hand, when it can be obtained only at excessively high costs or is unavailable because of monopoly supply restrictions, or when the size of its potential market is large enough to justify the cost of developing it locally because of the gains from successive applications. Efforts to acquire substantial innovation capability may pay off by reducing future costs and providing greater flexibility to adapt to changing circumstances. The difficulty of assessing these returns, together with differences in sensitivity to technological considerations, may' explain why firms in the same industry exhibit vastly different levels of technological effort.

Carl Dahlman and Larry Westphal, *Finance and Development*, 1983.

 # *The Caribbean Example*

Research has identified a number of problems associated with the transfer of technology in the Caribbean, for example, various types of restrictive business practices, transfer pricing, excessive payments for technology, various tie-in clauses and inappropriate technology. These problems were found to be widespread in such industries as food processing, bauxite, motor vehicle assembly, petrochemicals, construction, etc.

The impact of these practices is reflected in balance of payments problems stemming from excessive payments for imported technology, reduction in the export potential of the Caribbean as a result of clauses limiting export, problems of maldistribution of income between sectors using modern inappropriate technology and the traditional sector. Related to the question of choice of techniques is the lack of employment-creation possibilities since imported technology tends to be capital- rather than labour-intensive. Further, the use of restrictive business practices tends to limit the use of indigenous technology as well as the technological transformation possibilities.

These problems associated with technology transfer have also been documented in a wide cross section of developing countries. They assume particular importance in the Caribbean for a number of reasons. To begin with, there is the twofold problem of dependency and vulnerability of the Caribbean that has always preoccupied economists.

Dependency here is taken to mean economic dependency on the industrialized world for capital and consumer goods and technology in the form of proprietary know-how. Given the prevailing level of development in the Caribbean there are no capital goods industries worth speaking of. The capital goods sector, if it exists, tends to be basically service-oriented rather than production-oriented. The relative shortage of indigenous entrepreneurial talent means that foreign capital is often regarded as a source of 'resource tapping'. Further, although the Caribbean is basically an agricultural economy, most of its food is obtained from abroad because of the lack of adequate regional food processing capability. The situation has led the Heads of Government of the 12-Nation Caribbean Community (CARICOM) to establish a Regional Food Plan to bring about regional self-sufficiency in food. Recent estimates have put the region's food import bill at $400 million or about $80 per capita. Some economies, such as Guyana and Jamaica, have launched programmes on import substitution in food to cope with this problem.

The same applies to technology in the form of proprietary know-how. Separate studies by the CARICOM Secretariat on trademarks and patents show, for example that in some cases as much as 97 per cent of proprietary know-how in trademark is controlled from abroad. A similar situation exists for patents. Both trademarks and patents are important sources of the transfer of proprietary know-how. If the position is taken that small economies are more 'open' than larger ones at a similar level of development, then that structural openness would seem to mean that they are more dependent on foreign technology, all things being equal.

The vulnerability of the Caribbean could be interpreted to mean inability to exercise control over the macro-economic system given the structural links existing between the domestic economy and the international one. For example, in terms of development planning, several difficulties arise. First, most of the development plans aim at rapid transformation of economies, alleviation of major employment and underemployment problems, reduction of economic dependence and the satisfaction of basic needs. While it is true that some degree of economic diversification has taken place over the years, it has occurred without any major transformation of the technological capacity of respective economic sectors, whether primary agriculture and mining, secondary agro-industrial, light manufacturing and assembly-type operations or services.

Creation of employment poses a large problem in view of the highly capital-intensive nature of imported technology and its inability to absorb much surplus labour from the agricultural sector. It might be argued that, given the nature of these economies, capital-intensive technology should not pose a major problem for unemployment given the small absolute size of national labour force markets. Yet high unemployment rates are common throughout the region, in many cases as high as 20 per cent. This is often accompanies by high rates of under-employment, though this is very hard to measure.

Because small economies tend to be quite open, they encounter an inherent structural constraint to some degree of economic independence. In principle, however, the reduction of economic dependence is made all the more difficult by control of foreign technology over the region's economic life. The way of reducing this control is actively to encourage the use of indigenous technology in the region. For example, in Guyana a considerable part of the country's food processing technology is now supplied locally by an engineering firm. The country has also begun to export intermediate food processing technology to the Caribbean economies. Another way is to set up public institutions to import foreign technology that is consistent with development objectives.

The satisfaction of basic needs continues to be a problem largely because foreign technology is not geared to increasing the bundle of goods and services for rural and urban poor. Instead, such technology caters largely to satisfying the preferences of the relatively well-to-do social classes. Also, the satisfaction of basic needs is partly related to employment-generation effects of technology. If these effects are limited, people do not have access to income-earning possibilities. This affects their welfare levels. It is thus not surprising that a recent study has voiced concern over the problem of basic needs in the Caribbean, especially in health and nutrition. For example, some 44 per cent of the population are without minimum levels of protein and 56 per cent without recommended levels of calories.

Foreign technology is concentrated largely in urban centres. Inability to control it, as well as the terms and conditions of its acquisition, implies an inability to bring about balanced regional development, namely, reducing welfare disparities between town and country, a serious public-policy concern throughout the Caribbean. Where such technology is found in rural areas, i.e., plantation, it has brought about a destruction of traditional agricultural technologies to the detriment of rural development, as well as undue specialization to plantation crops.

With reference to investment projects, when local investors are ready to invest, the lack of an indigenous technological capability has meant that such an investment potential is unable to materialize into an actual project unless the technologies necessary to start production (equipment processes, various types of skilled expertise) are forthcoming from abroad. Industrial Development Corporations exist in some countries to assist local businesses, but seldom are they involved in the procurement of foreign technology.

When confronted by transnational corporations whose annual sales typically run into hundreds of millions of dollars, small Caribbean economies tend to be at a disadvantage in so far as bargaining for fair terms and conditions of technology transfer are concerned. This is aggravated by two other considerations. One is a lack of information concerning the operations of transnationals; the other is lack of joint bargaining strategies in the Caribbean to obtain foreign technology. Countervailing power has been advocated by a number of international organizations in an effort to equalize bargaining power between developed and developing countries.

To minimize problems associated with technology transfer, institutional considerations are quite important. In the Caribbean, there is no selection procedure (as, for example, in the Andean Pact) to ensure that technology transfer from abroad satisfies development requirements of respective economies. As a rule, most of the technology transferred is done on an *ad hoc* basis, with private

firms (subsidiaries of transnationals or independent local firms), or in a few instances public ones, taking the major decisions. Private selection criteria, however, are not the same as social ones and the former tend to be the prevailing elements in the selection process. Also, public firms often do not have market intelligence services for the procurement of foreign technology. In other words, scrutiny of the technology market is limited.

Further, research and development facilities scarcely exist in the Caribbean. In such traditional sectors as mining and agriculture, control by transnationals has meant that research and development were not actively encouraged locally. As a result, R&D activities relevant to the Caribbean economy took place in the parent headquarters of transnationals such as ALCAN, Reynolds, Booker McConnell and Tate & Lyle.

Frank Long, *Ceres*, 1982.

Transnationals and the Third World

In the current tide of interest in the development of technological capability in the Third World, the discussion has focused mainly on two objectives: the transfer and assimilation of imported technology, and the building up of an indigenous research and development (R & D) structure. The role of transnational companies (TNCs) has been considered mainly in the former context. On the positive side, TNCs have been taken to be the prime source of advanced technology, both as direct investors and as licensers and sellers of equip-

ment. On the negative side, they have been accused of a multitude of sins, from over-charging for the technology they provide to distorting the whole pattern of economic development.

While it is not the intention to enter into the larger debate, I would like to explore briefly the possibility that TNCs can contribute to the second objective of technological development, that of strengthening the R & D capability of developing countries. Assuming that the type of technology in which TNCs specialize

— usually modern, complex and capital-intensive — is beneficial to developing countries, or at least for those which have attained some degree of industrialization, can TNCs help countries in the latter category by relocating their R & D activities there?

The argument is mainly speculative. TNCs are starting to relocate R & D abroad, but very little of this is going to the Third World. Certainly, in relation to the relocation of production by TNCs from developed to developing countries, the relocation to technological activity is miniscule. However, it is only the start and the signs are that it will grow rapidly in the future. The economic forces which make for a shift in comparative advantage in production also make, though the degree and nature of the influence differs, for a shift in comparative advantage in R & D.

Expenditure on R & D overseas (developed and developing countries) by US TNCs totalled $1.3 billion in 1975. It had registered an increase of nearly 150 per cent in the period 1966-75, a growth rate far exceeding the 50 per cent increase recorded by the same TNCs in their R & D spending within the US. The years 1966-72, in particular, witnessed a dramatic difference in the growth of R & D overseas and at home: foreign R & D grew at 21 per cent per annum on average, as compared to 5 per cent for the TNCs domestic R & D, and only 4 per cent for all US industry. While foreign R & D was growing from a much smaller base, there clearly existed a propensity on the part of TNCs to relocate research acvitity abroad. In the recessionary period 1973-5, the rate of growth of R & D fell sharply both at home and abroad, with foreign expenditures being cut more severely (TNCs generally tend to retrench all their activities more abroad than at home). With a return to more 'normal' economic conditions, we may well see a resumption of the pre-1973 trend.

Foreign R & D constituted about 8 per cent of domestic R & D by US TNCs in the 1970s, having risen from just under 5 per cent in 1966 and the evidence for 1966-75 shows that over 90 per cent of US overseas R & D was located in the industrialized world. The Third World accounted for a small but increasing proportion of the total, from about 3 per cent in 1966 to about 9 per cent in 1975. As US foreign investment in the Third World is predominantly based in Latin America and, within it, in Mexico and Brazil, the location of overseas R & D is also similarly concentrated. If anything, the level of concentration is higher. Mexico and Brazil account, at a rough guess, for 6 per cent of total US foreign R & D, or two-thirds of R & D in the developing world, while they account for less than a third of the stock of direct US investment in the Third World. Outside Latin America, US TNCs appear to undertake little R & D in developing countries.

Data from other major capital exporting countries on overseas R & D is very scanty and anecdotal, but what there is suggests that European TNCs are following the same path as US firms. In some sectors they are ahead of US TNCs: Unilever runs the largest private sector R & D establishment in India; Ciba-Geigy has a large laboratory there; Hoechst

356

has laboratories in India, Brazil and Egypt; and Volkswagen has substantial technological work in progress in Brazil. In others they seem to be somewhat behind, but there is little reason to believe that the pattern is significantly different from that of US transnationals.

R & D location is the outcome of several complex forces. There are, on the one hand, various economic and historical reasons why TNCs began conducting most of their research at home. First, technological activity at home has provided the most important advantage which firms can exploit abroad, and so has preceded transnational production activity. Secondly, the home country (in the case of the US) has the most developed technological infrastructure, the largest and richest markets for testing new products, and the main production facilities for implementing new techniques. Thirdly, technological activity is highly 'communication intensive', requiring the close interaction of different scientific disciplines as well as of different production, innovational, managerial and marketing functions. Thus, there are inherent problems in breaking off one part of this activity and placing it abroad, in a different cultural, economic and geographical milieu.

To counterbalance these forces of inertia, on the other hand, there are factors which compel the relocation of R & D. First, for a number of industries, a certain amount of technological work is inherent to every production centre: food products and medicines have to be adapted to local markets; industrial machinery tailored to specific requirements; furniture designed to suit local climates,

and so on. Secondly, R & D is clearly much cheaper abroad than in the US; and other countries, especially in Europe, have well developed scientific and technological structures. Thirdly, certain types of testing, mainly clinical testing of drugs, can be done more easily or cheaply in countries with more lax controls than the US. Fourthly, the communication gap is becoming less important as TNCs establish a more effective international network of management, production and marketing. Fifthly, certain foreign locations, for example Germany or the UK, may be more advanced than the US in certain specific technologies and may attract R&D in those to gain from the 'fallout effects' of their own innovation.

There are, thus, several plausible reasons for expecting TNCs to diversify their R & D activity geographically. In essence, the process is determined by the interplay of the communication and scale economy factors which tend to keep R & D at home, and the growing scale of foreign operations and lower costs abroad which tend to force R & D to overseas locations. Over the longer term, we may expect that industries with innovational processes which do not require a very close relationship between the R & D and other managerial and marketing functions will continue to relocate research, while those which like electronics, instruments or sophisticated electrical machinery do require continuous co-ordination between them will tend to concentrate their activity at home. Naturally, industries where substantial product adaptation is required will relocate R & D *pari passu* with their growth of overseas production.

The benefits of R & D conducted by TNCs to developing countries are as follows. As far as applied research is concerned the attraction of TNC R & D to support their local production may mean that they receive more 'appropriate' products and technology than they would otherwise. They may gain, for instance, from medicines for local diseases, vehicles for local roads, machinery employing more labour and processes using more local materials. Furthermore, their own enterprises may be stimulated into adopting more modern technology, conducting more research and looking more closely at local needs. As far as basic research is concerned developing countries may gain from the 'fallout' effects noted above, from the employment opportunities, better training and stimulus given to scientific and technical manpower, and from a line of contact with the scientific community abroad.

The costs of TNC R & D in developing countries may arise in the following manner. First, TNCs may attract the best qualified manpower to themselves or what has been called 'internal brain drain', thus depriving local establishments of talent. Secondly, TNCs may use R & D to produce, not more appropriate products or techniques, but less appropriate ones (for example, more fancy packaging or small but unnecessary model changes) to bolster their market position, or they may exploit lax regulations to test potentially harmful drugs. Thirdly, the benefits of R & D done locally by TNCs may be less than those of R & D done by indigenous enterprises, because the former would be internalized by TNCs and so benefit their global operations while the latter would be internalized by the home country and so benefit its trading position or technological capability. Fourthly, the existence of large and powerful R & D activities by TNCs may suppress 'infant' technological activity by local firms rather than promote it, since the latter may never be able to bear the expense and risk of launching R & D at all. And, finally, TNC laboratories may be used to monitor and tap research done locally, though this is not a real danger as far as developing countries are concerned.

The final judgement on the net effects of attracting foreign R & D must depend on the benefits of getting appropriate technology and its 'fallout' effects as against the costs of stifling innovation in indigenous enterprises. This potential cost is not to be treated lightly: some semi-industrialized countries in the Third World have exhibited considerable technological progress when given some protection against foreign technology. Much more, however, needs to be known before we can pronounce on these important matters with any degree of certainty. As matters stand, developing countries can clearly derive some benefit by attracting TNC research activity into sectors where local technological activity is either well established (as with various kinds of equipment design in India) or is non-existent (as with basic medicinal research in most of the Third World) and with little chance of inependent success. The potential for advantage is greatest where the technology is susceptible to adaptation to suit local needs and resources, and especially when such adaptation leads to establishing a new export base for manufactured products. But some benefits exist even when 'basic' research is conducted, and these are greatest when local institutions are engaged in similar (rather than 'non-competitive') research. Needless to say, host governments must temper their welcome to exclude research work that threatens the physical well being of their populations — and this danger is important enough to have merited international attention in recent years.

Sanjaya Lall, *Third World Quarterly*, 1979.

Multinationals Revisited

The multinational is no longer so multifashionable. It is true that much is still being written about it, but, in spite of the continuing controversy, some of the steam has gone out of the debate. There is no longer the sharp separation between those who think that what is good for General Motors is good for humanity and those who see in the multinational corporations the devil incorporated.

The reasons for this lowering of the temperature are to be found in five recent trends that suggest that the role of multinational corporations in development has to be reassessed.

First, there has been a shift in bargaining power between multinationals and their host countries, greater restrictions on the inflow of packaged technology, a change in emphasis from production to research and development and marketing, among other fators, that have increased the uncertainties of direct foreign investment. As a result, there is some evidence that it has become the policy of multinational companies to shift from equity investment, ownership of capital, and managerial control of overseas facilities to the sale of technology, management services, and marketing as a means to earn returns on corporate assets, at least in those countries that have policies against inflows of packaged technology.

Second, many more nations are now competing with US multinationals in setting up foreign activities, which means that the controversy is no longer dominated by nationalistic considerations. Japanese and European firms figure prominently among the new multinationals. The number of US companies among the world's top twelve multinationals declined in all of the thirteen major industry groups except aerospace between 1959 and 1976, whereas continental European companies increased their representatives among the top twelve multinationals in nine of the thirteen industries, and the Japanese scored gains in eight. The reasons for this are to be found in the decline of US predominance in technology transfer; in the fact that foreign production follows exports, and exports from these countries steadily rose; in the steady growth of European and Japanese capacity to innovate; and in the greater adaptability — both politically and economically — of these companies to the needs of host countries. For example, Michelin's radial tyres, Bosch's fuel injection equipment, and French, German, and Japanese locomotives, aircraft, and automobiles are more energy saving than their American counterparts.

Third, developing countries themselves are now establishing multinationals. In addition to companies from the Organization of Petroleum Exporting Countries (OPEC), and firms established in tax-haven countries, the leading countries where multinationals are being established are Argentina, Brazil, Colombia, Hong Kong, India, the Republic of Korea, Peru, the Philippines, Singapore, and Taiwan. It may well be that these firms use more appropriate technology and are better adapted and more adaptable to local conditions. There is a strong preference in the developing countries for multinational corporations from similar countries. Korean companies put up buildings in Kuwait, pave roads in Ecuador, and have applied to Portugal for permission to set up an electronics plant; Taiwanese companies build steel mills in Nigeria; and Filipino companies restore shrines in Indonesia. Hindustan Machine Tools (India) is helping Algeria to develop a machine tool industry; Tata (India) is beating Mercedes trucks in Malaysia; and Stelux, a Hong Kong-based company with interests in manufacturing, banking, and real estate, bought into the Bulova Watch Company in the United States. C.P. Wong of Stelux

improved the performance of the US company.

The data on the extent of developing countries' foreign investment are inadequate and the evidence is anecdotal. A partial listing of major Third World multinationals contains thirty-three corporations with estimated sales in 1977 ranging from $500 million to over $22,000 million, totalling $80,000 million.

If there is a challenge, it is no longer uniquely American; and if multinationals are instruments of neocolonialism, the instrument has been adopted by some ex-colonies, and at least one colony (Hong Kong), and is used against others. Neither developed nor developing countries are any longer predominantly recipients of multinationals from a single home country.

Fourth, not only do host countries deal with a greater variety of foreign companies, comparing their political and economic attractions, weighing them against their costs, and playing them off against one another, but also the large multinationals are being replaced by smaller and more flexible firms. And increasingly alternative organizations to the traditional form of multinational enterprise are becoming available: banks, retailers, consulting firms, and trading companies are acting as instruments of technology transfer.

Fifth, some multinationals from developed countries have accommodated themselves more to the needs of the developing countries, although IBM and Coca-Cola left India rather than permit joint ownership. Centrally planned economies increasingly welcome the multinationals, which in turn like investing there, partly because you cannot be nationalized.

The nation state has shown considerable resilience in the face of multinationals; its demise has been somewhat exaggerated. The Colombians succeeded in extracting substantial sums from their multinationals. The Indians dealt successfully with firms that introduced inappropriate technologies and products. The Andean Group and OPEC showed that solidarity among groups of developing countries in dealing with multinationals is possible and can pay.

Paul Streeten, *Finance and Development*, 1979.

What Went Wrong?

'We are poorer now than we were in 1972,' said President Julius Nyerere recently, on the twentieth anniversary of Tanzanian independence. This, despite the fact that Western governments and organizations have poured about $3 billion in aid into the country in the past two decades. What went wrong seems to have been a combination of bad luck (global recession, oil costs, a war in Uganda) and, according to the Wall Street Journal, 'a home-grown crop of blunders that foreign aid programmes of the West helped cultivate.' According to the Journal, the West's contribution to Tanzanian distress included the following:

—Foreign assistance helped build schools, roads and other facilities but didn't provide sufficient ongoing aid to maintain them;
—Western aid-giving sometimes had commerce, not aid, in mind. Much of the aid received flows back to the donor country in the form of contracts that keep its industries humming — not necessarily those of Tanzania;
—Foreign aid was used to finance large numbers of 'jumbo projects that failed. We were all looking for our own monuments,' said the head of the British aid programme.

World Development Forum, 1982.

I sit on a man's back, choking him and making him carry me, and yet assure myself and others that I am very sorry for him and wish to ease his lot by all possible means — except by getting off his back. — Leo Tolstoy

The Mythology Of Aid

Every age contrives tranquilizing myths to justify its inequities. One of the most pervasive of our time is the mythology of foreign aid. It was conjured up thirty years ago, when the imperial inning was coming to an end, as an earnest of goodwill between the white sahibs on their way out to the pavilion and the brown sahibs on their way in to bat. The idea was that the newly independent nations were lacking in the technical expertise they needed and, since such skills were only available in the West, they would be provided through the generosity of Western loans and grants.

There was indeed substance to that notion. Some of the new countries found that they were desperately short of the indigenous skills needed to develop their material resources. But others, notably in subcontinental Asia, had most — if not all — the help they could use.

Disenchantment with the imported expertise became vocal before long. A Sri-Lanka Prime Minister expressed it succinctly as long ago as 1954: 'A foreign expert is a man who comes to find out and leaves before he is found out.' Sometimes the experts cost as much as $50,000 a year to maintain, and often they had barely completed their training or grown accustomed to local needs before they were reassigned to help another needy country. But the charade of foreign expertise continued to be played out — usually long after countries had trained their own specialists, and thus eliminated the need

for the outsiders.

Another persistent myth was the widespread belief, particularly in rich countries, that they were contributing lavishly to the development of poor nations. The truth is very different, unless its basis is that a dollar doled out by a millionaire to a stranger is as lavish a measure of philanthropy as a bowl of rice shared by one needy man with another. If the proportion of a rich nation's gross national product is taken as the measure, the most generous countries in the West are Sweden, the Netherlands and Norway, who have allotted nearly one per cent of their GNP to development aid. France and Denmark are next with about .7 per cent — the minimum agreed upon by the United Nations General Assembly. Well behind are Britain and West Germany (.41 and .33 per cent, respectively), Japan (.24 per cent) and the United States (.22 per cent). Although in monetary terms those percentages amount to many millions of dollars, they are minuscule when compared to the sums needed to lift poor countries to a level of self-sustaining development.

What function, then, does foreign aid serve? A third myth must be mentioned here before that question can be answered. The popular notion is that the rich nations' financial contributions to development efforts constitute a major portion of the money poor nations spend on development and, thus, that these foreign funds are absolutely crucial. In fact, nothing could be more absurd.

Rarely have foreign-aid funds to a specific country been more than 6 per cent of the recipient nation's total development budget. Indeed, the only area where this foreign contribution has regularly exceeded 6 per cent is in the case of 'investments' in arms and military hardware. But precisely because foreign aid to economic development is marginal to a poor nation's own efforts, it should be applied to the most sensitive parts of the over-all development programme. Only in that way can the modestly funded foreign-aid programmes achieve an impact that is at all significant in the context of the entire local development effort.

Following an old Asian saying, 'Give a boy some fish to eat and he will assuage his hunger for a day, but teach a boy to catch a fish, and he will never be hungry,' some developing countries have used foreign-aid funds to train their personnel. Others import their experts from neighbouring countries, where conditions are similar. This exchange of personnel between developing nations has even been institutionalized under the rubric of TCDC (Technical Co-operation Among Developing Countries).

A few nations, however, have been profligate with their funds, spending precious foreign exchange on glamorous but completely sterile monuments, or stuffing their new supermarkets with everything from Beluga caviar to pickled quail eggs. The Western press, quite rightly, has picked on those extravagances and other misuses of foreign aid. The slow nurturing of local talent, the building up of institutions, the gradual growth of industry and the

(GATE, 1983)

remarkable improvements in agriculture, education and health were, by and large, not regarded as salable copy. Apart from this, as the new nations matured, they became increasingly obdurate, refusing to dance on aid strings. A few rejected the very concept of aid dependency, demanding bigger transfers of wealth from rich to poor nations, instead of philanthropy, as a matter of economic justice. And beyond that, some nations even had the temerity to insist that such transfers be automatic, thus removing the demeaning connotations of charity from the process.

All of this has caused some of the aid-giving nations to harden their attitudes toward the developing world. The result, so noticeable at all jamborees of the international community, is that the North-South dialogue has degenerated into a cacophony of monologues often taking place simultaneously in the same room.

It usually seems quite earnest and may even be fun. But it will not do. We can now see the signs of a terminal disease afflicting our epoch, an age that has been based on the belief that it was right for a small number of people to accumulate wealth and power at the expense of billions beyond the pale. With the demise of the epoch, its web of self-serving myths ('survival of the fittest,' 'population growth is the cause of poverty') are also waning.

It is no longer a question of bridging poverty gaps by calculated dollops of 'foreign' aid, however generous. Rather it is a question of the poor world recognizing that its social structure, which keeps the benefits of growth from penetrating below the top-most layers, must be changed. Developing nations must respond to the needs of an increasingly vocal population that can no longer be suppresed by terror or tranquilized by promises. Poor countries must also look anew at their goals, and realize that they will never be able to develop the kinds of consumer societies that Europe and America have become.

The rich world must do some self-analysis as well. It must realize that the energy and minerals it needs from the Third World are not mere trade goods, but powerful arguments in the negotiating process between two mutually dependent areas of the world. Unless these simple realities are acknowledged in fresh new policies, we shall all be limping into the twenty-first century, still beguiled by the mythology of aid, and gibbering pathetically about new economic orders.

Varindra Tarzie Vittachi, *Newsweek*, 1979.

The best aid to give is intellectual aid, a gift of useful knowledge.
— E. F. Schumacher

Choice of Technology: Donor's Approach

Bilateral donors tend to associate appropriate technology with technologies for the poor, with village technologies, with technologies for basic needs. Projects which result from this approach tend to be small and often experimental: allocations under an 'appropriate technology' heading therefore form a negligible proportion of total project financing. The emphasis is usually on promoting research and development of technologies which are believed to be particularly appropriate, or on the collection and dissemination of information on technological alternatives. In a number of developed countries, special institutions now exist to promote appropriate technologies and to strengthen the liaison between their national institutes and those of developing countries. It is not easy to ascertain the impact of these appropriate technology institutes, centres or groups. It is fair to state, however, that in a number of developing countries they have usefully contributed to solving specific technical problems. Perhaps more importantly, they have contributed to a greater awareness in developing countries of technological alternatives. Today in many developing countries the appropriate technology concept is no longer regarded with suspicion, it having been demonstrated that technological alternatives are often a viable proposition. Finally, appropriate technology institutes, centres or groups in developed countries play an important domestic, didactic role. They depend on and address themselves to the unofficial aid lobby; they succeed in making the development problem — or at least part of it — tangible for the people 'back home'. The narrow approach thus becomes reinforcing.

Only a few bilateral donors are very explicit about their support for appropriate technology activities. In a statement of Dutch aid policy, certain types of aid activities are listed which are to be given more emphasis than others. These are activities in the health care, agriculture and education sectors. In addition, four cross-sectoral priority areas are specified, one of which concerns activities to promote appropriate technology which is less capital-intensive and whose prime aim is to create work and hence a means of subsistence. All too often in developing countries, influenced by experience in rich industrialized countries where the relationship between the availability of capital and labour is different, a technology is decided on which is very heavily capital-intensive and which does little, if anything, to help create employment. The direction of scientific research, and also the form taken by concrete projects, will be decided in accordance with the established fact that nothing is more conducive to an equitable distribution of wealth and genuine self-reliance than employment.

In the case to the US, legislation was adopted, authorizing the Agency for International Development to use up to $20 million over the three-year period covering fiscal years 1976-78 for activities in the field of intermediate technology, through grants in support of an extended and co-ordinated private effort to promote the development and dissemination of technologies appropriate for the developing countries. An official report on the new law stated that: the experience of more that a quarter century of development assistance programmes overseas has clearly demonstrated that much of the technology used in the United States and other industrialized countries is not well suited to the economies of developing countries. It is too big, it is too expensive and it does not create the jobs needed to absorb rapidly expanding labour forces in countries which already have an abundance of labour. It is not appropriate for use on the very small farms and in the very small business enterprises that make up so much of the activity in the developing world. If the poor are to participate in development they must have access to tools and machines that are suited to labour-intensive production methods and fit their small farms, small businesses and small incomes. They must have access to technology which is neither so primitive that it offers no escape from low production and low income nor so highly sophisticated that it is out of reach for poor people and ultimately uneconomic for poor countries — in short, intermediate technology. A proposal for a programme in appropriate technology was transmitted by the Agency for International Development to the Senate Committee on Foreign Relations and the House International Relations Committee early in 1977.

In response to a Congressional mandate to promote the development and dissemination of technologies appropriate for developing countries, Appropriate Technology International (ATI) was created. A private, non-profit organization largely funded by AID, ATInternational began operations in 1978 and focussed in its first two years on efforts to strengthen local delivery systems in order to increase the access of the poor to the technologies needed to sustain their own development. ATInternational has an active programme to encourage small enterprises and community groups to undertake productive ventures with a potential for commercialization.

In the United Kingdom, a Ministry of Overseas Development Working Party met in 1977 to consider how Britain could do more through its aid programme to promote the use in developing countries of technologies appropriate to conditions there. The main conclusion of this Working Party on Appropriate Technology was that in view of the importance now attached to intermediate technologies by the developing countries, of the role aid donors can play, of the direct relevance of the subject to Britain's aid policy of doing more to help the poor, particularly in the rural areas, and of the possible advantage to British industry, the current modest level of assistance to intermediate technologies within the aid programme should be increased. The Working Party's report recommended that existing aid activities to promote intermediate technology should be intensified and that a number of new activities be supported, initially by at least £500,000 a year from Britain's aid programme. The recommendations were accepted by the Minister for Overseas Development.

As a result of the recommendation that a portion of the new funds should be applied to the strengthening of the Intermediate Technology Development

Group (ITDG) a new unit, the Intermediate Technology Industrial Services Unit (ITIS) was set up to provide technical and financial assistance to meet the needs of developing countries for unfamiliar or new technologies, primarily in the small-scale industry sector. To meet the costs of such assistance, a project fund was established to support the dissemination of information, prototype development and testing, market surveys and campaigns and the like. Funds were also set aside to strengthen the headquarters staff of ITDG, and to enable ITDG to administer a small technical assistance fund for the benefit of developing country intermediate technology institutions. All these measures, however laudable, seem to fit a narrow approach perfectly well. In the case of the UK, however, the limitations of the narrow approach appear to have been recognized. An internal policy guidance note of the Ministry of Overseas Development suggests that the Ministry's field structure should look out for and respond favourably to requests for assistance in intermediate technologies. It suggested it is of particular importance that technology choice should be considered when projects are being identified, for example by writing into consultants' terms of reference explicitly that alternative technological options be appraised, since by the time a project comes forward for capital aid the technology may (though not always) be fixed.

The broad approach to choice of technology, which is particularly favoured by multilateral agencies, is harder to put into practice. Special measures are required to ensure that the appropriateness of technology is not considered as just another item on a long checklist. Such measures could be of two kinds. First, the standard routines for processing project proposals would contain tripwires to ensure that generalists in geographical departments, which in most agencies carry the main operational responsibility, seek specialist advice whenever necessary. Second, there would be some central unit or group within the agency to advise on all questions of technological choice.

Among the proponents of a broad approach, the World Bank stands out as the agency most aware of the relevance of choice of technology to all its operations. The introduction of a 1978 World Bank paper entitled 'Appropriate technology and World Bank assistance to the poor', speaks of 'the general organization and procedures of the Bank (which) ensure that appropriate technology is used in Bank-financed projects as a matter of usual and normal practice.' In the background document which the World Bank presented to the UN Conference on Science and Technology for Development, the first of four basic technological objectives of the Bank is described as to ensure that the most appropriate technologies are used in the projects it finances. Those technologies must be appropriate to the objectives of the project; to the broader development objectives of the country; to the local social, cultural, economic and environmental situation; the local raw and semi-finished materials; to the local and grass roots capacity to plan, operate and manage; and, if this is a principal project objective, to creating opportunities for productive employment and to alleviating poverty. The appropriateness of technology to specific factor endowments and local conditions will ensure efficiency in the allocation of resources. Elsewhere in the paper it is concluded that if government policy (in developing countries) were directed toward promoting a price structure that reflected the scarcity values of labour and capital more realistically, technological choices would be different. The results would be greater employment, broader income distribution, and more competitive patterns of production of precisely the labour-intensive goods that labour-scarce affluent countries need but cannot produce inexpensively. Therefore, 'The Bank has encouraged governments to change policies that may have led to distorted prices, subsidized capital equipment, overvalued exchange rates, subsidized interest rates, and discriminatory access to credit. It has also encouraged governments to change regulatory measures that may have promoted undue capital-intensity in investments and inhibited the ability of small-scale enterprises to function.'

These are undoubtedly statements which fit what we have called a broad approach to choice of technology. Yet, when the Bank reports on its activities, it appears to make a distinction between projects in which 'appropriate technologies' are used and other projects. Thus, appropriate technologies seem particularly called for in sectors such as urban development (e.g. for the upgrading of squatter housing through self-help), rural development (to improve the productivity of poor peasants), highway construction (building rural access roads in labour-intensive ways) and industry (notably for small-scale industries). In education and health, appropriate technologies are believed to be embodied in rural non-formal education and training, and in community health workers respectively. The fact that the Bank identifies appropriate technologies with cheap tools and small machines, to be used by the poor for the poor, would rank it among the followers of a narrow approach. Possibly, however, one is dealing with a combination of approaches designed to accomodate views on policy of various executive directors of the Bank or of the governments they represent.

Fred Fluitman and John White, *External Development Finance and Choice of Technology*, 1981.

Do you remember the game, Spillikins, where you must take a stick from a randomly dropped pile without jogging any of the others. There is a Spillikins effect which ensures that it's almost impossible to change anything important without starting some quite unexpected repercussions elsewhere . . . But when they tried [the Green Revolution] in India and Mexico, the results weren't quite what they'd hoped. . . . Does this mean that it's simply pointless to try? . . . Far from it . . . small projects, which grow up through the paving stones of old patterns, have a far better chance of long-term success than grandiose panaceas imposed from above; and the Spillikins effect is certainly one reason why. — Katharine Whitehorn, *The Observer*.

IX

The Dissemination of Technology

As PRECEDING chapters have shown, there is now an increasingly wide range of small-scale technologies avalable for use in developing countries — many of which have been designed and developed by local agencies and engineers. The case study chapters (III to VII) also show that many of these technologies are actually being used in the Third World to the benefit of the rural and urban poor. What is less clear is the extent to which such technologies have come to be used on a widespread basis.

One of the problems facing the AT movement lies in its 'project' orientation. Pilot projects are very necessary for developing and testing prototypes, and for proving to potential users, producers, donors and governments that small-scale technologies exist and that they are technically feasible and economically viable. They are of little help, however, if the next step is not taken and if the technology fails to disseminate beyond the project boundaries. Successfully introducing an improved plough, mill, oil-press or water or sanitation system to a few villages, or introducing a new construction or industrial process to a few co-operatives or entrepreneurs is all very well. What is needed, however, is that such products and processes should be adopted, and come into use, in thousands, if not millions, of villages throughout the Third World. A handful is not enough, and at the moment, a handful of people is all that is benefiting in many of the examples cited in Chapters III to VII.

Herein lies a crisis for the AT movement. Although the AT concept gained much support and credibility in the 1970s, it stands to lose much of the progress made unless practical proof of its effectiveness can be offered. For this to happen, there need to be more and more examples of small-scale, appropriate technologies being used efficiently on a very widespread basis. It is with the problem of getting beyond the individual project that this chapter concerns itself.

Appropriate technologies can be (and have been) introduced and can be proved to be technically and economically viable, but their dissemination throughout a country or region will be severely limited unless the socio-economic climate is such that it will encourage both the manufacure and the use of the new technologies. Government attitudes and policies obviously have an important part to play here, since they can and do influence the investment decisions made by producers and consumers in the private sector.

The first section of the chapter examines the role of government in influencing the adoption of appropriate technologies on a widespread basis. It starts with a general extract which examines the influence governments can have in creating the right environment through changing fiscal, monetary, import/export, industrial licensing and other economic policies. This is followed by an extract which summarizes the stated objectives in their national development plans towards rural industrialization and appropriate technology of forty-one Third World governments. This reveals that many governments are in fact now talking persuasively about rural industry and appropriate technology. But the extent to which they have implemented the policy measures mentioned in their plans, and the impact these have had in practice, is questionable. Much more research is needed on these issues if appropriate technology is to come into its own. Bearing this in mind, details are given of favourable government development plans in three countries — China, India and Kenya — and contrasted with the unfavourable environment prevailing in Indonesia in the 1970s.

A favourable economic environment is a necessary but not sufficient condition for the widespread dissemination of appropriate technology. Also needed are channels or mechanisms through which the dissemination process can occur. This is the subject matter of Section 2 of the chapter. It starts with a very useful account by A.K.N. Reddy of the complex and difficult nature of the dissemination process which, he argues, has to be a multi-institutional effort involving governments, R & D organizations, NGOs, commercial firms, financial institutions and potential beneficiaries. This means that the dissemination process involves a very different institutional framework to the technology generation process. While R & D organizations can take the lead and play a major part in the latter process, they are not suited to do so in the former. Development agencies need to co-ordinate the dissemination effort, involving R & D organizations as necessary.

This is followed by an extract by Charles and Olle Edquist which introduces the idea of the need for social carriers of techniques — people or enterprises which will implement and thus lead to the spread of technology. One type of agency particularly interested in the dissemination of appropriate technologies is, of course, the AT Institutions themselves. A recent OECD survey of the routes (carriers) through which such institutions have sought to disseminate their prototypes concluded that the financial system and big industrial firms play a surprisingly small role — perhaps because of the reluctance of these institutions to involve themselves in the sort of production system to which they are seeking an alternative.[1]

This attitude has had a noticeable affect in biasing AT organizations against commercial routes to the dissemination of technology (including the route through small enterprises) and in favour of the more welfare-orientated project route. Given that the AT movement grew out of a concern for people with little or no cash, who must be helped to gain access to products and services in ways other than through the market place, the bias is understandable. This does not mean however that the mass market and commercialization should

---

[1] Nicholas Jéquier and Gerard Blanc, *The World of Appropriate Technology*, 1983.

be completely ignored by those wishing to disseminate appropriate technologies to millions of producers and households. On the contrary, if a technology such as an improved stove or a more efficient brick-moulding press is to reach the masses of poor people who could benefit financially from its use, then commercial channels will almost certainly need to be utilized — ideally in combination with consumer credit.

The advantages of, and the constraints on the commercial route to dissemination are dealt with in Section 3. This starts with a general extract by Malcolm Harper on the ins and outs of appropriate technology marketing and is followed by three extracts on support mechanisms for small firms — one on the nature of small industries and the problems they face; a second on the policies implemented by governments (particularly the Indian Government) to support such industries; and a third on credit programmes established to assist small industries in the developing and developed countries.

The final section of the chapter examines the extension route to technology diffusion — the route characterized by projects, rural development agencies and self-reliance. This, of course, is an essential part of any poverty-focused rural development strategy. People who are without assets need some help in moving themselves to a situation where they begin to have control over assets and have cash to buy basic goods and services through normal commercial channels. Poverty-focused projects should therefore be seen as a means to an end rather than an end in themselves: the latter perpetuating a welfare approach and a dependence on project personnel and resources.

Having said this, there is a need to ensure that projects aimed at introducing improved technologies to the poor are properly designed and implemented. It is with this issue that the extracts in this section are concerned. Those by Marilyn Carr and David French offer suggestions to project managers about the questions they need to ask when designing and implementing projects. The former emphasizes the special problems involved in disseminating technologies to rural women who, as was seen in Chapters III to VII, are the potential beneficiaries of many (if not most) AT and small industry projects.

Examples are also given of project design and implementation. The extract by M.K. Garg describes the steps through which many of the successful ATDA projects have progressed. That by Martin Greeley describes the attempts to disseminate improved storage technologies through the extension route, as opposed to the commercial route. Both extracts emphasize the importance of the commercial sector.

We should not feel ashamed to ask and learn from people below.

— Mao Tse-Tung

Creating the Right Climate

The identification of technologies which are appropriate to conditions prevailing in the developing countries will be of litttle use unless they are applied on a widespread basis. Also, there is little point in simply advocating the increased use of these technologies without offering suggestions as to how this can be accomplished.

What might appear to be desirable technologies can be introduced and proved to be technically and economically viable, but their diffusion throughout the country will be severely limited unless the socio-economic climate is such that it will encourage both the manufacture and the use of the new technologies.

In theory, it should be possible to devise a comprehensive package of incentives and disincentives to fit any circumstances. Thus, if a government is concerned with reducing unemployment and poverty, it should be possible to devise a package of policies which lead to the adoption of technologies which are appropriate in the context of achieving these objectives. In this particular case, the policies involved will be ones which induce existing producers (e.g. large-scale manufacturers and land-owners) to adopt — on economic grounds — those technologies which benefit the poor and unemployed. They may also be policies which increase the purchasing power of the rural and urban poor so that an effective demand is created for goods which lend themselves more readily to production by low-income producers who are more likely to use 'appropriate' technologies. In practice, the extent of government intervention depends not so much on what the government might wish to achieve, but how far it can go without losing the support of such powerful groups as land-owners, large industrialists and the urban elite. As far as the government is concerned, the extent to which it can achieve its stated development objectives (e.g. reduction of unemployment and poverty) without losing the support of those who — in the short term — control the economy, will determine what is and what is not appropriate.

Government policy with respect to import licensing, industrial licensing, price control and many other issues can and often does stop the spread of profitable, small-scale technologies. One example of this is the development in Ghana of an intermediate technology for producing animal feed from brewer's spent grain. This technology, which consists of simple hand-pressing and solar drying, produces animal feed at a much lower cost and with less use of scarce foreign exchange than the advanced technology normally used by the breweries. However, the decision of the government to allow the breweries to import sophisticated drying equipment looks as if it may put existing small-scale entrepreneurs using the intermediate technology out of business (by eliminating the supply of spent grain), eliminate the possibility of other small entrepreneurs joining the market, and lead to a rise in the cost of animal feed.

Since the intermediate technology results in lower unit costs and more employment than the advanced technology, and also provides a source of income for small entrepreneurs, it would have been more appropriate from the point of view of the consumers of animal feed, the unemployed and the small entrepreneurs to encourage this by refusing to grant an import license to the breweries. It would also have been desirable from the government's point of view in saving foreign exchange. This, however, would not, in the existing circumstances, have been the most appropriate policy from the point of view of the breweries. The questions to ask here are whether indirect government policies such as price controls could alter circumstances so that the breweries no longer found it desirable to use the imported technology, or if direct measures such as import or industrial licensing would be necessary; and also to what extent would such policy measures be feasible from the government's point of view?

Those who argue for the need to create the right social-economic environment before desired changes in the lot of poor people can take place, often do so in a negative fashion, arguing that nothing can be done unless 'the system' is changed. That argument is not the basis upon which the above considerations are put forward. It is merely that in planning the introduction and diffusion of appropriate technologies, there is a need for governments to look at the different interventions they can make and to examine their feasibility as well as their possible value and effect in achieving desired results. By the same token, outside agencies advocating the use of appropriate technologies must be aware of the need for appropriate interventions and must be able to assess the extent to which such interventions are possible in the circumstances.

Institutions, including those dealing with appropriate technology, are part of the infrastructure package in development plans and the package is, itself, one instrument to be used in creating the right climate. Institution building must be seen within the context of the whole infrastructural framework. For example, how is the need to create specific AT institutions balanced against the need to instil AT concepts and capability into the institutional fabric as a whole?

Although in the long term it may be ideal to have the concepts of appropriate technology incorporated into the institutional framework as a whole, there is probably a need for specific AT institutions in the short term until the concepts become part of the conventional wisdom. If these specialist institutions are successful, there should eventually be no need for their existence. But, in creating them, to what extent is the concept that appropriate technology is different from technologies which are appropriate being perpetuated?

Marilyn Carr in J. de Schutter and G. Berner (ed.), *Fundamental Aspects of Appropriate Technology*, 1980.

Planning for Rural Industrialization

A national development plan has been described as a technocratic conception of a feasible programme to attain certain policy goals. A study of national plans should, therefore, provide a good indication of the policy goals of particular countries as well as the programmes considered feasible by policy makers in support of those plans.

This review of the latest available development plans of forty-one developing countries is based on plan documents specifying goals and programmes for sub-periods within the decade 1975-85. In the table, these countries are categorized according to the importance and encouragement to rural industrialization

implied by statements in the plans. As the table indicates, most governments now recognize the importance of rural industrialization as part of the strategy of development. A large majority have backed up this recognition with specific policy measures for rural industrialization in their plan documents and nearly half have explicitly emphasized the importance of appropriate technology development, dissemination and adoption for the purpose of rural (or decentralized) industrialization. In all of these latter countries, measures aimed at improving the process of development and transfer of appropriate technology have actually been implemented.

Categorization of Countries According to Encouragement/Importance given to Rural Industries

I No specific encouragement to small rural industries	II Described as important. Some encouragement but no specific policies stated	III Encourage rural industries but not much attention to technology development and dissemination	IV Encourage rural industries. Programmes include development and dissemination of ATs
CUBA	ALGERIA	AFGHANISTAN	BANGLADESH
IRAN	ECUADOR	BARBADOS	BOTSWANA
OMAN	LIBERIA	BURMA	BRAZIL
SAUDI ARABIA	MALAWI	CAMEROON	CHINA (mainland)
YEMEN ARAB	MALAYSIA	MAURITIUS	THE GAMBIA
REPUBLIC	MEXICO	SOMALIA	GHANA
	PHILIPPINES	SUDAN	INDIA
	SENEGAL		INDONESIA
	SEYCHELLES		IVORY COAST
	SYRIA		JAMAICA
	THAILAND		KENYA
			KOREA (Republic of)
			NIGERIA
			PAKISTAN
			PAPUA NEW GUINEA
			TAIWAN
			TANZANIA
			ZAMBIA

Category 1 countries — Cuba, Iran, Oman, Saudi Arabia, Yemen Arab Republic — appear to have undergone no shift in their commitment to conventional large-scale capital-intensive development strategies. Those in Category II recognize the importance of rural development and (at least by implication) rural industrialization, but set out, in their development plans, no specific policies for rural industrial growth. Of these, Algeria, Ecuador, Liberia, Syria, Senegal and Seychelles couch their encouragement in rather general terms relying, presumably, on agricultural development alone to foster the growth of rural industries. In Malawi, the expectation of interlinked and balanced agricultural-industrial growth is clear, with the planning document stating explicitly that rising output in primary production will give rise to increasing opportunities for processing industries; equally, rising incomes derived from the sale of this output will open up the market for industries producing consumer goods and agricultural inputs. The governments in Malaysia, the Philippines, Thailand and Mexico provide considerable support for small industries for reasons of employment generation and income

(FAO)

distribution; but the emphasis stated in their plans is on sub-contracting and export-oriented industries rather than on creating specific links with the rural economy. References to decentralization are limited to directing industrial investment away from metropolitan areas to other towns with reasonably developed infrastructure rather than to rural areas. These four national plans, while recognizing the importance of rural development, appear to be heavily influenced by the development experiences of Japan (in the case of Malaysia, Philippines and Thailand) and the USA (Mexico).

Category III countries give considerable importance to rural development and specifically decentralized industrial units in rural areas. Barbados, Burma, Cameroon, Somalia and the Sudan have plans which concentrate on rural artisan activity, building on skills available in rural areas to develop productive manufacturing enterprises. Modern small industries catering to local demand or utilizing local resources are also encouraged. In the Mauritius Plan, emphasis is placed on small industries utilizing local resources both to meet local demand and for export purposes; while in Afghanistan there is a marked tendency to encourage cooperative effort for agricultural processing. None of the category III country plans makes more than a passing reference to the development and dissemination of appropriate technologies for rural industrialization. Judging from the size and stage of development of most of these countries, it is unclear whether this is a reflection of the inadequacy of the local science and technology infrastructure or of the failure officially to recognize the importance of appropriate technology as part of the development strategy.

Category IV consists of a wide range of developing countries of various sizes and at various stages of development. The national development plans of these countries aim to promote rural industrialization as part of a comprehensive strategy of rural development. Programmes incorporate the development and dissemination of appropriate technolo-gies as part of the effort to bring about balanced economic growth within the framework of a dynamic and equitable rural society. As might be expected, the programmes undertaken in these countries vary considerably in both conception and detail. The variations reflect not only differential capacities to undertake the requisite programmes but also conceptual and methodological differences based on the degree of understanding of the strategy and on considerations of political economy.

Sanjay Sinha, *Planning for Rural Industrialization,* 1983.

 # Technology Policy in The People's Republic of China

A coherent general approach to locally initiated industrial development was articulated in China shortly after the establishment of the People's Republic, and was incorporated into the First Five Year Plan (1953-57). According to this approach, while the central government was busy building a modern large-scale industrial sector to serve as the backbone of the industrialization effort, the various localities (provinces, municipalities, districts and counties) were to undertake an effort of their own. Their task was primarily to produce industrial inputs for agriculture and consumer goods required by the peasantry, and secondarily to serve as adjuncts to the emerging modern industries in the cities. The technologies to be adopted by the local industries were dictated by a constraint central to the entire approach — that these industries use only such resources as were not required by the modern, large-scale sector. Under no circumstances were they to compete with the modern sector for raw materials, fuels or other inputs. Hence, they were to be confined to waste and scrap materials, second-hand machinery and equipment, small, scattered or low-quality ore deposits, and locally available skills and financial resources. The technologies consistent with this constraint were relatively small in scale of output and relatively labour-intensive. Hence the administrative division of Chinese industry into centrally and locally operated components conformed quite closely to the technological division between relatively large-scale and capital-intensive units, on the one hand, and relatively small-scale and labour-intensive ones, on the other, except that a large part of the latter category strictly speaking consisted of handicrafts and was under collective rather than local state jurisdiction.

During this period, the peasantry depended for a major portion of its production and consumption needs upon the local industrial and handicrafts sectors. The First Plan called for most investment in agriculture to be self-financed by the agricultural co-operatives and individual peasant households, and local industry and handicrafts were the natural source of the real goods counterparts to local savings. But with its leadership resources stretched thin by a variety of massive economic and social programmes of higher priority, the Government did not vigorously promote local industrialization at this time, and, with neither the central nor the local industrial sectors providing agriculture with sufficient increments of needed equipment and inputs, farm production lagged.

The Government became aware of these problems by the middle of the First Plan period, and began to encourage greater attention to local industries. In 1957 the central ministries worked out a set of more than a hundred designs for small-scale plants capable of being maintained and operated locally. Their importance was to emerge only later, for shortly after their appearance they were caught and temporarily inundated by the advancing wave of the great leap forward movement (1958-60).

This movement had as one of its chief characteristics, of course, the rapid establishment (within a year or two) of literally tens of thousands of small enterprises, many no more than workshops or sheds, under the administration of the newly formed people's communes and their sub-units (production brigades) and of the local governmental bodies directly above them. A certain amount of consolidation of the earliest such shops took place, but by mid-1960, according to one report, there were about 60,000 industrial units run by the counties (hsien), for an average of about thirty such enterprises per county, and some 200,000 units run by the rural communes, not including the even smaller shops of the brigades. These industries included iron and steel, non-ferrous metals, coal mining and refining, electric power generation, agricultural tools and machinery, chemical fertilizers, construction materials, light industries of various kinds and transport and communications.

Outside of China, today's evaluation of the great leap's small industry experiment tends to be harsh. The movement was indeed laced with flaws that provide ample reason for a critical evaluation. Local resource bases and market conditions were not adequately investigated before construction, objective technological constraints were not observed, necessary production of small tools, utensils and household articles fell away as handicaraft equipment and raw materials were commandeered by the new factories to make iron and steel and other heavy industrial goods; commune enterprises used their substantial financial freedom to make high profits by gearing production to the demands of materials-short urban factories, ignoring the pressing needs of their own peasants who could not meet such competition. The resulting conditions of acute shortage of small commodities, inferior quality, great waste of materials and equipment and an unsound drain of labour away from other activities — especially farming — called for rapid rectification.

Despite these grave weaknesses, the small-scale industry movement of the great leap period had some substantial positive effects on the country's devel-

377

opment. Many of the thriving regional industries of today had their origin in a primitive workshop established in 1958. The idea of dispersed, small-scale, locally operated and indigenously equipped industries responsive to the needs of the localities was in many areas so good, and the enthusiasm so great, that the enterprises established transcended their obstacles and became viable. In Maoist terms, first the question of presence or absence was solved, and later that of meeting acceptable standards of efficiency. Even where the shops established during the great leap were forced to close, however, the initial experience with industrial methods they had afforded the peasants and the lessons, both positive and negative, to which they gave rise, proved invaluable later when local industrialization was again pushed vigorously.

The internal problems of the small industry movement contributed to, and were themselves aggravated by, more generaal errors in planning and organization that characterized the great leap forward, by three successive years of harvest failure, and by the emergence into the open of the Sino-Soviet dispute with the withdrawal of Soviet experts from China in June 1960. These conditions, which caused a sharp downturn in industrial and agricultural production in 1959-60, necessitated a consolidation of local industries in which the majority of the small-scale enterprises were dismantled.

It is clear that during this period policy toward small-scale local industry became the focal point of vigorous political debates between those who regarded such industry as unruly, uncontrollable and a hindrance to effective central planning, and those who saw it as a means of mobilizing local resources, arousing local initiative and combating bureaucratism. The former charged that 'those people claiming reliance on their own resources actually impair the State's interests and thwart the completion of the planned projects and the smooth progress of production.' Later, during the polemics of the cultural revolution in the late 1960s, the mass closure of small industries of the early part of the decade was blamed exclusively on the political motivations of officials who opposed local control as contradictory to their plan to esablish state monopoly 'trusts' in various lines of industry. There is indeed evidence of political opposition to local industrialization at this time, but it is impossible to establish from afar its responsibility, relative to that of objective economic constraints, for the mass closures that took place.

Nevertheless, despite the closures, a good many of the leap-originated enterprises survived or were restored when economic conditions began to improve in 1962. The planning chief of the time, Po I-po, called attention to the 'by no means negligible' role of local small and medium industries, as early as 1963, in an article that warned equally against the tendencies 'to put a one-sided emphasis on the development of national industry and large enterprises, or to follow the decentralized method of blindly developing local industry and small enterprises in desregard of the unified state plan.' The Chinese press of this period is replete with examples of localities with flourishing small and medium industries even in backward areas, such as Kansu Province in north-west China, which had not been able to make its own nails in 1949, but which had some seventy farm equipment factories in 1963 (compared with five in 1957, before the great leap) producing chemical fertilizer, cement, waterpumps, diesel engines and insecticide sprayers. In the economically advanced regions of the country, development of small industry was more extensive. In the northeastern heavy industrial centre, Liaoning Province, small rural machinery plants and grain processing enterprises run by means of small thermal power plants were operating in remote towns in 1963, in which year the province had some 400 small and medium units in metallurgical, coal, chemical, agricultural machine and light industrial production. Therefore, it is clear that the view expressed during the cultural revolution that such industries lay dormant until that movement bestowed its revitalizing kiss, is something of an exaggeration. While the impetus of the great leap years was certainly lost, the level and growth of small industrial activity in the early sixties was undoubtedly far greater than in the years preceding the leap. Just as important, these industries were on a much firmer basis, producing goods that were locally needed at costs not unreasonably high and without the waste that had characterized such great leap phenomena as the 'backyard iron and steel' units.

It was from this relatively sound foundation that rural industrialization entered a dramatic take-off phase during the cultural revolution. Under the stumulus of that movement's encouragement of local, decentralized initiative, and with the more cautious central planners under attack, various provinces and hsien began to report growth rates for local industries of a magnitude that had not been seen since the great leap. The new industries were closely linked to agriculture. For example, the production capacity of small nitrogenous fertilizer plants — a sector that had already grown considerably during the early sixties — increased by five times between 1965 and 1969. The share of local plants in national fertilizer output rose from 12 per cent in 1965, to over one-third in 1969, to 43 per cent in 1970, to 60 per cent in 1971.

Local industrialization in the years preceding the cultural revolution centred on chemical fertilizer and cement production, the manufacture and repair of farm equipment and machinery, and farm products processing industries. These continued to be stressed in the spurt that began in the late sixties, but they were joined by a rapid spread of local power industries (hydroelectricity and coal) and of local iron and steel production. The latter industries provided a producer goods base for local self-reliant development far more extensive than had previously existed. The number of small hydroelectric stations built in the two years 1970-71 was said to exceed the total for the previous twenty years; by early 1972 there were some 35,000 such installations in China providing 16 per cent of the total national installed hydroelectric generating capacity.

The rapid extension of electricity to areas previously without mechanical motive power was of great importance in 'forcefully stimulating the development of local industry'. For example, in the much publicized Lin hsien of Jonan Province, the extensive development of some 244 country- and commune-run industries had to await the coming of electricity with the completion of the famous Red Flag Canal in the late sixties.

Iron and steel had been perhaps the major victim of the closures of small plants in the early sixties, and there is no evidence of any significant revival of this local industry until the end of the decade, when county-run blast furnaces began to go up in numbers, turning out three times as much pig iron in 1970 as in 1969. By 1971, all provinces, municipalities and autonomous regions except Tibet had their own small and medium iron mines and iron and steel plants, which were responsible for a quarter of the iron ore and one- fifth of the pig iron produced in the nation. The erection of numbers of small- and medium-scale oxygen top-blown converters brought a rapid increase in steel production in many provinces. The development of this industry on a local basis was significant in that it released the local machine-building and other industries to some extent from the constraints of scarce centrally produced iron and steel.

Thus, from the late 1960s on, rapid development of the 'five small industries' (iron and steel, cement, chemical fertilizer, machinery and power) provided the basis for relatively comprehensive 'local industrial systems', whose components had tight backward and forward linkages with each other and collectively with agriculture. These 'small but complete' systems were most often at the level of the hsien. Their completeness was relative, for, while striving for self-reliance, they sought and received aid from higher levels in obtaining a certain irreduceable minimum of sophisticated equipment, technical assistance, and sometimes investment funds. Their erection was aided by the politics of the cultural revolution and by a run of consecutive good harvests that provided raw materials as well as local savings to mobilize.

The speed of local industrialization decreased after 1972, as the local networks filled out in those areas of the country most advantageously situated with respect to raw materials and markets, and perhaps as a more central control-orientated leadership exercised authority. Thus, there is today a fairly stable if still technically dynamic local industrial system in place, whose principal economic characteristics and whose relations with other sectors of the economy and society have been available to study for several years.

Carl Riskin in Austin Robinson (ed.), *Appropriate Technologies for Third World Development*, 1979.

 # India's Five Year Development Plan 1978-83

The sectoral strategies, targets and outlays closely reflect the main strategy and the basic objectives of the Plan. Thus in drawing up the Plan:

— the highest priority has been given to the sectors which generate the maximum employment and which have a significant impact on the standard of living of the poorest, like agriculture and allied activities, village, cottage and small industries and inputs like irrigation; fertilizers and power, which are required to sustain them;

— upgraded norms have been adopted in the Revised Minimum Needs Programme and, to the extent possible, programmes in sectors like communications, science and technology, housing, health and family welfare, education, social welfare and nutrition have been oriented to benefit the rural and urban poor; and

— the objective of achieving self-reliance both technologically and by investment in sectors which will ensure that when necessary we can do without foreign aid, has been kept in view.

The employment objective depends crucially on increased labour absorption in agriculture and allied activities. This means increasing the productivity of available land through irrigation, multiple cropping and improved technology. The main thrust of the planning strategy, therefore, would be to expand the area under irrigation as rapidly as may be possible, and to develop cropping patterns and agricultural practices which optimize the use of land and water resources. Detailed agricultural plans would need to be drawn by regions and sub-regions, based on the full exploitation of the water resources in the command areas of irrigation projects, and on the principle of water conservation and management in rainfed areas which would enable us to break out of the constricting historical trend rate of growth of around two per cent per annum. Fortunately the agricultural development potential today is greater than at any time in the past, in terms of availability of improved seed, modern cultural practices, applicable research results and farmers' awareness of and access to all physical inputs and credit. The new Plan would provide for massive

India's development plans have emphasized appropriate technologies

381

investments in expanding the rural infrastructure, covering not only irrigation and supply of seed and fertilizer but also expansion of credit, storage and marketing.

To maximize employment in agriculture it is necessary not only to provide for the infrastructure and inputs which will increase physical productivity, but also to push forward the implementation of land re-distribution programmes and schemes for the consolidation of holdings; and to regulate the growth of farm mechanization to ensure maximum labour use consistent with optimum land and water utilization.

Improved productivity and employment intensity can be achieved not merely in the production cereals and cash crops but in animal husbandry, horticulture, forestry and fisheries, where the scope for expansion is even higher.

A marked increase in agricultural employment should lead to significant growth in secondary employment in rural areas in distribution and transport, and in tertiary employment in the economic activities generated by the growth in rural income.

After agriculture, household and small-scale industries producing consumer goods for mass consumption hold out the greatest potential for employment. This is a sector which has received inadequate attention in earlier plans. The planning strategy would aim at protection of existing livelihood of rural artisans and a substantial increase in the employment content of the rural industrial sector. It would seek to improve the quality of production, increase productivity, reduce costs and expand the market.

Some rural industries like handlooms and handicrafts can be viable and even competitive in export markets, given organization, credit-supply, small improvements in techniques, design and marketing assistance. The output and employment of these industries can be substantially expanded. The employment in a number of other rural trades can be stabilized or expanded.

The share of the small-scale sector, excluding household industries, fell from 19.5 per cent of the income arising from all industrial production in 1968 to 16 per cent in 1976. This trend has to be arrested and reversed by a vigorous programme of promotion of small industries.

Government of India Planning Commission, *Draft Five Year Plan 1978-83*, 1978.

 # Kenya's Four Year Development Plan, 1979-83

The majority of the rapidly growing younger population must find employment and self-employment in the rural areas. Hence, the emphasis on rural development in the new Plan is closely tied to the need for employment creation. Idle and underutilized land is a serious waste of resources when many have inadequate land to farm. The Plan specifies means for ensuring better land use and provision of essential services, such as

382

credit, extension, inputs, markets and transport, to small farmers; it addresses the problems and opportunities for those on arid and semi-arid lands; and it presents guide-lines for the pricing of agricultural products intended to ensure a steady advance in the farmer's income *vis-à-vis* those in the non-agricultural sectors.

With respect to agriculture, the Plan focuses on small-scale agriculture and arid and semi-arid lands. Small farms will be given every opportunity to increase their participation in the monetary economy. Among small-scale farmers, greater attention will be given to those who have been lagging behind up to now. Rural access roads will be built so that the flow of inputs to such farms can be expanded and produce can more easily be brought to market; extension services and credit will be increasingly directed to the small farmer; consumer goods and social services will be brought within reasonable distances; water and power will be increasingly extended into the rural areas; and the means and resources devoted to identifying inexpensive and easily repairable technologies which will enhance small farm productivity and ease the problems of off-road transport will be increased.

The development of arid and semi-arid lands will also receive special attention from Government. More than 20 per cent of our population lives in these areas comprising 80 per cent of the land of Kenya. Because of the complexity of the problems — economic, environmental and social — our approach to their development must be integrated, requiring a high degree of co-ordination among Ministries. The Ministry of Agriculture will play an important role in developing programmes and in establishing co-ordinated procedures. While planning will proceed on the basis of regional anylysis considering watershed and agro-climatic areas, implementation will be on a district basis, relying on established administrative systems and ensuring local-level participation in assessing needs and priorities.

But while rural development must feed on and supply agriculture, it cannot be limited to agriculture if it is to be successful. Hence, the Plan will provide incentives for the dispersion of industry, and rural non-farm activities in the informal sector will be accorded high priority. Such activities are even now an important source of rural employment and income, with about 43 per cent of smallholder income being generated in this way. In addition, such activities lend themselves to the use of labour-intensive methods of production. Their production processes are characterized by a low capital-labour ratio. For our economy where capital is scarce and labour abundant this fact is an important consideration. Secondly, small-scale production is often the only means of meeting demand when the size of the market for any given item is small. This is particularly the case for relatively isolated local markets in small towns and rural areas. These enterprises can, therefore, play a useful role in programmes of industrial decentralization. Thirdly, they help in the tapping of resources such as entrepreneurship, capital, and raw materials which otherwise would remain unused. They generally mobilize family or community savings which might have remained idle or been spent on unproductive activities.

Developing technologies for small farm and non-farm activities in the rural areas is also a matter of high priority. It is recognized that the currently available hand tools, like the hoe, are inefficient. On the other hand, imported farming machineries, like tractors, are either inaccessible or inappropriate for the majority of the small-scale farmers. Similarly, rural industries for processing locally available crops and other resources on small scale basis will require technologies appropriate to their scale and location. Hence, research and development institutions will be given the encouragment and resources they require to address themselves effectively to the transfer, adaptation and development of specific appropriate tools and technologies for the rural areas.

An essential ingredient of successful rural development is increased participation in the decision-making process at the district level. Some programmes such as the Rural Access Roads, the Rural Development Fund, and Rural Water Supplies already encourage local-level participation in helping determine needs and priorities. More effort will be made to involve local-level technical staff, elected representatives and members of target groups themselves in programme decision-making. The district is seen as the basic unit for development planning and implementation. In this regard, the District Development Committees will be strengthened and revitalized.

Republic of Kenya Development Plan, 1979-83.

An Unfavourable Environment

Increasing attention on the part of a number of economic specialists has brought to light serious underlying problems in Indonesian Government policies as well as market signals facing entrepreneurs that lead to an anti-employment bias in the choice of technique. There seem to be five basic problem areas.

Firstly, the Indonesian tariff structure, if fully administered, would be seriously distorting. The direct distortion is brought about by allowing duty-free imports of capital goods while levying substantial duties on consumer goods and some raw materials and intermediate goods. Firms are thus encouraged to use capital instead of labour and imported capital equipment rather than domestic equipment. An indirect distortion also exists because duties on some commodities permit a lower price for foreign exchange than would occur otherwise. This hampers exports, promotes imports, and tends to raise the capital intensity of investment. A more general consideration, possibly the most important, is that high tariffs on goods that are also domestically produced allow the economy to become 'inward looking' rather than forcing it to be 'outward looking'. Empirical evidence is accumulating that the latter strategy generates greater growth in output and employment (and in employment relative to output) than the former. These tariffs may not have had a seriously distorting impact on the present structure of the Indonesian economy because they are alleviated by substantial smuggling. But the tariff structure is distorting current decision-making by foreign and domestic firms.

Secondly, the price and availability of credit is subject to distorting factors. The medium/long-term (up to five years) investment programme makes a limited amount of money available at 12 per cent per year, well below other government and private bank rates. One purpose of this programme is to promote investment by Indonesian entrepreneurs who are at a significant disadvantage relative to the Indonesian-Chinese entrepreneurs and foreign investors who frequently have access to capital from outside Indonesia (especially Hong Kong and Singapore) at substantially lower rates than the market rates in Indonesia.

Perhaps not unexpectedly, indigenous entrepreneurs have benefited little from the programme. The below-market-rate credit must be rationed, and it goes either to very good credit risks (mostly non-indigenous entrepreneurs) or is allocated on the basis of personal or political favour. Apart from who gets the credit, the distorting effects of the subsidized rate of interest can be significant. Rice milling is an interesting example. A number of large modern rice mills (by Indonesian, not world standards) have been built under the medium-term investment programme. With a few exceptions they have been built by Chinese or Chinese-backed investors (who were the only investors able to qualify for the 12 per cent money). At the same time, a very rapid investment in much smaller facilities — hullers and small rubber-roll rice mills — has been made by indigenous entrepreneurs paying the market rate for their capital, between 2 per cent and 3 per cent per month. . . . These smaller facilities, much more labour-intensive, are far more appropriate for the Indonesian economy than the larger rice mills. . . . Few of the large mills are able to operate near their planned capacity because the smaller units have outbid them for supplies.

A second major distortion in the credit market is the greater ease in obtaining investment capital relative to working capital. The cheapest working capital available, and then only for firms or individuals able to post 150 per cent of the loan as collateral, is at 24 per cent interest per year. Private rates, e.g. to the milling industry, run about 5 per cent per month. Limits on the availability of working capital seem to be an effective constraint on output for several industries. It follows that raising the amount of working capital available from the banking system for a given loan of investment capital would raise the employment content of any fixed investment significantly.

Thirdly, several aspects of present labour laws and regulations (which are now under extensive government review) tend to convert labour from a variable to a fixed cost and make it easier to use machinery rather than

385

men. Act No. 12 of 1964 prohibits the firing of a single worker without the prior consent of a Regional Committee for the Settlement of Labour Disputes. If ten or more employees are to be laid off, Central Committee authorization is required.

Act No. 1 of 1951 limits working hours to seven hours during the day and six hours at night. An eight-hour day is granted in most cases upon application to the Government, but Act No. 1 discourages multiple shift operations. Rice mills, for instance, cite this law as the reason for not running night shifts during the harvest season. (An additional factor affecting those mills which use electricity is that the night rate for industrial electricity is almost prohibitive.)

Indonesia has a very protective legislation for female employees: paid leave for the first two days of their menstrual period and up to three months paid leave for childbirth.

These laws and regulations are obviously of no consequence in the traditional sectors. But they do have a significant impact on investments in the modern sector, especially for the highly visible foreign and government firms. It is often said but worth repeating that it is the welfare of the total work force that the laws should protect, not just the privileged segment that receives jobs in protected areas of the economy.

Fourthly, indigenous management talent is at least as scarce as capital in Indonesia. There is no good empirical evidence yet on whether capital-saving projects require less management talent than labour-saving projects. In rice milling the management requirements of the small enterprises that are also capital-saving are so low that ample supplies of indigenous managers have been forthcoming. This would almost certainly not have been the case if much larger, more complex facilities had been built. On the other hand, there are no doubt industries where capital can substitute for management, and for these investment in management training is likely to have a high payoff by making it possible to substitute managerial expertise for capital.

There is some feeling that Indonesian management training also introduces an anti-employment bias. The focus of this training is process-orientated rather than organization-orientated. The effect is that Indonesian management trainees know how to run factories with machines but not with people. Since the chief bottleneck to highly labour-intensive projects is usually administrative and organizational, this is a serious problem indeed.

And lastly is the problem of red tape. Foreign investors in Indonesia must contend with complex forms and procedures and be prepared to wait for months or years for processing before their project can be started. If the firm and the total size of the investment are large enough, these essentially fixed-cost barriers to entry are worth crossing. But these firms tend to be the most capital-intensive in structure and oulook (and frequently carry overtones of economic imperialism). The very firms that are most desirable — small labour-intensive electronics, clothing, furniture-making concerns primarily interested in export — have little political impact and so cannot afford to overcome the obstacles mentioned above. This systematic exclusion of smaller firms substantially reduces the labour intensity of foreign investment. The same argument applies, in lesser degree, to domestic investment as well.

C. Peter Timmer in C. Peter Timmer et al., *The Choice of Technology in Developing Countries*,1975.

2 CHANNELS OF TECHNOLOGY DIFFUSION

Dissemination of Appropriate Technologies

The dissemination of conventional Western technologies is a process the modalities of which have been established over several decades. The beneficiaries of these technologies are usually powerful and articulate groups expressing themselves through clear-cut market mechanisms. As a result, commercial enterprises can, through profit-seeking efforts alone, disseminate the technologies quite successfully. In contrast, the dissemination of appropriate technologies is a relatively more recent process and challenge. Also, its prospective beneficiaries are invariably weak and inarticulate sections of society, e.g. the urban and rural poor. These sections can rarely back up their needs with purchasing power, they do not constitute a significant market, and, therefore, the task of responding to their needs cannot be left solely fo industry. Catalytic assistance from external sources is often essential and inescapable. The purpose of this external assistance should be to facilitate the technology implementation process with technological know-how, with credit for equipment and working capital, with input deliveries

and output off-take, with managerial help and training programmes, and entrepreneurial leadership. In addition, the beneficiaries must themselves play an active role if the whole exercise is not to peter out for the lack of popular participation.

It follows, therefore, that the dissemination of appropriate technologies must be based on a multi-institutional effort involving development agencies (either government or voluntary agencies), R & D organizations, industry, financial and credit institutions, input (e.g. raw materials) delivery and product off-take (e.g. marketing) organizations, management and personnel training institutions — and, of course, organizations of the beneficiaries (e.g. co-operatives of the urban or rural poor).

This multi-institutional effort, which is so necessary for the dissemination of appropriate technologies, implies that a host of structures and procedures must be worked out for each appropriate technology. In particular, attention must be focused on the procedures for the procurement of inputs and credit, for the off-take of outputs, and for the management

387

of organizations, training, manpower and finances. In short, an entire hardware and software package must be worked out in detail for each appropriate technology, bearing in mind its specific features. Thus, the package for appropriate road-building technology may be completely different from that for mini-cement plants.

Too often inadequate attention is directed towards the elaboration of these total packages, the general tendency being to assume that if the hardware (machinery, equipment or process) has been developed, the appropriate technology will diffuse under its own steam. In all except a few cases, even this hardware is rarely worked out with the same turn-key, engineered finesse as the technologies of the industrialized countries — in short, the hardware development is rarely thorough. But even when this is the case, successful technology diffusion depends on the elaboration of the software. It is this shortcoming that has proved to be one of the major obstacles to the dissemination of appropriate technologies, and until this inadequacy is overcome, the process is unlikely to gain much momentum.

The insufficient emphasis on the development of the software aspects of appropriate technologies is what may be termed an internal constraint on the successful diffusion of these technologies. In many circumstances, however, it is the external constraints which are of far greater significance. Of these external constraints, the most important one arises from the fact that the partisan vested interests of the elites (or powerful groups within elites) in the dual societies of developing countries are often inimical to the adoption and diffusion of appropriate technologies. In such an unfavourable environment, inadequacies in the software aspects of these technologies are only amplified, and used against them in decision-making. The above discussion of the dissemination of appropriate technologies shows that, though this process must be coupled with that of technology development, there are crucial differences between the two processes. Unfortunately, a blurring of these differences takes place too often, and it is therefore necessary to make them explicit.

Firstly, the agents for the two processes are usually quite different — whereas R & D institutions (at the macro-, meso- and micro-levels) are mainly responsible for technology generation, technology diffusion is usually the responsibility of a development agencey, acting in co-ordination with the people, local self-government organs, R & D institutions, financial and credit institutions, and marketing organizations. Thus, technology generation can be achieved by the sole effort of R & D institutions, but technology diffusion must be a multi-institutional effort. Secondly, the power structure need not necessarily be disturbed by the generation of technology, but it cannot but be affected by technology diffusion. Thirdly, the levels of operation of the two processes are quite different — technology generation can be achieved at the institutional level; technology diffusion must be accomplished at the level of society. . . .

Fourthly, and as a consequence of the above two differences, the technology generation process is much more autonomous than the technology diffusion process, in that, given funds for R & D; sufficient awareness and commitment among those doing R & D; and the absence of direct political hostility towards the R & D, the generation of technology appropriate for the development can be accomplished successfully.

In contrast, technology diffusion cannot be achieved against the wishes of the ruling groups in society. And, where the technologies to be diffused are against the vested interests of the privileged — which in dual societies they often are, if they are indeed technologies appropriate for weaker sections — then the success of the diffusion depends on the particular balance of power between various groups in society. The ruling group is rarely homogenous, and if, within this group, some powerful sections, e.g. the urban elite, are not against the diffusion of appropriate technology for the rural poor, then the process stands a favourable chance. If, on the other hand, all the privileged sections are unitedly opposed to the technology, then the attempt to diffuse it is almost certain to fail; nevertheless the attempt must be made as an essential and integral component of the struggle of the under-privileged and its allies for a more just and equitable society. Thus, a necessary condition for the successful diffusion of technologies appropriate for the urban and rural poor is a large measure of active political support from the rulers of society.

Finally, the role of the people in the two processes is quite different. Though close consultation with the people is vital for obtaining better insights into felt needs, traditional solutions, local conditions, local materials and local skills, and though these insights are quite essential for ensuring the appropriateness of technology, an R & D institution can in fact generate technology without the active participation of the people in the designs, calculations, experiments, fabrications, etc. In other words, appropriate (including socially acceptable) technology is unlikely to be generated by R & D institutions without close consultations with the people, but their active participation in the technology generation *per se* is not necessary. This is not to deny that widespread popular participation can raise the efficiencey and appropriateness of technological innovation to a qualitatively higher level. Such popular participation should therefore be the objective, since an intimate interplay between institutional and popular innovators is an ideal state of affairs.

In contrast, the active participation and involvement of the people is a necessary condition for technology diffusion. These distinctions between technology generation and diffusion, particularly between social consultation and single-institutional work for technology generation as distinct from social participation and multi-institutional work for technology diffusion, lead to some important perspectives and conclusions, with regard to the role and scope for appropriate technology institutions.

For instance, it is clear that institutions of education, science and technology can assume — and successfully discharge — the responsibility of generating technologies. If, however, these institutions also assume the responsibility for diffusion of technology, they must realize that they will have to lead, co-ordinate and manage the concerted action of a large number of institutions, viz., development agencies, local self-government organs, financial and credit institutions, marketing outlets, etc., and that they are almost sure to deviate from their charters of education, science and technology.

Whether they are structured and competent to discharge this onerous responsibility is a moot question. In general, it would be unwise for educational, scientific and technological institutions to assume this responsibility for technology diffusion without being aware of all the implications and consequences. On the other hand, technology-generating institutions must be an essential part of the technology diffusion process — the vital need for their active participation in the process follows logically from the linkage between the technology generation and diffusion processes.

A.K.N.Reddy in A.S.Bhalla (ed), *Towards Global Action for Appropriate Technology*, 1979.

 ## *Social Carriers of Techniques*

A social carrier of a technique is a social entity which chooses and implements a certain technique. The carrier may be a company, an agricultural co-operative or an agricultural extension agency which is introducing new tools or agricultural machinery in a specific context. Individuals sometimes play an important role as introducers of improved techniques; farmers and artisans may serve as social carriers of techniques. For a large-scale industrial technique in an underdeveloped country, the social carrier may be a government agency and a transnational corporation — alone or in different combinations.

For a technique to be chosen and implememted in a specific context or situation, the technique must, of course, actually exist somewhere in the world. But in addition some necessary conditions must be fulfilled:

— A social entity that has an interest in choosing and implementing the technique must exist;
— This entity must be organized to be able to make a decision;
— It must have the necessary social, economic and political power to materialize its interest; i.e. to be able to implement the technique chosen;
— The social entity must have information about the existence of the technique;
— It must have access to the technique in question;
— Finally, it must have, or be able to acquire, the needed knowledge about how to handle the technique.

If the six conditions in the definition are all fulfilled the social entity is an actual social carrier of a technique. Every technique must have an actual social carrier in order to be chosen and implemented. If the six conditions are fulfilled the technique will actually be introduced. In other words, the conditions are not only necessary but, taken together, they are also sufficient for implementation to take place.

If, on the other hand, the social entity has an interest in choosing and implementing a technique but not enough power to materialize that interest, it is only a *potential* social carrier of the technique. . . . In various national and international forums a large number of proposals for suitable techniques are proposed. Everything from solar stoves to cook food and hovercrafts for transporting heavy equipment in the deserts are suggested. Many of these ideas are sound and solid from a technical point of view. There is no lack of ideas in the UN system and elsewhere, but often nothing comes out of those ideas. The problem is that the proposals are not implemented.

Evidently there are a great many 'problems' which can be immediately identified: the technique is not sufficiently tested and adapted to local conditions, it is too risky and no one wants to pay, there is no organization for spreading it or instructing the users, etc. Behind these problems one can, in many cases, identify one single 'obstacle' of a fundamental nature; there is no actual social carrier of the technique.

If a new agricultural technique is to be introduced there must be peasants or peasant organizations which can acquire the inputs needed (machines, seeds, implements, fertilizers, pesticides etc.), organize the labour (own or hired) and distribute the products. If these requirements are not met there is not much use in proposing new techniques. Even minor improvements require carriers. For example, agricultural improvements occurred rapidly after the land reforms in the liberated areas in China during the Sino-Japanese war and afterwards. The reason why this was possible was that the landlords were forced to give up land to the poorest peasants and the small peasants formed mutual aid teams and later small co-operatives. This made it possible to plough more land, to utilize draught animals better, and to improve irrigation, drainage and harvesting. By creating carriers for these simple changes, techniques could be improved and productivity was slowly raised (and the income was more evenly distributed). In a later stage (1957/58) large-scale people's communes were formed out of the advanced co-operatives and an organizational network was created for vastly improved production techniques both in agriculture and in small-scale industry. New social carriers for these more advanced techniques had emerged. Potential carriers had been transformed into actual ones.

Sometimes less drastic social changes may make it possible to introduce new techniques or to improve existing ones. But the basic necessary condition is that an actual social carrier for the technique in question exists or emerges. If, for example, a government wants to raise the technical level in the dispersed, small-scale industries and, in

China's communes have provided an effective route for disseminating technologies

this way, improve the production of simple commodities, it is necessary to support the local, already existing, entrepreneurs and to increase their ranks. Alternatively new organizations to handle the tasks have to emerge as happened in China through the establishment of the people's communes. . . .

It is important to point out that it is not enough for a developing country to import new techniques (by paying for patents, licences, etc.). It is also necessary to organize carriers inside the country for the effective diffusion and use of these new techniquese. This may be relatively simple for large-scale industrial production; managers and technical experts can be imported together with the technique. But for techniques for small-scale production it is necessary to create these managers and technical experts in great number inside the country from its own ranks. And this must be done in opposition to the interests of other influential power groups.

Charles Edquist and Olle Edqvist, *Social Carriers of Techniques,* 1979.

3 THE COMMERCIAL ROUTE: PROMOTING SMALL ENTERPRISES

AT and the Entrepreneur

A residual contempt for the mass market, for the popular product and for commercial success, has subtly affected the Appropriate Technology movement. Some engineers who work on Appropriate Technology projects are not only unaware of how many people have used the things they have designed, but are not even interested in finding out. After all, they argue, the product works, in the laboratory and the field experiments, the manual has been written, what more is there to be done? It might be argued that the marketing task should be carried out by somebody else; the difficulty is that there is nobody else to do it, and unless the initiator of the idea is fully involved in marketing it, it may not be a genuinely appropriate idea at all. In some developing countries, in fact, there seem to be two sorts of small-scale technologies; one kind is found in every town and village vigorously and profitably applied in small enterprises, but the other kind is only to be found in demonstration units and university courtyards. People may fail to make use of the second kind because it is not suitable, since its designers saw their job as developing technologies, not satisfying needs.

Nobody ever looks on Appropriate Technology institutions as money-making affairs, but rather as some form of public service or charity, which must necessarily be subsidized because they could not, and probably should not be expected to be self-supporting. Those who work in such institutions are unlikely to get rich themselves; they are not unnaturally reluctant to help others to do what they themselves have voluntarily or otherwise failed to do, and they prefer to make their ideas available to co-operatives, state enterprises or other 'non-exploitative' forms of business, rather than to private entrepreneurs whose motive is to make money. Again, this would not matter if co-operatives or state enterprises were likely to be as successful as private businesses in using, diffusing and, in fact, 'exploiting' the new technology. Unfortunately, they are not, and the appropriate technologies are forced into fundamentally ineffective marketing channels. It is easy to criticize and to ignore the many individual successes which run counter to the general trend. It is harder to suggest what might be done to improve the marketing of appropriate technologies, so that they are used as well as developed.

There are some lessons to be learnt from three appropriate technology marketing case studies which have been published by Intermediate Technology Publications. These are about the Small Industries Research, Training and Development Organization at the Birla Institute of Technology in India, the Technology Consultancy Centre at Kumasi University, Ghana, and the development of the Egg Tray Machine initially sponsored by ITDG. In these three very different situations, there is one common thread: the technologies have been developed in close collaboration with, and mainly at the instigation of individuals or businesses which stand to make a profit from using them. The student entrepreneurs from the Birla Institute themselves develop the products and processes on which their businesses will be based, at the end of their university course. The successful products of the Technology Consultancy Centre have all been adopted and put forward by business people at a fairly early stage in their development, and the Egg Tray Machine, although the initial idea and funding came from ITDG, was substantially developed by the company which now manufactures and sells it. The last case has a further lesson; of all the machines which have been installed, only the privately managed units, in Zambia and Nigeria, have shown consistent profits, in spite of their being the least favoured in terms of investment incentives and concessionary finance.

In many situations of this sort, private profit-motivated support may only be accepted as a last resort. The entrepreneur, who dares to admit that his intention is first to make a profit and only incidentally to use or to sell a socially appropriate piece of equipment, is all too often mistrusted, rather than regarded as the unsung hero of development. In this, as in so many cases, he or she employs people, saves imports or generates exports, uses local materials and otherwise epitomizes successful development. The main differences between the efforts of the entrepreneur and of most co-operative or public agencies are first that they do achieve their own society's objectives, and second that they do it

A successful example of the marketing of appropriate technology

394

at no cost to the public purse.

Entrepreneurs, unlike co-operators or civil servants, are used to being 'sold to'; they understand and respect others who want to make a profit, and who will appeal to their own motives for doing the same. Agencies whose aim is to promote appropriate technologies to those who are most likely to make effective use of them must therefore be prepared to market and sell their ideas as aggressively as if the survival of their institution depended on their success or failure. The best way of ensuring that they do this, of course, is to try to work with and through organizations whose survival does indeed depend on the adoption of their product. This is not too difficult if the product is a large and sophisticated piece of machinery like the Egg Tray Machine. It can however be done with less expensive equipment, which can be made locally in large quantities. For example, simple agricultural tools have been widely manufactured and sold in the Philippines and elsewhere, because their designers worked with local entrepreneurs from the outset, and devolved responsibility for the later stages of development and manufacture to these enterprises as soon as possible.

Malcolm Harper, *Appropriate Technology*, 1983.

Creating the Right Environment for Small Firms

Small enterprises (defined here as firms employing up to 100 workers) account for more than half the industrial employment and contribute a large proportion of total output in the developing world. Their relative importance tends to vary inversely with the level of development, but their contribution remains significant even in the most advanced economies. In Indonesia, for example, 87 per cent and in Colombia 70 per cent of manufacturing employment were provided by small enterprises in 1975. In Japan, their share of employment and of value added was still as high as 53 per cent and 37 per cent in 1965.

Small-scale enterprise could play a more significant role than it does in creating balanced economic and social development of developing countries. Its potential advantages are well known. Small firms tend to use less capital per worker than their larger counterparts. . . . Their labour-intensive character is consistent with the relative abundance of labour and the shortages of capital and foreign exchange characteristic of developing economies (although wide artificially induced disparities in capital/labour ratios between small and large firms can have negative effects). Small enterprises also have the capability to use capital productively.

Perhaps most important, through that elusive characteristic, entrepreneurship, they make use of resources that may otherwise not be drawn into the development process. They generally employ workers with limited formal training, who then learn skills on the job. They are able to mobilize the small savings of proprietors who would not use the banking system but who will invest in their own firms. Experience with small firms in Indonesia shows that firm owners have a surprisingly high propensity to save and to reinvest, even at quite low income levels. They are able to utilize scattered local raw materials (such as straw and clay) that would otherwise be wasted. They are often effective in subcontracting arrangements with larger firms. And they add flexibility to the industrial structure, by engaging in small batch production and made-to-order or other types of 'finishing' operations complementary to the activities of large-scale industry.

The disappointing results of small enterprise development programmes in many countries show that if such enterprises are to flourish, the right conditions must exist. The failure to create these conditions often goes undocumented because small enterprises are frequently excluded from official statistics. But persistent poverty among those who depend upon small-scale activities for a livelihood provides indirect evidence. . . . In most developing countries small-scale enterprises are either left to their own devices and are unable to realize their full potential because of distorted and overregulated markets, or help is extended to them by governments in ways that are often largely ineffective or may even hinder their growth. Significantly, where small-scale enterprises have flourished, it has been in those countries where governments have allowed markets to operate with a considerable degree of freedom. In such an environment, firms in the same line of business compete with each other on a more equal footing. Just as important, in this environment firms in different fields can form complementary, mutually beneficial relationships. Where interdependence between suppliers and customers is allowed to develop freely, many of the supporting services will be provided by small firms and no longer need to be provided by governments.

It is almost impossible to make valid generalizations about the activities of small enterprises in developing countries, simply because these enterprises are so diverse. They seem to emerge in almost any sector, in any country, in response to unmet needs of the public, where there is ease of entry, an absence of pronounced economies of scale in that line of activity, or a positive advantage in being small.

Yet if they are born easily, their mortality rates are high. To some extent, the failures of small firms are a healthy sign in a dynamic economy: uneconomic enterprises should find it difficult to survive. But in many cases failures, or retarded development, of firms are caused by an unfavourable economic climate. Starved of capital, cold-shouldered by the development finance companies, overlooked in development plans, and, until recently, ignored by most foreign capital aid programmes, small-scale firms have had to rely largely on internally generated funds for expansion and modernization. To compound the problem, their profitability and incentive to invest is often undermined by both overt and hidden subsidies to large-scale industry.

Moreover, the numerous regulations that affect markets in developing

396

countries often seem to encourage small firms to develop uneconomic ratios of capital to labour. Small-scale enterprise is typically labour-intensive, and this is usually a desirable characteristic in developing countries. Some countries have even promoted small firms as proxies for government-sponsored programmes to provide employment. The low capital intensity is often, however, accentuated by government-induced distortions, with undesirable effects on efficiency and income levels.

In many developing countries the markets for capital, foreign exchange and labour are affected by numerous types of regulation, subsidy and taxation. Most small firms have very limited access to foreign exchange and to bank and institutional credit, whereas they can usually draw upon a large reservoir of cheap unemployed or underemployed labour. The result may be capital/labour ratios so low as to result in very low labour productivity in many small firms. Large firms, on the other hand, may be allocated investment funds at negative real interest rates and be provided foreign currency at artificially low exchange rates. This encourages them to be profligate in their use of capital. In addition, trade union pressure and labour legislation (which usually cover only the modern sector) may require large firms to pay relatively high wages which, in turn, may induce excessive substitution of capital for labour.

The net effect can be the worst of both worlds — technically backward production methods and depressed incomes in small firms, where the majority of the labour force is employed, and excessive use of capital in the modern sector, which receives the bulk of the investment funds but creates few jobs. A dualistic structure results, inimical to both efficiency and equity.

While inappropriate regulation may distort their structure, it is also clear that specific government measures of support to small firms may, unless carefully designed, actually discourage or even prevent their efficient operation. Public support for small enterprises is frequently justified, for example, on the basis that they act as a 'breeding ground' for entrepreneurs. This argument needs to be looked at closely, however. Clearly, entrepreneurship is an essential element in economic development: risks must be taken in introducing innovations; businessmen are more likely to take risks, and usually better at gauging them, than public servants. They are able to 'think small' and, as 'economic lubricants', to search for, identify, and fill needs likely to be overlooked by the public sector. Moreover, businessmen are more likely to terminate activities which the public will not pay for.

Nevertheless, certain limitations must be recognized if programmes to assist small firms are to be appropriately designed. Most fundamentally, entrepreneurs are by definition self-reliant, energetic and innovative, and do not generally need to be coddled by promotion programmes. It has, in any event, proven extremely difficult for the public sector to devise programmes that can develop entrpreneurship. Not all — or even most — artisans and petty traders are capable of developing the capacity to innovate by taking the special courses that have sometimes been provided under government auspices.

In general, the skills and aptitudes of most proprietors of small firms are best utilized if concentrated on a narrow range of activities, leaving the more innovative entrepreneurial functions to other organizations such as trading companies or larger firms which then subcontract to smaller ones. As economies develop and their structures change, many independent artisans may, in fact, be absorbed into larger organizations as workers and supervisors. The exceptionally entrepreneurial small businessman will usually survive (and eventually stop being small) if he is left free to operate, subject only to basic ground rules of public safety. . . .

To overcome the shortage of capital available to small enterprises, many governments earmark resources — both from the budget and from foreign assistance — for loans to small-scale enterprises, usually through one or two specialized and subsidized public institutions and subject to relatively low interest rate ceilings. Often the result is excessive demand for this cheap credit and a need for credit rationing, which often leads to abuse in the allocation process. In some countries, cheap foreign currency loans have encouraged the adoption of imported, capital-intensive technology.

Given the relatively high costs and risks of lending to small firms, low interest rate ceilings deter other financial institutions (such as commercial banks), which cannot draw upon subsidized funds, from providing loans to small firms. So the total supply of credit available to small firms tends to contract rather than expand.

Another common measure is to create — often with foreign assistance — a central small-industry institute, to provide services such as technical training, advice and information, management consultancy, legal couselling, common facilities, procurement, product design, quality control, marketing and display, and assistance in obtaining loans from financial institutions. These services are usually supplied either free or at a nominal charge to the recipients.

Experience with such institutions has often been disappointing. There have been numerous obstacles in practice. The heterogeneity of small enterprises has made it impossible to encompass the range of expertise they need in a single institution. Salaries have usually been too low to attract or hold experienced staff. The rapport between staff and client has often been weak, stemming from the civil servant's traditional distrust of the 'free-wheeling' entrepreneur and the businessman's fear of a government official prying too closely into his affairs.

A third approach has been to introduce selective controls to protect small firms from competition. Such controls are difficult to design well and have sometimes backfired. For example, in some countries licences for the production of products such as cotton textiles are confined to small firms using traditional handicraft techniques; technological development and adaptation have thus been discouraged, and consumer demand has switched to more competitively priced, or more 'modern', synthetic textiles made in medium- and large-scale industry not subject to controls. Incomes in the protected sectors have consequently declined or remained stagnant. Similarly, attempts to protect the livelihood of taxi and rickshaw operators or small traders by restricting the number of licences issued have sometimes led — through the corrruption of public officials — to these activities being controlled by larger operators or by particular ethnic groups.

Keith Marsden, *Finance and Development*, 1981.

Small is Appropriate

Appropriate technology needs its own appropriate institutional software. The cheapest and quickest way of spreading appropriate technology widely is to encourage small-scale enterprise, which is rapidly joining basic health care, participation and redistribution as one of the new received ideas among the development agencies.

Small-scale enterprises use capital much more efficiently in two key objectives of development: employment creation and expanding production. In a survey of four countries, large-scale enterprises had between 3.9 times (Mexico) and 8.8 times (India) as much capital in fixed assets per job than did small-scale. In some branches, the discrepancy can be considerable. In India, small-scale textile firms had $1631 of capital per employee against $18,130 for large, while in iron and steel the ratio was $2,522 to $39,917. In other words, for the same investment, small enterprises create many times more jobs. Moreover, small-scale enterprises tend to employ more unskilled workers — two-thirds of their labour force, against half for large firms — hence providing more work for the urban poor, who have few or no educational qualifications.

More unexpectedly, small firms achieved a higher return on capital. The same four-country survey showed that small firms produced between 80 and 300 per cent more output per unit of fixed capital than large firms. Large firms usually achieve a higher productivity per worker — but not always. In Indian sawmills and car repair, the value added per employee is actually higher in small than in large firms.

Indirectly, too, small firms create more employment, because thay are far more likely to use local raw materials and other inputs, while large firms are more prone to import, creating jobs in other countries (usually Western). Small firms in other words create a more egalitarian style of development by spreading the benefits more widely. Small firms in Colombia used only 11 per cent of imports in their inputs, while large firms used 18 per cent. Of the large firms' fixed investment 45 per cent was accounted for by imports, against only 24 per cent for small firms.

There are other advantages in smallness. Small firms need less infrastructure (power, water, roads, etc.) and so do not need to be concentrated in the biggest cities. They provide a cheap on-the-job school for vocational and entrepreneurial skills. They encourage and mobilize family savings among the poorer groups.

Small-scale enterprises employ a considerable part of the workforce in developing countries — between one-third, in the most industrialized nations of Southeast Asia and Latin America, and four-fifths in South Asia and Africa. Yet the majority of small firms belong to the so-called informal or unorganized sector, operating largely outside official regulations, and almost totally without official help. Government policy usually favours large firms, giving them cheap credit (hence encouraging them to be capital intensive), access to foreign exchange and import licences (hence encouraging capital imports) and protecting their often uneconomical production with tariff walls ranging up to several hundred per cent. By contrast, small firms are discriminated against. They find it difficult to get official credit. An International Labour Office survey in Africa found that only 4 per cent of unorganized sector enterprises in Kumasi (Ghana) has official credit, and only 2 per cent in Freetown (Sierra Leone). Small firms have no political pull and cannot get hold of foreign exchange or import licences. In most cases they are located in slum or squatter areas and may be without water and power and sometimes even a roof over their heads. Some governments and municipal authorities, using inappropriate Western regulations, actively harass them, demolishing their premises, demanding licences they cannot afford and documentation they have not the education to prepare.

Yet precisely because of their smallness, small firms tend to need more official help than large. They lack the manpower or the finance for separate departments dealing with accounting, engineering, design, quality control, marketing, and in all these fields they tend to be deficient, and less able to compete with the giants.

Fortunately, the climate of opinion is now changing in favour of smallness. The world's largest development finance organization, the World Bank, has been lending for small-scale enterprises since 1972, though initially only on a very modest scale — in the five years to 1976, out of $2,200 million loaned to development finance companies (national bodies which then relend the money to companies) only $100 million was earmarked for small-scale enterprises. In September 1975 the Bank's president spoke of the need to alleviate urban poverty and the importance of helping informal and small-scale enterprises. Since 1976, the Bank's lending for small-scale enterprises has increased considerably, and it is now official Bank policy to foster appropriate technology, to encourage governments to correct policies that encourage undue capital intensity (for example, too low interest rates or currencies overvalued) and hamper small industry.

Recent Bank loans to development finance companies in Cameroon, Colombia, India and the Philippines have earmarked significant sums to be reserved for small firms. Large bank-financed urban development projects in Madras and Calcutta have made sure that premises are provided for small workshops, and set aside sums to finance loans to small entrepreneurs.

Several developing countries now have programmes to encourage small-scale industry, . . . but the country that has the largest and most ambitious small-scale enterprise programme is undoubtedly India, which also has the world's biggest employment problem. The Sixth Five Year Plan outlines its dimensions. In 1978, out of a total labour force of 265 million, unemployment and underemployment amounted to the equivalent of 20.6 million full-time unemployed. Every year 6 million new workers are entering the labour market. The task ahead then is to create 50 million new jobs by 1983. There is no hope, the plan states, of large-scale industry absorbing more than a tiny fraction of the total. Organized factories employed only 4.8 million people in 1971 — only 22 per cent of the total in manufacturing and repair. Employment in the modern sector tends to grow much more slowly than investment or output from the sector: between 1961 and 1976, investment rose by 139 per cent and output by 161 per cent, but employment grew by only 71 per cent. Large-scale industry, in other words, was getting more capital-intensive all the time.

Of the total extra jobs, India plans to create around 23 million in agriculture through expanded irrigation facilities; 17 million in services, including labour-intensive construction of infrastructure (roads, etc); and 9.4 million would be in small-scale and village industries.

India's small-scale enterprise sector was already booming. In 1977, the average firm employed eleven workers, with a capital of around $10,000. In the five years from 1972, the number of small firms nearly doubled, from 140,000 to 269,000, while the value of their production trebled. In 1976, though they represented only one-tenth of the capital investment in industry, they provided two-fifths of the jobs and two-fifths of the production: in other words, for one unit of scarce capital they created six times as many jobs and six times as much output as large-scale units.

In December 1977 India's industry minister tabled a statement on industrial policy which was nothing short of revolutionary. It proposed that henceforth employment in consumer goods industries would grow faster than production — reversing the previous trend. In other words, industry would grow more labour-intensive, and use machinery producing less output per worker, than before. This would be brought about in large part by reversing the traditional discrimination in favour of large-scale firms. In future, official credit would be reserved for small-scale and village industry, while large-scale firms would be expected to finance expansion from their own profits. The move to smallness would be speeded up by encouraging large firms — including publicly owned ones — to hive off parts of their business as independent ancillary units run by small entrepreneurs.

Still more drastic, a long list of products . . . was reserved for production in small-scale enterprises, and no large-scale firm would be allowed to expand its capacity in these goods. The practice of reservation began in 1967, when 46 items were listed. By 1976 this had grown to 180 products, but in 1977 another 324 were added. The reserved list includes a considerable number of fairly sophisticated products, including, for example, most laboratory chemicals from ammonium carbonate to zinc oxide, plastic compression moulded products, diesel engines up to 15hp, electric motors up to 10hp, TV antennae, TV games, digital clocks and cheap radios. All the products have been chosen because they are suited to small-scale production — for example, they may involve labour-intensive assembly or production in small batches. As an additional booster, the largest single buyer in India, the government, has a separate list of 241 products which it will purchase only from the small-scale sector. In other products it will accept tenders from small firms preferentially, as long as their price is not more than 15 per cent above the cheapest large-scale tender. Reservation of products acknowledges an important factor that most government programmes ignore: small-scale industry is highly vulnerable to competition from large firms, who enjoy all the economies of scale in production and marketing. If they are to prosper, they need protection against the big boys in the same way the budding national industries need tariff protection against foreign competition.

But protection can easily turn into mollycoddling of unsuccessful firms producing over-costly products, unless it is paralleled by technical assistance aimed at boosting efficiency. To provide its positive help to small-scale industry India has had since 1954 a Small Industries Development Organization (SIDO). SIDO's most visible function is to run a network of Small Industry Service Institutes throughout the country. Typical of the help these offer is the work of the Institute for Haryana State and Delhi, on the southern outskirts of the capital. Its 220 staff run training courses for managers of small firms in a whole range of subjects including production, finance, cost control, personnel management, work study, marketing and market research. They run the Government's adventurous Entrepreneurship Development Scheme, which trains unemployed engineering graduates, helps them plan and set up a viable industrial project and provides them with unsecured loans of up to £12,500 to start up in business.

They provide management consultancy, including feasibility studies and technical consultancy, advising on available and suitable machinery, liaising with suppliers, where necessary designing tools, machines and even products from scratch. It gives help in improving quality control and matching products with Indian or foreign standards, providing facilities for chemical and electrical testing. Finally, it has its own workshops for specialized processes that individual small firms cannot afford to do themselves: heat treatment, lens finishing, making of sophisticated tools and dies requiring expensive equipment to produce. These last activities are the only ones for which the Institute makes any charge.

Until this year, assistance to small industry was still fragmented in India. Besides the Service Institutes, a small entrepreneur seeking machinery or raw materials would have to go to the National Small Industries Corporation. For credit he would go to the state banks, for help with marketing to the State Trading Corporations. This was complex and time consuming, and a small business might easily miss out on some important service. Earlier this year India started up its first District Industry Centres, of which 200 are planned by the end of 1979. These will provide all the services needed by small industries under a single roof; including all the functions of the Small Industry Service Institutes.

To these they will add one important new function which could radically alter the impact of research into technology. Based on their assessment of the technical needs of their client firms, they will contract out research and development work to local higher education institutes. This will ensure that research into appropriate technology — so often conducted in an economic vacuum — will tie in very firmly with the real needs of practising entrepreneurs. This type of liaison is so important that it ought to be emulated by every developing country.

India's success has had other ingredients. Industry has an ancient pedigree here, and India was the world's largest exporter of textiles until Britain came along and dismantled her cottage industry. Her education system was developed very early on to meet the needs for administrators of the British colony, and has a hundred year start on many African countries. As a result, India now has the world's third largest body of trained technical and scientific personnel. Her fiercely protectionist tariff barriers have developed in close harmony with her abilities to make products at home.

But her experience does show that small-scale entrepreses, quite apart from the social advantages outlined above, can and must form the basis of any truly indigenous technological revolution in the Third World. The technological development of the major Western industrial powers started small and built the large on that foundation.

Paul Harrison, *New Scientist*, 1979.

4 THE EXTENSION ROUTE

Asking Appropriate Questions

There is considerable debate about the nature of the dissemination process. Technologists tend to argue that the problems are of a purely institutional nature and that there is nothing inherently wrong with the hardware devices themselves. Others have argued that the difficulties are often technical as well as institutional, and that it is the tendency to try separating these two inter-related aspects of technology projects which contribute to their lack of impact. In the absence of concrete data based on experiences with introducing and using new devices, it has always been difficult to pin-point the nature and causes of the problems encountered in disseminating technologies to rural families. However, several recent studies aimed at investigating this issue give some useful indications of the kind of problems being encountered and raise many questions which should be asked by anyone involved in technology projects.

In an article based on work in Upper Volta, the point is made that there is no universal 'intermediate' level of technology, and that attempts to solve villagers' problems by developing small-scale low-cost devices (in the belief that all such devices are automatically appropriate) can often be counter-productive. This is well-illustrated by the example of a small peanut oil-press which was designed specifically to augment the earnings of rural women through the increased production and sale of peanut oil. Although small, low-cost and locally developed, this 'improved' device proved to be more time consuming and less profitable than the traditional method. Besides doubling the processing time and reducing profits by changing the nature of the by-products, the new press could only take about 2 kilos of nuts at a time — too small a quantity to make the time and effort worthwhile. The press's main benefit was to eliminate the need for skimming oil, but this is one of the easier steps in the traditional process. By contrast, some women had already identified ways of overcoming more difficult and time-consuming processes; for example, some were taking nuts to a local grain mill to be ground into paste. The grinding fee paid by the women was more than covered by their extra sales. Despite its larger size and greater cost, in the circumstances the grinding mill turned out to be a much better example of an appropriate technology than did the oil-press.

Similar problems were identified in a recent evaluation study of small palm oil-presses introduced elsewhere in West Africa. Again, although developed locally with the specific aim of helping women to earn more income from a traditional industry, the presses failed to gain acceptance because they were too small, saved no time and were less profitable than the traditional method. One design presented particular problems in that it made it necessary for the women to separate manually the nuts from the pounded fruit before pressing. This extremely tedious and time-consuming operation was not traditionally required and was an obvious factor limiting the acceptance of a device introduced partly on the basis of being effort- and time-saving. Many similar instances describing the rejection of new devices have been recorded; women have failed to see the benefit of purchasing hand-held maize shellers (however cheap) when they can shell maize more quickly with their bare hands; they have rejected improved stoves which require finely chopped wood, creating rather than saving work for women.

The message that comes across here is fairly clear. Any new device which results in more work without adequate economic return, or involves additional expenditure without measurable benefits is unlikely to gain acceptance with rural people. That such devices are developed in the first place seems to indicate a lack of understanding of the traditional way of doing things and a lack of appreciation of the wider implications of technical change. This would be less likely to happen if engineers and rural development staff worked more closely together on the collection at village level of data aimed specifically at identifying where problems exist and at specifying the nature of the technology package best able to overcome them. Strong as the arguments are for making profits or saving time, there are many instances recorded of the rejection by women of technologies which apparently meet these criteria. In these cases, the problem has often been one of lack of social and cultural acceptability. For example, a study of experiences in introducing new technologies in Ghana revealed that accustomed practices, tastes and beliefs were crucial variables in explaining the rejection of several technically efficient and apparently profitable technologies. An improved fish-smoking technique was spurned because it changed the texture and taste of the fish; a new hybrid maize was rejected because its taste differed from that of the local variety and proved unsuitable for the preparation of traditional dishes; and bullock ploughs were left unused because the farmers were women, and local customs forbade then from touching cattle.

Practices and customs vary considerably between and even within countries and a technology which is appropriate in one place may be less so somewhere else. Emotions tend to run high when projects run up against this kind of obstacle, with technologists complaining about the conservatism of rural people. However, traditions cannot be changed overnight and, in any case, it should be remembered that one of the basic premises of the appropriate technology movement is that technologies should be adapted to the needs of people, rather than people having to adapt to them. A more reasonable approach in situations of this kind would be for those charged with the responsibility of identifying, developing and introducing technological improvements to consult or work with those who have a thorough knowledge of local customs and beliefs.

A variation on the same theme is the failure of technologies which, although technically sound and culturally acceptable, relate to a low priority need as far as rural people are concerned. For example, a study to investigate the lack of co-operation in a self-help water supply scheme in Ethiopia revealed that although women thought that clean water was important, they were less concerned about this than with acquiring access roads and better marketing facilities.

Resources that are wasted in providing rural people with technologies they don't want could be far better utilized on pre-feasibility studies to help identify the priority needs of target groups and to plan the type of technological intervention best capable of meeting these needs. Developing technology packages in accordance with people's priorities makes more sense than introducing a new technology (for reasons which may be unrelated to village needs) and then implementing other interventions to persuade people to use it.

When the target group is rural women, and particularly in areas where it is difficult for men to communicate directly with women, strategies such as employing female interviewers or working through existing women's programmes have normally succeeded in identifying the women's needs; their husbands' interpretations of their needs often give a very biased view of the real situation.

Having sorted out problems of profitability, acceptability and need, there still remains the all-important question of accessibility. A study in Sierra Leone of the factors affecting uptake by women of technologies relating to their tasks revealed that the most crucial variable was one of access; evidence from many other countries supports this finding.

In many cases, rural women are completely unaware of the existence of improved technologies which could help them. When information does filter down to the village level, it is usually men who receive it. This is because most extension workers in rural areas are men who, by choice or custom, tend to communicate only with other men, even if the information relates to work carried out by the women. When the women do learn of the existence of certain technologies further obstacles are placed in their way. They are often denied access to credit facilities because the land and buildings which are needed as collateral are held in the men's names. They rarely have access to advice on how to form themselves into a co-operative so as to secure a loan through a co-operative scheme. Many rural women could not afford to purchase (even collectively) a grinding mill, an oil-press or a pedal thresher without the help of a loan, and their husbands may see little point in utilizing credit facilities to acquire such devices when there is plenty of 'free' female labour available to do the work. Basically, what all of this adds up to is that in most countries there is a lack of purchasing power for many of the technologies which could be used by women to promote rural development.

The appropriate technology movement is frequently criticized for concentrating on technology in its narrow sense — meaning tools, equipment and techniques — and for neglecting the system for acquisition, transmission and control which properly forms part of the definition of technology. If one of its objectives is to get improved technologies to the people who need them, the evidence reveals the importance of concentrating much more on entire hardware and software packages, rather than thinking solely in terms of technology hardware.

Technologists and technology centres obviously have a role to play in this, and that role consists of far more than simply assuming that participation ends with getting the device right and leaving dissemination to the rural extension channels. The experiences of many technology centres with modern small-scale industry schemes reveal that the processes of identification, adaptation, development and dissemination are inextricably linked; in the absence of adequate inputs from training, credit and advisory agencies, many technology centres have, of necessity, become fully involved in the process of securing financial and other assistance for their clients, and in providing necessary training and continuing advice.

Marilyn Carr, *Appropriate Technology*, 1982.

Necessity never made a good bargain. — Benjamin Franklin

The Ten Commandments of Project Planning

If the availability of technology were the only issue, there would be little reason for concern. But a new system must 'fit' socially and economically as well as technically if it is to do more good than harm. It can be difficult for us to judge this fit, since we and the poor inhabit such different energy worlds. For us, energy means electricity and petroleum for running our cars, stoves and air conditioners. For them, energy is mostly firewood and charcoal for cooking, along with muscle-power for planting, hauling water and grinding grain. If we are to help, we must somehow bridge this experiential divide. In so doing, we will do well to observe the analytical 'commandments' now set forth.

Commandment 1: First, Do No Harm
It is sometimes argued that development is necessarily disruptive, and that we should not worry unduly about our contributions to the process. The argument is a bit suspect on the face of it, since it is always we who disrupt and the poor who are disrupted. Moreover, disruption is often harmful by any measure. . . .

Commandment 2: Be Reality-Led, Not Technology-Driven
To be more useful than harmful requires us to understand the environment in which we are working. Our line of inquiry should be this: how do we open channels of communication between us and the poor? Given what we can find of their sense of needs and priorities, is there anything we can do to help?

Commandment 3: Filter Your Data Well
One important way of measuring reality is the survey, a data-gathering exercise which can help lead us to correct action at the local level. . . . Central to this process are the analytical 'filters' through which we pass raw data to determine whether systems being considered are socially and economically appropriate. For example, the correct economic filter is benefit-cost analysis, which allows us to judge whether we will get as much out of a system as we put into it. Since different analytical filters require different kinds of data, no survey should be set in motion until we know which filters will be used. Only then can we know which data we need to collect.

Commandment 4: Find Net Present Value
The best model we have for the way individuals view economic decisions is benefit-cost analysis. Such analysis rests on two intuitively obvious assumptions. First, we take for granted that people who buy a system will expect to receive more resources from it (benefits) than they put into it (costs). Second, we assume that people will attribute different values to given flows of benefits (or costs) depending on when these take place: today, next year, or ten years

from now, with future values discounted because present values are closer and more certain.

Properly handled, information on these points tells us the system's 'net present value' to the individual. If net present value is positive, the system is worth its cost; if net present value is negative, it is not. With appropriate adjustments, the same approach can be used to find the system's net present value from the perspective of society as a whole.

Although benefit-cost analysis is widely appled in the modern sectors of developing countries, this experience can be transferred only with the most extreme care to the poor. Point by point in our analysis, we must work to un-fold their reality, rather than express our own. . . .

Our tendency in estimating these values is to bend them in favour of the technologies under review. In the absence of local data, for example, we tend to extract numbers from our own reality for purposes of analysis. In key instances (benefits, discount rates, operating life) our numbers will be syste-matically more favourable to new technologies than would local values. The problem is compounded by the fact that the people carrying out such analysis (manufacturers, consultants, energy bureaucrats) often have a strong vested interest in showing that the systems they prefer are sound. Together, these distorting forces can easily encourage us to support systems with no prospects for success in the real world.

Commandment 5: Set your Calendar to Local Time
When we travel across the world, we adjust our watches to local time. When we try to cross the psychic distances between us and the poor, we must do the same to our calendars. The issue here is not what year it is, but how far away the future seems. For us, next year may appear imminent, the year after close behind. The consciousness of an impoverished villager, on the other hand, is likely to be focused more closely than our own on a precarious present, the years ahead receding rapidly from view.

This matters most when we try to decide how heavily the potential buyers of a technology system will 'discount' the portion of its benefits which will only be gained in future years. For example, we may assume that buyers con-sider having $1 a year from now to be the same as having $0.91 today. This implies a 'discount' rate of 10 per cent, a figure often used in benefit-cost analysis because it is a round number somewhere within the range of discount rates we seem to apply to our own economic decisions. The poor, however, may have discount rates far higher than this. For example, $1 next year might be viewed as having the same value as only $0.77 (or less) in hand today. This would imply a local discount rate of 30 per cent (or more).

Although the correct discount rates for the poor are not easy to establish, economic theory suggests they will be roughly equivalent to interest rates on local loans. Outside of urban areas, the relevant rates would be those charged by nonofficial sources of credit, like merchants and money lenders. . . . A reality-led approach will start by finding what discount rates are among the poor. While a wide variety of systems may prove justifiable using these rates, we are likely to be forced into considering systems much cheaper and more labour-intensive than those to which we are accustomed in our own world.

Commandment 6: Master the Unquantifiable
Technology systems may have effects which are not readily measured or expressed in numbers. Examples might include ecology and health. A renewable energy system which reduces demand for firewood will save trees. This will keep watersheds healthy, which will conserve topsoil, which will help maintain agricultural productivity. Where dung is used for cooking, improved stoves will leave more dung in the fields, allowing nutrients to be restored to the soil. Although such effects may be an important aspect of energy systems, they are very hard to isolate and weigh. Improved stoves often incorporate chimneys, meaning that smoke need not accumulate inside the home. Especially where dung is used for cooking, the smoke-free environment which results may significantly reduce respiratory and eye disease. If human wastes were fed into biogas systems, pathogens (germs) would be destroyed which could otherwise pose a threat to public health. . . .

Where unmeasurable benefits are real, benefit-cost analysis can at least indicate how great their value would have to be for a system to be worthwhile. For example, consider a hypothetical $60 stove which burns dung more efficiently and vents the smoke outside the dwelling. The stove's most obvious benefit from the investor's point of view might be labour worth $15 annually which is freed from dung collection. An added, unquantifiable benefit would be improvements in health from no longer breathing dung smoke. Assuming a local discount rate of 30 per cent, it involves a relatively simple calculation to find that health benefits would have to equal at least $4 annually for the stove to have positive present value. The potential buyers can then be asked whether clearer eyes and lungs are worth more or less than $4 a year.

The approach we select for dealing with a system's unmeasurable qualities will vary both with the system and with the place where it is to be tried. Nonetheless, there will almost always be some way in which the unmeasurable can be mastered — or at least better understood — for purposes of analysis.

Commandment 7: Be Replicable
A single technology system — a pump, a stove, a woodlot or a solar cell array — will in itself make but a trivial contribution to a country's needs. It is only when a system is adopted by the thousands (or hundreds of thousands) that significant impact will be felt. From the earliest stages of conceiving and testing such systems, we must therefore consider what would happen if they were widely replicated. There are two major tests of replicability, net present value and claims on national resources. An energy system should produce more resources over its lifetime than it consumes — that is, its net present value shoud be positive. When this criterion is met, the more of such systems the better. If the system fails this test, on the other hand, a programme to promote its use will magnify the net present harm it does to society. At the pilot

stage, the testing of almost any new system will require an unsustainably heavy infusion of human and financial resources. Beyond this stage, however, systems will differ greatly in the amounts of personnel and money which governments would have to provide to support their use. In some cases, claims on scarce resources would simply be too great to consider, whatever the system's apparent advantages.

These principles are fairly obvious — and commonly ignored in development practice. In terms of claims on national resources, many pilot woodlots demonstrate how not to be replicable. At least until very recently, a common model for village woodlots involved ten hectares of land, a process for planting seedlings there, and a barbed wire fence to keep out the goats. Unfortunately, to extend this model broadly enough to meet the fuelwood needs of a representative developing country could cost so much for the barbed wire that it would not be replicable in terms of available resources, a point that was often neglected until demonstration woodlots of this sort had been planted around the world.

Commandment 8: Provide No Subsidies
Subsidies are the opiate of the technology programmer. It is possible to sustain a bad system almost indefinitely with subsidies sufficiently large that the system need never be judged on its merits. . . .

Formally speaking, we can outline conditions where subsidies for a technology system would be warranted. The criteria are: the system must have positive economic benefits for society as a whole, and these must be greater than benefits provided by competing systems; the system must depend on subsidies to lower its cost (or increase its returns) in order to appeal to investors; and

It's what happens afterwards that matters

410

the necessary subsidies must be more attractive to the government involved than spending the same money on other development activities.

In the real world, however, almost no technology system will meet all these tests. If the system has value for the society, it is likely also to appeal to private investors. If it does not sell commercially, it will probably be also unsound from the national point of view. There may be instances where subsidies are warranted, but these are very, very rare.

None of this is to argue that governments should provide no support at all for new technology systems. 'Provide no subsidies' simply means that the systems which are the focus of all this activity should be expected quickly to stand on their own, without the reality-evading props that subsidies provide.

Nor does anything here argue against the provision of credit for the purchase of energy systems. However, credit should not be a disguised form of subsidy. The World Bank estimates that 'real' costs of providing rural credit (adjusted to eliminate inflation) may be 22 per cent annually even for efficient lending institutions. There is no reason whatever that potential investors in new systems should not be expected to cover these costs. Credit should simply be a way of enabling capital-poor investors to buy inherently attractive devices; it should not be used to give false sheen to the nonviable. . . .

Commandment 9: Keep Track of What You Do
A great many technologies are now being tested in developing countries. Unfortunately, such tests almost always address strictly technical questions: do the woodlots grow wood? do the pumps pump? do the grinders grind? . . . It is not enough to know if our systems work, however. We need to find if they make sense to users, and that is more complex. A system makes economic sense if it has positive net present value. To determine whether this criterion is met, we need various kinds of local information, ranging from wage rates and unemployment levels to materials costs and interest rates. As necessary, we can adjust test data to suggest what costs and benefits would be if the system were more widely distributed. If it is not to be shed when the technicians depart, a system should be in rough accord with local sex roles, land tenure patterns, forms of co-operation and authority, work habits and the like. The implementing ministry for the test should be one with established extension links to the poor. By involving trained extension workers, we will help ensure that the right information is fed into the monitoring process. We will also find whether the ministry has the technical skills which would be required of it if the decision was ultimately made to spread the system generally through the land.

Real information on these matters can be acquired only if the system being tested is woven fully into the lives of real people. This requires that people, government and (where relevant) donor agree in advance on key elements of the test. . . . Once the experiment's basic structure is established, progress should be closely monitored. . . . As the experiment proceeds, it may prove desirable to redesign various elements of the system, adjust the ways in which it is being managed, correct inequities in the distribution of benefits, or otherwise respond to the unforeseen. Monitoring can ensure that problems are caught early, that responses are based on current local information, and that the results are recorded. More than incidentally, the process also ensures that all the parties concerned are involved in doing these things.

The monitoring process may seem too demanding for government officials already stretched thinly across their work. . . . But if we are unable to keep track of what we do, we are better off doing nothing.

Commandment 10: Think Small

Scattered throughout the earlier commandments is the admonition to be 'reality-led'. The relevant realities, we might add, are seen only when our vision is narrowly focused. To 'filter your data well', for example, requires us to understand the specific outlook of the particular people we survey. Similarly, to 'set your calendar to local time' or to 'master the unquantifiable' implies a very concrete sense of local realities. Large thoughts may follow about policies and programmes, but such thoughts will be correct only if their origins are small.

We should also think small in terms of the energy technologies we support. When these devices are analyzed using numbers from the real world, there appears to be a direct relationship between their simplicity and their economic merit for the poor. Small systems also seem to have greater developmental impact, in terms of those unquantifiables which register an area's capacity for healthy future growth. Complex devices will sometimes be appropriate, of course. Nonetheless, relative to a technology-driven view of things, a reality-led spectrum of energy systems will be weighted much more heavily toward the simple and the cheap.

David French, *Development Digest*, 1983.

 # A Successful Methodology from India

A technology programme must consider:

— *Product preference.* Manufactured products must be acceptable to the society in and for which they are produced. Production must be able to adapt to changing needs and tastes;

— *Technology.* Plant, machinery and processes must operate on as small a scale as possible consonant with keeping product quality and cost competitive with large-scale production;

— *Organization.* Ownership, financing and marketing must be organized so that surplus capital remains available at the lowest level and can be used to further local development.

Unless these factors are well integrated, the chances of the programme's success are slight. An appropriate technology for producing cloth, for example, may be available, but unless the quality of the cloth is acceptable to the society, the technology is useless. On the other hand, an appropriate technology may not work if suitable entrepreneurs and organizations to own and use it are not available in a particular location. The Appropriate Technology Development Association (ATDA) has developed a pilot-project plan which takes into account these factors: proper product selection; specification of technically feasible and economically viable production processes; and identification of the appropriate type of organization for ownership and operation.

ATDA methodology for developing such technology is as follows:

Stage I: Survey and analytical studies. Surveys and studies identify areas in which appropriate technology can be developed and tested on the basis of a pilot project. The studies explore the potential of the technology; define its present status in the country (both as a large-scale modern industry and as an indigenous, local or artisan-type of industry); recommend improvements and developments to scale up artisan or indigenous units or to scale down large units; identify technological gaps and other problems to be solved; and identify factors which would ensure the success of a smaller unit.

The studies reveal the state of the technology in question. The data and information collected are also useful to other research workers even if they do not justify recommending establishment of a pilot project.

Stage II: The Pilot project. The pilot-project plan covers these points: background need; objectives; product specification; organizational pattern; technology (including sources and availability of operational advice, process details, machinery specifications, manpower needs, capital cost structure and total funds required, operational details); and timing and extension methodology.

Pilot-projects proposals are published and circulated to interested agencies and organizations and may lead to organization and operation of the pilot unit by one or several collaborating agencies within the context of existing operations or organization and operation of the pilot unit as a new enterprise under realistic field conditions.

Stage III: Operational and development planning. Depending on the results of the pilot project, the decision may be taken to design integrated plants; to offer them to entrepreneurs guaranteeing the availability of technological information and assistance on installation, erection and operation; begin manufacture of machinery by making drawings and designs available to technical personnel and machinery manufacturers; publicize the technology by publishing technical and operational reports, by holding seminars, and by arranging technical training programmes. Reports on the results of pilot-plant operations are published as case studies.

Stage IV: Research. Large-scale industry is supported by intensive R & D efforts. If appropriate technologies are to remain competitive, they also require support through R& D. ATDA therefore attempts to encourage continuing and dynamic R & D on appropriate technology. Accordingly, in collaboration with owners, on a cost-sharing basis, it studies ways and means to promote higher efficiency. It refers fundamental or long-term problems to universities and research institutes for solution. It helps to plan and set up independent R & D institutions concerned with the problems of appropriate technology. By these means ATDA hopes to ensure that small-scale technologies will continue to develop and remain competitive with the technological advances of large-scale industry.

M. K. Garg, UNIDO, *Monographs on Appropriate Industrial Technology,* 1979.

 # The Organization of an Appropriate Storage Technology Programme

The returns to a programme designed to popularize improvements to traditional structures depends ultimately upon the rate of diffusion of the technology through the government extension services. However, the type of extension required in the case of improvements to traditional structures is a more complex skill than that required with a programme to popularize metal bins. It involves the organization of village artisans to work in collaboration with farmers, who must provide the raw materials and access to their store.

The use of non-market inputs requires that the extension officer motivates his target group to provide these inputs itself. This has to be timed for a period when the existing store is empty to permit the improvement to be carried out. But to popularize the metal bin requires demonstration of an easier kind. It is an attractive product, and the major constraint to its use is transport. Demand for bins at harvest time at the market, when farmers have both cash in hand and transport available, is high despite the economics involved. Possession of a modern manufactured product is a mark of affluence — which we all recognize in some form or another — and this provides an important motivation to store-owners to purchase a metal bin. The monetary involvement is small compared with many crop production expenses. Sales figures in the pilot project indicate that the extension services have enjoyed some success in marketing the metal bin to more affluent farmers. No choice of product has been available and the increased economic benefits described by the extension agents through purchasing the bins, if not illusory, are at least not well-founded on thorough research results. It is quite probable that the attractions of the product listed above and the selling techniques employed are more relevant in explaining sales figures than actual economic returns. This does not in any way reflect upon the rationality of the purchase, rather it demonstrates that a narrow economic rationale is not the only determinant of the human process of decision-taking.

On the part of the policy-maker, the selection of a programme based on improvement to traditional stores limits the impetus of the programme: e.g. there are no improvements for many indoor stores, either because the cost of providing them is prohibitive or, more commonly, because they are simply not possible without structural alteration to the farmer's house. This sort of limitation is almost inevitable where situation-specific technologies are concerned. The metal bin can, however, be adopted without extra costs.

The organization of the improvements to traditional stores causes more upheaval within an extension department. It requires a greater knowledge of the causes of storage losses on the part of the extension officer — it necessitates provision of education for knowledge of both storage science and construction techniques. The continuing motivation of individuals to learn these skills and apply them is difficult when easier tasks are available. Attempts to use specialist extension services have shown that priorities from other wings of the agriculture department tend to override even these specialist roles: e.g. involvement in fertilizer distribution.

The costs of labour management associated with particular technologies have not been subjected to empirical analysis in a sufficiently systematic manner to quantify them even in an industrial situation. However, when the task of labour management is to be performed by those inexperienced in the task, and at the same time as learning and dissemi-nating skills, as would be the case with an extension officer in an improvements programme, the possibilities of a fast rate of diffusion are slim.

A counter-argument is that if the improvements to traditional stores are genuinely economically viable, then they will spread eventually, albeit more slowly than the alternative, a metal bin, would spread to more affluent farmers. The problems are seen to be a knowledge gap and an organization gap on the part of the farmer, and a motivation gap and organization gap on the part of the extension agent. These factors are not quantifiable. In a perfect market a narrow sense of economic viability may be highly significant in explaining decisions, but in this more complex non-market area the factors affecting decision-making are considerably wider.

The complexities of an improvements programme for the individual extension officer (or the bureaucracy establishing his function) and for the farmer he is 'selling' to, may limit the rate of diffusion dramatically. In a large part these complexities are intangibles which are not reflected in usual methods of evaluation of technology choice.

A simple cost-benefit comparison has shown that the improvement of traditional structures is a more desirable technology on private economic grounds than the metal bin. The advantages of the improvement in terms of employment, use of local raw materials and potential distribution suggests that most social cost-benefit comparisons would support this result in the absence of a consideration of the issues discussed

415

above. However, the problems associated with the non-representation of the poor in the market economy has led to the generation and application of a technology (metal bins) that does not benefit the poor either absolutely or compared with other economic groups. In an economy with unevenly distributed and scarce capital and plentiful labour this has meant that the direction of technology development has been away from the most appropriate technologies which would optimize use of factor endowments: i.e. they have been insufficiently labour-intensive.

Relieving poverty depends upon developing technologies that provide employment and therefore purchasing power. This, in the context of a scarce-capital/plentiful-labour economy is the essence of the appropriate technology philosophy. The example of farm-level storage indicates that there are major obstacles to the practical operation of this philosophy where successful operation depends upon new forms of organization and operation of development agents.

The successful examples to date of the operation of this sort of 'appropriate' technology programme in South Asia have all been small-scale and without any attempt to utilize the conventional market and government channels for the introduction of innovations. This is likely to continue to be the only method in which the introduction of 'appropriate' rural technologies will be operationally successful.

The key factors which limit the operational feasibility of a storage improvements programme are all associated with a technology choice based on local resources, skills and knowledge. These are precisely the features which are most commonly advocated in the literature on appropriate technology, and, in that they will use actual resource endowments more efficiently and meet distributional objectives, they are highly 'appropriate'. However, there are serious constraints on their policy viability for institutional reasons of organization and motivation. These are inherent biases which appropriate research methods and the development of appropriate technologies as they are presently conceived do not circumvent. Indeed the need for the development of an appropriate technology approach itself reflects the existence of these biases, and whilst this may be as much as research can contribute, there are more fundamental factors affecting the operational feasibility of its results than their 'appropriateness'.

Martin Greeley, *Food Policy*, 1978.

You may, by accident, snatch the market; or, by energy, command it; you may obtain the confidence of the public, and cause the ruin of opponent houses; or you may, with equal justice of fortune, be ruined by them. But whatever happens to you, this, at least, is certain, that the whole of your life will have been spent in corrupting public taste and encouraging public extravagance. — John Ruskin

X

Education, Training and Communication

Education has long been recognized as a central element in development and, more than ever before, the development of human resources through increased access to improved education and training is seen as a key factor in alleviating poverty, and in raising levels of national productivity and income. Yet vast numbers of people in developing countries, and especially those in rural areas, still have no access to any type of formal education.

In December 1948, the United Nations adopted its Universal Declaration of Human Rights. Article 26 states: 'Everyone has the right to education. Education shall be free, at least in the elementary and fundamental stages . . . Technical and professional education shall be made generally available and higher education shall be equally accessible to all on the basis of merit.' However, according to a recent World Bank Report[1], this objective is far from having been achieved three decades later. In the developing countries, less than 65 per cent of children between the ages of 6 and 11 are enrolled in school, and of them, only about 50 per cent reach the fourth grade. The enrolment ratios of the 12-to-17 and 18-to-23 age groups are about 38 per cent and 9 per cent respectively.

The situation varies significantly among regions and countries. In addition, there are unequal education opportunities within countries based on sex, socio-economic status, and different regional, urban, and sometimes, ethnic background. Of all these, none is of greater hindrance to development than that based on sex. One of the greatest obstacles to increased living standards is continuing population growth, and the educational, social and economic status of women is clearly correlated with fertility levels. In addition, women — as mothers — have an important influence on the education level of the future generation. In view of this, the disparity rate between the sexes in enrolments for primary education (72 per cent for boys and 56 per cent for girls) is an issue of concern. Factors contributing to this are the 'Western' concept of identifying the man as the bread-winner and thus more in need of formal, employment-oriented education than females; and the tendency to keep girls away from school to help with 'female' domestic chores.

Provision of basic education to the adult population target group (15-to-45 years) is likewise inadequate and uneven. While the percentage of adult illiterates in developing countries declined from 44 per cent to 32 per cent between 1950 and 1975, the absolute number is steadily increasing; from 544

[1] World Bank, *Education Sector Policy Paper*, April 1980.

417

million in 1970, it had reached about 600 million by 1978 and is not expected to turn downward before the year 2000. Again, statistics show a disparity between men and women. In Pakistan, for example, 36 per cent of males are literate as opposed to only 11 per cent of females. A major factor contributing to this divergence (apart from disparity in school enrolment rates) is the inabiity of women to participate in adult literacy courses because their farm and domestic chores simply leave them no time to do so. And yet, literacy for women would seem to be vitally important given their heavy involvement in agriculture, health and productive non-farm activities.

In recent years, two clear, inter-related trends have emerged in the education field, which if followed through, promise to do much to meet the educational needs of the masses of the rural poor and the development needs of the developing countries. First is the questioning of the appropriateness of existing education systems in terms of individual needs (for productive employment) and national needs (for relevantly skilled people). Formal schooling has come to be identified with a passport to a job in the formal 'modern' sector, but this sector has proved incapable of providing jobs on anything like the scale needed. The result has been a flood of unemployable school and unversity graduates — educated to expect jobs which don't exist

"Look—if you have five pocket calculators and I take two away, how many have you got left?" (Punch)

and unable or unwilling to contribute to national development through farming or non-farm work in rural areas. Questions have obviously been raised as to the possibilities of gearing education — at all levels — more towards the development needs of the country, and to equipping school leavers and university graduates with qualitifications and skills which will enable them to obtain productive, sustained employment.

Second, while increased funds are obviously required to meet the pupil explosion caused by the growing proportion of youth in the population and the increased emphasis on achieving universal primary education, demand will overwhelm the supply if existing standards and costs per capita are adhered to. The pressure of educational expenditure on national budgets is extreme; and may even be approaching a limit in many countries. Thus, unless more economical methods can be found, it looks as though the extent of present inequalities between rich and poor countries, between men and women, and between urban and rural people in developing countries may actually grow for the rest of this century. Hence, questions have been raised as to whether education, especially in rural communities, could be provided at a lower cost.

The extracts in this chapter have been selected in such a way a way as to illustrate either or both of these two themes. In the first section, on formal education, the appropriateness of the formal education system is discussed, and examples are given of appropriate and inappropriate systems in various parts of the world. Particularly interesting here is the extract by Ronald Dore on the Diploma Disease in which the 'for its own sake' educational system of Sri Lanka is compared and contrasted with the more appropriate, vocationally oriented system in Mexico. Also useful is Ken Darrow's account of attempts to make school-level science education in various countries more appropriate in terms of community development needs.

For the lucky few who reach university-level education in developing countries, the quality and content of training received is particularly important. As Nicolas Jéquier puts it in his extract on the higher education system — 'the students of today are the engineers, technologists, industrial entrepreneurs and political leaders of tomorrow, and the type of education they are now getting will determine to a large extent the type of society that will exist twenty or thirty years hence'. In particular, if appropriate technology is to play an increasing role in the development process, there will be an increasing need to encourage the training of engineers and technologists more towards the needs of rural areas and the levels of technology discussed in earlier chapters. The difficulties in bringing such changes about and the experiences of some Third World universities in working on appropriate technology and influencing future generations are discussed here.

Many Third World students must still look to developed countries for their university training in scientific and technical subjects. Thus, it is very important that some courses overseas should be conscious of, and oriented towards, the 'appropriate technology' approach. The extract by McPhun is illustrative of the change in emphasis occurring in many European universities — a change

419

which is useful not only for developing counries, but also for the developed countries which increasingly need more 'appropriate' technologists themselves.

For the main part, the twin problems of reaching the vast masses of the rural poor with some form of education or training, and of providing them with skills and understanding appropriate to the needs of their communities, involves fairly radical thinking and action. A variety of approaches has been attempted, to meet this challenge. Section 2 of this chapter has extracts which describe many of these approaches ranging from the now famous Brigades in Botswana and Village Polytechnics in Kenya to the lesser known 'rural university' scheme for training technicians and engineers in Colombia. Many schemes such as these have been started in a small way by a non-governmental agency or group of individuals in an attempt to prove (to governments and donors) that alternative edcuation systems are viable. Some have been adopted and expanded by governments which see them as a way of providing technical traiing to a larger number of school leavers within a limited budget. Both the Brigades and Village Polytechnics have been 'adopted' and expanded by national government.

A frequent problem with 'formal' technical training of this type is that graduates still feel inclined to migrate from the rural areas in search of wage-employment in urban areas. Another problem is the overwhelming bias of such schemes against women — not only are far fewer women involved, but the training they receive is also less market-oriented than that given to men. Some schemes show that, with the correct orientation and support, it is possible to persuade technical graduates to stay and serve rural communities. There are far fewer examples of schemes which provide women with more marketable skills.

One way of ensuring that local communities have the technical services they require is to follow the 'barefoot' approach and train local people (usually with little previous education) in a particular skill. The extracts given in this section include the training of pump mechanics in India and community promoters in Peru. Other chapters have also included extracts on various aspects of this principle: barefoot vets in Chapter III; barefoot doctors in Chapter IV; and barefoot foresters in Chapter V. An important point touched upon elsewhere, but elaborated upon in this section by Clare Oxby, is the need for women as well as men to be trained in this type of village-service work if the needs of the whole community are to be catered for successfully.

The inter-relationship between community needs, appropriate technology, skills training and adult education are pointed out in the final extracts in this section. Particularly important is the extract by Brenda McSweeney in which she points out how the introduction of labour-saving devices such as grinding mills and carts has increased the participation of girls and women in education programmes.

The last section of the chapter looks at the means of transferring knowledge (education) — especially to rural communities. A variety of communication techniques are examined including low-cost basic education literacy materials, radio and T.V. programmes, video and puppet shows.

With transistors and electronic equipment becoming smaller, cheaper and easier to run and maintain, they are providing new opportunities for reaching larger numbers of rural people, and people in remote communities. The extracts in this section illustrate some of the educational achievements of the new media techniques. Colin Frazer's extract on rural radio broadcasting gives some particularly impressive statistics such as the increase in the number of households using iodized salt from 5 per cent to 98 per cent after broadcast messages about the prevention of goitre. In the circumstances, it is unfortunate that so little air time in developing countries is devoted to rural broadcasting. There is, of course, a danger here. Cheap transistors or even cheap solar TVs will do little for rural development (and may even harm it) if the messages they broadcast are inappropriate in terms of the needs and means of rural people. As Colin Frazer points out 'cheap and appropriate communication technology does virtually nothing to ensure its use for rural development'. Similarly, although the 'barefoot' microchip is an appealing idea — keeping villagers informed of vital weather and marketing information; linking teachers with the taught through satellite communication, etc., the dangers of increased control and exploitation are just as real (if not more so) as the potential benefits. The extract from *Development Forum* explores some of the issues involved in the 'information' revolution.

Finally in the midst of the excitement over new communication techniques, many of which are visual in nature, a warning note is sounded in an extract from *World Development Forum* in which it is pointed out that some villagers are visually illiterate — they canot 'read' pictures. In circumstances such as these, perhaps the older traditional media techniques such as theatre and puppet shows are still the most appropriate. Martin Byram's extract of People's Theatre in Botswana gives an interesting example of how this type of media technique works.

In any strategy aimed at bringing about the use of more appropriate technologies, the role of education and training is crucial. Education shapes the attitudes and techniques of tomorrow's politicians, planners, architects, engineers and teachers. The way they run tomorrow's world will depend very much on the quality of the education they receive today. Further, dissemination of technology on a widespread basis is closely linked with access to training — in production, use, maintenance, repair, healing, management and business skills. These in their turn are based on some minimum level of literacy and numeracy.

Appropriate education and training must be keys to successful development, for without them, the chances of achieving the widespread and successful use of appropriate technologies are slim.

The serious problem is the education of the peasantry. — Mao Tse-Tung

Education for What?

No matter how high the percentage of national resources pumped into the education sector, no matter how earnest and well-motivated the effort, it is impossible for conventional schooling to meet even the basic learning needs of the majority of the world's poor. Yet the comfortable myth that depicts schooling as the only possible type of education still persists and is almost universally adhered to. People by the million get lost between the lines of fancily-packaged manpower-planning reports, academic discourses and politicians' manifestos. There are those for whom there will never be enough schools or enough teachers. And there are others who participate daily in an empty ritual of attendance at institutions where what they are taught is so divorced from the reality of their lives in the shanty towns and villages, and so out of register with the jobs open to them, that they exist in perennial confusion, anxiety and false hope.

Of course, educational planners are aware of the disadvantages to the poor that are built into conventional schooling. But, inevitably, the solutions that have been proposed from within the system, and in some cases implemented, are piecemeal. Slogans like 'equality of opportunity' ring down the corridors of most schools today, but the words are largely without meaning. Certainly, there have been some minor adjustments that make it more difficult for the rich to buy a better education than the poor; but educational achievement can still be predicted fairly precisely on social background.

Schooling has killed education and replaced it with a steeple-chase, which has a university degree at the finishing post and two or three qualification hurdles along the way. Meanwhile, down on the track, society bestows a head start on the wealthy, allowing them to perpetuate their good luck down the generations. The rhetoric of equality of opportunity only serves to delude the poor into believing that they set off from the same starting line as their more privileged brethren and that their failure is their own fault and they had better accept it gracefully.

The real problem, the problem which has hardly been tackled at all, lies in the structure of the educational system, its institutional framework, its competitive ethos, and the very fact that it is a steeple-chase. Qualification-getting dominates the view of all participants — rich and poor, teachers and pupils alike. Chasms of achievement measured by paper qualifications segregate winners from losers, failures from successes and whole societies come to be structured around this dualism. As one critic has put it: 'It is too much to expect educational administrators to contemplate the alternative. It is on the bread-and-butter importance of general education certificates — on their universal use in job allocation — that the survival of the whole educational industry depends. One could hardly expect that industry to devalue their importance'.

So, education for what? Education for qualification-getting, for supplying industries and state bureaucracies with suitably regimented personnel, for docile acceptance of pervasive injustice, for a world in which the great mass of people don't have a chance of fully experiencing their own humanity? Or education for liberation, personal development, and a full awareness of self within society? In practice the choice can never be as immediately clear-cut as that. Certainly, there is no justifiable defence for the further proliferation of existing systems. But, at the other extreme, the language of alternatives has become inflated with a good deal too much pomp at the expense of far too little circumstance. Ivan Illich is over-optimistic when he asserts: 'We are witnessing the end of the age of schooling'.

School may well be shot full of holes but it is far from dead. One particularly depressing development has been the extent to which the conventional school system, dying or not, has managed to reproduce itself with new vigour in the Third World. But it is in the Third World also that some of the freshest new thinking and practice has emerged.

New Internationalist, 1977.

Climbing Jacob's Ladder

For the great majority of Kenyans, education looks like Jacob's ladder. It seems to go up and up to a good job, a nice car, a brick house and a suit and tie. In fact, Jacob's ladder goes to cloud cuckoo land. In 1982, out of 400,000 primary school leavers, well under 100,000 will get a secondary school place. Some of the rest will join the ranks of the unemployed in Kenya's cities. Others will try to farm on small family plots or hire themselves out to bigger farmers at rock-bottom hourly rates.

Practically nothing that they will have learnt in primary school will be of any value to them. Indeed, if anything, their school experiences will disqualify them from finding happiness and moderate prosperity. These experiences will make them envious and restless and fill them with the poignant sorrow of paradise lost — a paradise that they will spend the rest of their days fruitlessly, pitifully, trying to regain. This is because the Kenyan school system exists solely to distil just one tiny drop of educational excellence — the 2,000 undergraduates at Nairobi University.

New Internationalist, 1982.

423

The Diploma Disease

There is a striking difference between Mexico and Sri Lanka. Mexican education is much more vocationally oriented at all levels than the Sri Lanka system. The latter is much more influenced by the aristocratic emphasis on 'for its own sake' education (humanities and 'pure' science, plus social science as bastard latter-day humanities rather than as disciplines-for-use) which was derived from the British model. At the junior secondary level in Mexico, only a little over 50 per cent of students were taking general courses in 1973, many of the remainder being enrolled in those of the 8,000 private institutions in Mexico offering training in all kinds of occupations — technical and commercial — that accept primary school graduates. At the senior secondary level there are a variety of technical institutions, most notably the vocational school which also prepares students for entrance to the National Polytechnic. In the National University, the country's largest, in 1974 fewer than 10 per cent of the students took courses in general arts and humanities. In Sri Lanka, on the other hand, although the very brightest students are creamed off into the two prestige faculties of engineering and medicine, the bulk of students at the tertiary level (some 86 per cent in 1967) are taking general arts or science degrees, and at the secondary level vocational training is practically nonexistent.

These differences are clearly reflected in the institutional rules which govern recruitment in the two countries. Mexican employers seem to take note only of what a person has studied and to what level — secondary, senior secondary or university — he or she has studied it. Technical and business studies are preferred. General education is rarely sufficient for clerical posts, and, at executive

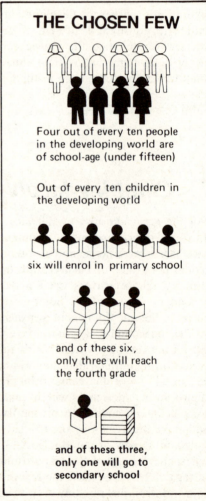

THE CHOSEN FEW

Four out of every ten people in the developing world are of school-age (under fifteen)

Out of every ten children in the developing world

six will enrol in primary school

and of these six, only three will reach the fourth grade

and of these three, only one will go to secondary school

(*New Internationalist, 1982*)

levels, is likely to be interpreted as showing a lack of motivation and orientation towards work. In Sri Lanka, on the other hand, one frequently comes across rules which discriminate between people who have been educated to the same level, favouring, for instance, honours graduates over general degree holders, science specialists over arts specialists (only the highest scorers are allowed into science streams in secondary schools), those with credits and distinctions over those with passes (at the same level of education), those with passes or credits in notoriously 'difficult' as opposed to 'easy' subjects, or those who have accumulated a certain set of subject passes in fewer rather than more sittings of the examination — all rules which, not being concerned with the substantive content of what has been learned, only make sense if the certificates are being used as some measure of 'general ability'.

This is a point of striking importance for the academic debate between the proponents of 'human capital theory' and the advocates of 'screening theory' which has enlivened discussion of the economics of education in recent years. These debates are too often conducted as if they were debates about the role of the school system in the economy — as if the whole world were the same. In the case of Mexico, assumptions of 'human capital' operates obviously. In the case of Sri Lanka, the screening theory offers a better explanation of what has happened.

Ronald Dore, *IDS Bulletin*, 1980.

Education: Designed for the Rural Poor

In most of the Third World, primary-school systems have been developed in the towns and suffer from 'urban bias'. The whole concept of the teaching, and very often the training of the teachers, has been along lines suited to the needs of a certain class of urban dwellers — or certainly more suited to their needs than those of rural dwellers. This arises from the conventional view that the principles underlying education must be the same regardless of whether it is provided in towns or the countryside. The belief has grown that a range of signals from an urban environment are more exciting than those from a rural locality, and so the school turns its back on the rural area in which it is situated and fails to help the child to grasp the reality of rural community life.

The components of the educational system designed for the rural poor in a given country should be considered from a number of aspects. Primary schools, for example, though a critical step in preparing young people for the future, are rarely considered in relation to the agricultural or rural development knowledge system. In some countries the primary system scarcely affects the rural poor. Even where primary education does exist, children of the rural poor often do not attend or attend only for a very short period. What can be done to change this? There is plenty of experience of primary-school teachers who have played important roles in rural development. In recent years, however, education authorities have turned the emphasis away from this. Agriculture is taught in some rural primary schools but rarely in a way that integrates it with rural life as it will be lived by the pupils. Yet, there is reason to suppose that, if the primary school aimed at developing rural aspirations rather than inculcating urban ones, rural youth would gain more of a real benefit from the schooling experience. Also worth examining is the extent to which the primary-school system relates to the other learning arrangements that may exist or be introduced to the community. What is the connection, for example, between the school and youth programmes, or the school and the extension service? How much can the services of the teacher be used for teaching other people in the community who have missed out on education? There is now quite a lot of experience of older people and parents attending school with children and expanding both their own and the children's learning opportunities. The school might become a centre of learning for the village or community rather than simply a place in which formal schooling takes place. The costs of this need not be very great once it is accepted that you do not need a teacher's qualification to take part in teaching. Obviously, every school needs a qualified teacher but his or her services can be greatly expanded by the use of local people with local wisdom. The main

cost for most countries would be creating schools where none exist or expanding schools that are inadequate.

Secondary agricultural schools embrace many kinds of institutions. Basically, they are high schools with an agriculturally-based curriculum. Sometimes they are designed to turn out people who will form the field level of the extension and related services, but others are simply secondary schools from which the pupils, in theory at least, go back to the rural community. In many cases, they are a stepping-stone to the third level of education. However, few children from the poor sector ever reach them, and if by chance some do, they probably will not return to their homes but rather seek their living in the urban sector or in government services. Because of the scattered nature of rural society, most of these schools are residential and result in the alienation of children from their environment. The agriculture they teach is scientific with scant regard for the socio-economic side of agricultural change.

More emphasis should be placed on producing graduates with a practical orientation and a deep knowledge of the problems of the smaller farmer. Such people could fill a critical gap in development services as local-level change agents. While it seems unlikely in the foreseeable future that more children of the rural poor will find their way into secondary education, these institutions could be utilized much more for short courses for young people who have some primary education. No heavy investment would be involved but rather a change in emphasis in the way of working. Staff might even be available from these schools to go out and give short courses in communities, working alongside their ex-pupils and with primary-school teachers.

Teacher- and extension-training institutions include all those institutions who turn out people who will teach and those who will staff the extension and development services. In most countries, teacher training and extension training are quite separate functions and the institutions are unrelated. There is a strong case for training together, using some common services, or at least in close proximity of all those who will be involved in rural development. Such an innovation might mean the creation of some new or joint institutions, but since both teacher training and extension training are always expanding, it might be more a relocation of new investment than actual additional investment. Those teachers who will work, at least initially, in rural areas would benefit greatly from contact with rural issues during their training. Extension training is in general one of the weakest links in preparing people to work with small farmers, the emphasis in most extension training being on preparing a man or woman to deliver technical information rather than as a change agent.

Faculties of agriculture produce the graduates who will staff the higher levels of the development and administrative services, and the research institutes and experimental stations. Most of the faculties were developed under colonial rule or have been founded since with much aid from abroad. Few of the older institutions are sufficiently closely associated with current needs in agricultural research or in rural development and few of them attempt, except in a very general way, to serve the community within which they exist. Of course, a standard of excellence is looked for in university education but that will be of little use if it is totally remote from the community. Thus graduates who staff administrative services have little idea of the small-farm problem; those who staff research stations do not understand the nature of the relationship between research and extension and so on. A number of new faculties have tried to overcome these problems but there are difficulties of a very practical nature. For example, most universities have a leadership role to play, which goes far beyond the provision of trained staff; in few cases are they doing so. An expansion of graduate output would achieve nothing in improving the understanding of the small-farm problem. It is true that more graduates may be required if there is to be a large programme aimed specifically at the small-farm sector. But, in the meantime, much can be done by reorienting the thinking of faculties to help train people to understand the problem and to give them enough practical experience to be able to become directly involved in it.

Whatever is done in the future to improve the agricultural knowledge system, there remains the problem of services staffed with people who, on the whole, are not trained to deal with small-farmer issues at any level. In recent years, there has been constant talk of retraining programmes and some countries have attempted them. In general, however, courses are needed for policy-makers (including politicians), planners and administrators, teachers and trainers of teachers, extension workers and others in the development services, research workers and the staff of faculties of agriculture. Where possible, such courses should be given to mixed groups of people from different areas of concern and at mixed levels so that responsibilities are more clearly understood. Some countries already have colleges at which such courses could be located but the courses would be of little value unless at least a part of them could be given in the field among the farming community, thus providing for direct problem-oriented contact and the exchange of experience.

John Higgs and David Moore, *Ceres*, 1980.

 # Reorienting Science Education

In a document prepared for the 1979 UN Conference on Science and Technology for Development, the Committee on the Teaching of Science (of the International Council of Scientific Unions) emphasized the importance of improving the effectiveness of science and mathematics education as a major part of efforts to use science and technology for development. . . .

A past director of UNESCO's division of science teaching concluded that the inquiry mode of science and the design mode of technology should both infuse the science education of the future. . . . An essential part of a new approach to science education is the fostering of creativity. If the importance of support for rural innovation is recognized, programmes of action in this area will depend heavily on practical, relevant science education through the primary grades and in non-formal education programmes. . . .

The following examples contain elements that appear to have key roles to play in putting rural innovation on a more firm scientific footing:

— The African Primary Science Programme (APSP) represents an unusual departure from conventional practice. The APSP focussed on the development of the child, rather than structuring the programme around the discipline of science. APSP science activities were designed to help children to develop cognitive and manipulative skills, as well as curiosity, persistence, self-reliance and a respect for scientific reasoning.

— The Palawan National Agricultural College, located on one of the outer islands of the Philippines, has sponsored a remarkable programme of community-based high schools on the island. These high schools have both a structure and a curriculum that is intended to provide an education that is relevant to the future farmers of Palawan. The first two years consist of practical agricultural work on small pieces of land provided by the parents of each student. The students apply the scientific knowledge of farming gained through class study, and the instructors visit students on their plots to help in the application of these concepts. The sale of produce from these plots goes to pay for the instructors (who are chosen by the community). The last two years of high school are used to teach students more abstract concepts and additional scientific skills. In this way, the schools remain relevant and responsive to the practical needs of the communities.

— In Brazil, the Foundation for Science Education Development (FUNBEC) has designed a series of fifty low-cost pocket-book size science education kits. These are sold thoughout the country on news-stands. They are intended to stimulate inquiry and an appreciation of the scientific method.

— In a village grain storage project in rural Tanzania, school children were involved in an experiment with small containers of stored grain mixed with different materials (insecticides, sand and other materials) to discover the effects of these different techniques. At the same time, their parents were engaged in a non-formal education process in an attempt to develop and choose some grain storage improvements.

Ken Darrow in UNESCO, *Technology for Rural Development*, 1981.

AT in the Higher Education System

The students of today are the engineers, technologists, industrial entrepreneurs and political leaders of tomorrow, and the type of education they are now getting will determine to a large extent the type of society that will exist twenty or thirty years hence. This time-lag or production cycle of the educational system suggests that if appropriate technology is to play an important part in the development process, the students of today need not only be familiar with it, but also to have a certain 'sympathy' for it. The shift in values and attitudes which this requires cannot be achieved in a simple and straightforward way, but a number of steps have been suggested. One for instance is the Indian idea of integrating intermediate technology in the university curricula and eventually having degrees in this field. This obviously poses a number of problems. Academic curricula, notably in the engineering field, are designed in such a way that it is difficult to add on new courses without either suppressing some existing courses which are essential, or lengthening by several semesters the total amount of time spent by students at the university. And for the time being at least, it does not seem possible to design a full curriculum focussing exclusively on appropriate technology.

Another approach is to accept intermediate technology as a legitimate subject for an engineering student's diploma work. Designing a windmill, a palmnut crusher or a water distribution system which uses only local resources and which can be manufactured by people with little technical education is just as challenging, difficult and instructive of the student's abilities as the design of a diesel engine, a steel bridge or an electronic circuit.

Encouraging universities to develop intermediate technology requires a modification of their reward and promotion system. In most institutions, the evaluation of a faculty member's performance — which determines his salary level and promotion prospects — is based to a large extent on publications. This yardstick is perfectly legitimate if research, preferably of a high quality, is accepted as one of the main missions of the university. It is, however, strongly biased against those who are very good teachers but poor researchers, and it does little justice to the other missions of a university, and notably to its educational role and its contribution to economic and social development.

Changes in the evaluation methods can and do take place, and there is no reason why in the universities of the developing countries, contributions to the development of intermediate technology might not also be taken into account. In fact, from a social point of view, such activities are probably much more useful and rewarding than a publication in a recognized foreign scientific journal, and the knowledge that they would also serve as a basis for the professional evaluation at the end of the academic year would provide a powerful motivation to do more.

The various steps amount in effect to an institutionalization of intermediate technology. That they could be useful is more than likely. However, even if they were carried out, and this is beginning to be done in some places, many problems would remain. One of the most important is that the university as such belongs essentially to the modern urbanized segment of society. Students and teachers, even if they come originally from a rural community, have for the most part lost contact with the farmers and craftsmen in the villages. If they are to help develop new technologies which are truly appropriate to the daily conditions of the farmers and poor people, they need some much closer contacts with those to whom these new technologies are addressed.

Universities in the developing countries may well be able to make an important contribution to the development of new intermediate or low-cost technologies, despite their primary commitment to education and their cultural bias towards modern technology. However, if one considers the ways in which their R&D activities could be mobilized for development, it is probably a mistake to look exclusively at innovations in low-cost or intermediate technology. Technologies of this type are no doubt important, but they cannot solve everything. There are a number of complex problems — in agriculture, natural resources, transportation or public health — which can only be solved through the most sophisticated modern technology. One of these, for instance, is the eradication of bilharzia, a parasitic disease which affects hundreds of millions of people in the poorer parts of the world. Another is the mapping of underground mineral resources by space satellites. In the transportation field, there are a number of alternatives to the traditional methods like ox-carts or the modern methods like trucks, but these parallel technologies (e.g. the airship) still require a certain amount of development work before they can become economically attractive.

Many of the problems which are still unsolved could probably be dealt with successfully in the next ten to twenty years through a large-scale R & D effort involving the universities of the developing countries. The technologies required to provide new tyes of crops, more efficient transportation systems, or a better utilization of natural resources are in many cases neither low-cost nor intermediate, but highly sophisticated and very modern. Yet they are particularly appropriate, in the sense that they try to solve problems which are crucially important to the developing countries but which have been neglected by the research community of the industrialized countries.

The new appropriate technologies required by the developing countries can either be low-cost and intermediate or highly sophisticated and very advanced. In both cases, the universities of the developing countries have an important part to play, both as educational centres and as R & D centres. Their educational mission should not only be to train scientists and engineers and give them the basic qualifications for a professional activity in the modern world, but also to familiarize them with the social and technical problems of the millions of people living in the rural areas. Their R & D activities should probably focus more on the development of technology, but this should not be at the expense of highly sophisticated research in the scientific and technological problems which are specific to the developing countries. Their 'modern' R & D activities might well in fact serve as the basis for the creation of new science-based industries focussing on the problems of development.

Nicolas Jéquier, *Appropriate Technology, Problems and Promises,* 1976.

Thoughts of a Field Worker

Enough has been said to indicate that the project is far from being secured — that it has not seized the hearts and minds of the people. The education system is no help: these notes would be incomplete if I failed to mention I taught in the mission primary school for a term . . . rediscovering that rocks may be classified as igneous, sedimentary and metamorphic . . . Some time ago, I was asked to prepare plots for the boarding children, which I did, but they were never sown or planted: the children were too busy, I guess, learning about the European Renaissance . . .

Stuart Mortished,
*Sebapala
Communal Garden
Project,* 1983.

University-based AT Groups (ATDCs)

Autonomous ATDCs (Appropriate Technology Development Centres) within academic or research institutes have started off by concentrating either on technologies appropriate for small industries and entrepreneurs or on technologies appropriate for the needs of rural communities. Experience indicates that they have had more success with the first category. Possibly, this is because it is much easier for a small group of university-based professionals to handle (or at least co-ordinate) the software elements involved in transferring technologies to urban and semi-urban small entrepreneurs than to widely-dispersed rural artisans and communities. This does not imply that small industry development and promotion is a simple matter. The experience of TCC in Ghana and SIRTDO in Bihar[1] show that it takes years of work with many setbacks before significant results are achieved. In both cases, the major problems were not so much in developing the technologies appropriate to small-scale production as with the commercialization of the products and processes once developed.

TCC found that small entrepreneurs were unwilling to invest in new processes in the absence of proof that they would form the basis of a profitable venture. The only way to overcome this was to set up and operate pilot production plants in the TCC workshops to demonstrate the technical and economic viability of the new processes. The commercial operation of these production units is a major undertaking, which requires much of the TCC staff's time. The production units do, however, provide income, accounting for nearly half of TCC's total income.

SIRTDO's approach has been somewhat different. Products needed by local heavy industry and capable of small-scale production were identified and laboratory models were developed by students and SIRTDO staff in the Institute's workshops. Industry, however, required commercial prototypes to test before giving any assurance of orders, but the workshops were not equipped (or allowed) to produce commercial prototypes. To resolve this, funds were obtained and technical personnel seconded from the Birla Institute of Scientific Research to set up and operate a workshop capable of developing commercial prototypes. Once markets were established, there was the problem of actual production, which was generally beyond the technical capabilities of existing small firms in the area. SIRTDO resolved this problem by helping some of the University graduates (many of whom had worked on developing the technologies as students) to set up small industries.

TCC and SIRTDO have found that entrepreneurs require assistance in acquiring loans, installing equip-

[1] Technology Consultancy Centre (TCC) at the University of Kumasi in Ghana and the Small Industry Research, Training and Development Organization (SIRTDO) at the Birla Institute of Technology in India.

ment, and training a workforce, and also need technical support to overcome initial operational problems. SIRTDO's task is made more difficult in that its entrepreneurs are inexperienced. To help with this problem, SIRTDO has established a 'nursery' industrial estate adjacent to the campus. TCC's task is easier, in that it is largely dealing with established entrepreneurs; the entrepreneurs are widely dispersed, however, which has led TCC to establish satellite units in key informal industrial areas.

The transfer of technologies in rural areas is more difficult, and requires greater time and resources. Both TCC and SIRTDO cite lack of resources as an explanation of their relative lack of success in the dissemination of rural technologies. However, both have recently started new rural development initiatives, and have established autonomous units to handle their rural development work.

In some countries, such as Indonesia, university students are required to spend part of their time on practical work in rural areas. As involvement in rural problems is thus an acceptable part of university work, the environment would appear favourable for an ATDC to be generating and transferring technologies to rural communities. However, the experience of the Development Technology Centre of the Bandung Institute of Technology[2] suggests that this is not sufficient. There appear to have been problems of inadequate interaction between hardware development and needs identification, insufficient effort on field testing prototypes, and an imbalance of staff and resources in favour of laboratory research work.

Similarly, university-based ATDCs in other countries have had little success in the generation and transfer of technologies to rural communities. In Papua New Guinea, evaluations of ATDI[1] have cited problems of insufficient capacity to undertake more than a few pilot schemes, concentration on academically-oriented projects, and an inability of faculty and students to become adequately involved in rural development work. An exception is the Universidad de Los Andes in Bogota, Columbia, which has developed an approach to apply its resources to rural development effectively. A formal agreement was established that allowed the staff and students of the Engineering Faculty to work in an integrated manner with the staff of a government-sponsored rural development programme at Las Gaviotas. A faculty member and a group of students move between the University and the rural development area, determining needs in the field, developing prototypes at the University, and testing the prototypes in the field: eventually, successful prototypes are produced in the rural deveopment area. This approach resolves the problem of how to direct sufficient numbers of technically-qualified staff to work in the rural areas for a sufficient time.

These experiences suggest that successful development and transfer of technologies requires a well-planned methodology and input of

[2] Development Technology Centre (DTC) in Indonesia and the Appropriate Technology Development Institute (ATDI) in Papua New Guinea.

substantial technological capacity over five years or more. Moreover, while technologies can be developed by small numbers of technologists within the university, the transfer of technologies to rural areas requires many more people with a variety of non-technical skills and requires the ATDCs to work through existing rural development programmes. The technologists must still spend substantial time in the rural areas, which they may be unable or unwilling to do. It is thus easier for university-based ATDCs to transfer technologies to small entrepreneurs, although this still requires a commitment by the staff to apply their skills for the benefit of others.

Richard Whitcombe and Marilyn Carr, *Appropriate Technology Institutions. A Review*, 1982.

 # *Appropriate Engineering Education*

Although many engineering departments are now researching into renewable energy technologies, there has been little change in the approach to the undergraduate degree course. This traditionally places the emphasis on theory and analysis rather than practice and synthesis. . . .

Thus, after graduation, students typically take up positions and gain their experience in large-scale industry or government establishments, sometimes through formal training schemes. They are not expected to be immediately productive, and 'design' as opposed to 'analysis' is learned on the job. Where the teaching of design exists within the degree course, it is usually limited to the technical details needed to meet specified requirements. Certain unwritten assumptions usually underly the teaching of engineers, including:

— after graduation, the student will work for a large firm or organization;

— the work will be concerned with high technology;

— it will consist of purely technical problems presented by someone else;

— new technology is necessarily best.

Hand-in-hand with these assumptions go the expectations of engineering graduates that they will:

— be employed by someone else;

— be technical decision-makers;

— receive guidance from commercial management;

— have back-up from other specialist engineers and services, e.g. skilled artisans.

The result is that the normal engineering graduate is not at all suited to undertake a range of activities absolutely vital to society, such as:

— creating employment as the graduate knows little about the social and commercial context of technology and does not expect to occupy multiple roles;

— identifying problems — the real ones which often underly apparent difficulties;
— making socio-technical decisions, for example deciding what to design and manufacture;
— making products themselves.

The appropriate technologist, however, must be prepared for these types of activities. Consider, for example, the creation of new employment in engineering in the UK. Most new jobs occur in small firms, but very few graduates of any kind run small firms. The small engineering firm may have only one or two engineers who have to do all types of work relevant to the firm's business. Traditional engineering education does not prepare its graduates for this role. A new approach is definitely called for.

Far more important than details of the syllabus is that teaching should be based on the right assumptions. These will affect the way that subjects are presented, the content of examples and examination questions, and the type of student applying for entry to the course. The main assumption is that technology should be appropriate.

Engineers should be primarily concerned with the social objectives: they then acquire and use technical skills in order to achieve them. Other assumptions which may follow are:

— the word 'appropriate' denotes fitness for purpose, taking into account all factors, not just the technical ones;
— an engineer may be called upon to decide what to design or make, or what not to. Sometimes the solution to a problem may not be technological;
— high or new technology may not be appropriate;
— the graduates' responsibility may be as much to create employment as to be employed;
— working in, or creating, a small firm, co-operative or community, or working in rural development overseas is a possible future activity for graduates;
— the graduates may have to turn their hands to anything;
— practical skill, as well as knowledge, will be required.

Thus, these graduates need skills not taught in normal engineering degree courses. Either a longer course is needed, or else some specialization must be accepted in order to work within the constraint of the traditional three-year course. To decide on the direction of specialization one must match likely fields of work for graduates with the expertise and interests of the available teaching staff.

M.K. McPhun, *Appropriate Technology*, 1981.

Towards what ultimate point is society tending by its industrial progress? When the progress ceases, in what condition are we to expect that it will leave mankind? — John Stuart Mill

2 FORMAL AND INFORMAL TRAINING

The Rural University

To those who think of universities primarily as urban institutions, the term 'rural university' may seem self-contradictory. But not to the inhabitants of the Cauca Valley, Colombia. For many of them, the FUNDAEC rural university, based in a village near Cali, is an agent of change, an institution in their midst that helps them use knowledge to improve their lives and livelihoods.

The Foundation for the Application and Teaching of the Sciences (FUNDAEC) was created in 1974 by a small group of professors from the Universidad del Valle, in Cali. The FUNDAEC philosophy holds that disadvantaged rural people can not only benefit from higher education but can also help create and exploit new technologies to improve their standard of living. This tenet, coupled with disappointment over the failure of national and international development efforts to improve the well-being of the rural poor, led to a novel approach to education and development.

'Usually when one thinks of Institutions working with peasants, they are not supposed to take a very high-powered approach intellectually,' says FUNDAEC director. 'What is usually taken to the peasants is information, not knowledge. As far as I can tell FUNDAEC is a rare kind of institution because we make the creation of knowledge the basic issue. A rural population needs a university, not just primary or technical schools,

to act as its learning institution.'

The three levels of the FUNDAEC learning system are the 'promoter' of rural well-being, the 'technician' and the 'engineer'. FUNDAEC sees promoter training not as sufficient qualifications for a rural job, but as a basic education to be gradually offered to the entire youth population of the Norte del Cauca region. In this sense, promoters are the base of the pyramid of workers.

Promoters study in their own village under a tutorial system called Sistema de Aprendizaje Tutorial (SAT), run by FUNDAEC engineers in the last stages of their formal training. At present, this training is available in twenty villages of the Cauca Valley, with a total enrolment of about 200 students. Between thirty and forty were expected to complete their training by September 1983.

Although promoter training ideally lasts one year, students learn at their own pace and, once finished, their training is considered by the Ministry of Education to be the equivalent of two years of high school. The FUNDAEC administrators point out that the cost of training a promoter using SAT is about three-quarters that of sending the student to a regular high school for two years.

The SAT curriculum has five components: service to the community, mathematics, science, technology and language. Service to the community is basically an exercise in getting to understand development at the family level. Students establish a personal dialogue with a number of families in the village, observing their problems as well as the opportunities and resources available to them. The students discuss their observations with other students to establish a clear picture of how the village operates. Subjects such as health, small industries, and marketing are studied throughout the course.

Students chosen to go on to the technician level continue their training in the same subjects for two years or more. But, academically, the level is higher. This is considered the equivalent of a four-year high school programme.

At this stage, the 'service to the community' component is more complex, organized into 'research-action-learning packages'. For example, in a package dealing with environmental issues, the students and professors make a detailed analysis of the state of sanitation in a village and relate it to the health of its inhabitants. In another package, 'small units of production', the students share a production project with a village family, such as a chicken-raising operation. Here they apply their technical skills and receive their first training in simple economics and community organization. Such projects are geared toward villagers who do not have access to much land.

The final three-year stage of training brings the student to the level of a university graduate, an engineer in rural well-being. The engineers-in-training tutor promoters, and continue both academic studies and joint production ventures with peasants, finishing with the one-year super-vised residence in a village.

When students reach the engineer stage, they are intimately involved in the creation of knowledge. They identify specific needs in the community, search for, and experiment with, technological alternatives, and participate in the dissemination of solutions, mainly through technical bulletins and the university's documentation centre.

What has worried FUNDAEC is whether these young professionals can find adequate sources of income once they graduate. More than a few small-scale production ventures with local farmers would be needed to make a living. To help solve the problem, an association of engineers has been created as a non-profit organization with capital of about CA$240,000. The association will ben an investor-partner in a number of large enterprises such as a plant for feed concentrate. Profits will not be divided among members but put into a fund administered by the association's elected board. The members will present proposals for village development and include their salaries in the budget.

'The engineers believe that 1983 and 1984 will witness the consolidation of their association', says the director, and that they are finally finding a reasonable answer to the question every visitor has asked them since the beginning of the programme 'How are you ever going to earn a living after you graduate?''

FUNDAEC's professors feel the curriculum, especially the service to the community component, has engendered a strong sense of commitment among the new engineers for rural well-being. Not everyone has been so optimistic. 'Many people made bets with me that none of them would be there after the first couple of years', recounts the director. 'It's turned out to be totally the opposite. I think we've broken the myth that getting an educational automatically means leaving the rural areas.'

While FUNDAEC goes on training promoters, technicians, and engineers, perfecting its curriculum, and finding technological solutions to villagers' problems, a nagging question remains. How far can people develop with such limited resources?

Gerry Toomey, *IDRC Reports*, 1984.

Rural development where there is no land available to the farmers is meaningless.

The pattern of land tenure is unfavourable to the villagers of the region. But the correction of such a structural problem is beyond the ken of the rural university, which is non-political. Perhaps FUNDAEC's contribution to Colombia's development is that it is providing the people of the Cauca Valley a way of improving their daily lives, however modestly and slowly, without resorting to violence or revolution.

Botswana's Brigades

Two aspects of education which cause particular concern in many countries are the high cost of schooling and the fact that many pupils see little relevance to their future lives in what they learn at school. . . .

In Botswana, a developing country with under a million inhabitants, a system of education which tackles both these problems is in operation. Appropriate education for many primary school-leavers is given in the Brigades, named after the Workers' Brigades in Ghana. Brigades provide on-the-job vocational training as well as giving supporting academic education. . . . There are now Brigades in sixteen large villages in Botswana, giving training in building, carpentry, dressmaking, auto-mechanics, welding and other trades. On-the-job training for Brigade members usually lasts three years, ending with a Government Trade Test: this aims to

provide trainees with a better chance of employment. Sometimes a Brigade forms a production unit specifically to employ the trained youths. The training is productive and cost-covering: for example, if a trainee carpenter is learning how to make a table, materials have to be bought and his instructor has to be paid. However, when the table is sold, the money earned pays for these expenses.

New Brigades are started by the village community, and overseas donors may help with initial expenses. The community elects a Trust to manage the Brigade Centre, where several trades may be taught and other activities chosen to suit the particular locality. Most Brigades offer training in building trades and grow vegetables for sale. As houses are increasingly required in Botswana, bricklayers can get a job locally rather than having to go to South Africa to work in the mines. Also, fresh, locally-available vegetables are far preferable to the community than imported ones.

Trainees are mainly made up of those who have left primary school then spent a few years working with their parents before joining the Brigades, although some will have spent a few years at secondary school. No fees are charged unless accommodation or meals are provided, and some Brigades even pay trainees an allowance from income earned.

The government recognizes the benefit of the Brigades in giving both academic education beyond primary school level, and much needed vocational training, as well as offering rural employment in non-training production units. It therefore subsidizes training which helps costs to be covered, and also offers advice on business and training. The government has also enabled different Brigades to discuss common problems by forming a National Brigades Co-ordinating Committee (NBCC).

Trainees spend about four days a week on money-earning, productive work ranging from sewing-up school uniforms to building a house. In order for good instruction to be given and a sufficiently professional job to be done, there should not be too many trainees to each instructor. About half-a-day a week is allotted to learning about the theory of the trade, and both the theory and the practical are tested in the Trade Test. Because Brigades operate as small businesses in that they must cover their costs, the marketing of their goods and services is very important. Often, the Brigades only make goods to order, but there may be a small showroom at the workshop or shop in the village where goods may be sold. . . .

About half-a-day a week is spent studying English, Mathematics, Science and Development Studies. All the textbooks are written in English, as are the Trade Tests, as it would be very difficult to translate into Setswana all the technical terms used in each trade. It is for this reason that English is taught, but it should be taught in an appropriate way to help trainees understand their trade's terminology and textbooks.

A knowledge of mathematics is useful in any trade, and indeed in general life. However, the instructor will need to apply to the subject examples and questions concerning the particular trade being taught to make it more relevant. This point particularly applies to science. For example, trainee bricklayers will be greatly helped by seeing just why a building needs foundations, and complex trades like auto-mechanics are heavily dependent upon scientific concepts.

Gerald Giffould, *Appropriate Technology*, 1981.

Give a man a fish . . . and you are helping him a little bit for a very short while; teach him the art of fishing, and he can help himself all his life.
— E. F. Schumacher

Training Village Entrepreneurs — Easier Said Than Done

The importance of training village entrepreneurs is evident. They play a vital role in the renovation of village economy.

— they make a living in a rural area, without depending on agriculture;

— they slow down the brain drain of educated youth from villages to towns;

— they help diversity the village economy;

— they bring badly needed skills and managerial competencies into the village;

— they give village youth a hero image, which some may want to imitate;

— they help the indigenous population to fulfil a business role which otherwise is appropriated by outsiders in an exploitative manner.

The EDP (Entrepreneur Development Programme) requires six months, at least, if one deals with persons from underprivileged communities, or groups which have had no previous exposure to business. The training comprises three components: motivation, management and skill training.

Success of an EDP depends on the motivation of the young entrepreneur. Motivation is the driving force within the individual, which urges him or her to strive after an objective, whatever be the obstructions on the way. It is in the individual, and cannot be given from outside. But it can be nurtured by the creation of a favourable environment. Motivation is something the candidate already has before he or she enters the portals of the training agency. It must be reinforced, clarified, legitimized, during and after the training. Experience shows that entrepreneurs may lose their determination to achieve their goal, if reinforcements are not adequate to maintain a positive balance in motivation against the formidable odds which the candidates face when setting up a small business. . . .

A village entrepreneur is a barefoot manager, and fulfils — albeit in embryonic manner — all the functions of a self–employed manager. He needs the following managerial abilities to do justice to his task:

— planning ability: setting objectives, breaking them down into targets and following a work plan to reach the latter;

— financial ability: the capacity and self–discipline to handle money, his own or borrowed, and to make it grow;

— social ability: ease and grace in dealing with others and facility in befriending people, remaining businesslike at the same time;

— achievement orientation: striving after one's goal with single-–minded determination, discipline in using one's time and other resources;

— technical competency: being familiar with the techniques, processes, raw materials, prices, customer expectations, and with the sources from where relevant information can be obtained;

— relation to village environment: the entrepreneur must bear a sense of responsibility towards his community.

Managerial training is best imparted in theoretical sessions, during afternoon hours, although again mere lecturing is to be shunned. The knowledge and skills that are imparted must be seen as relevant by the trainees. The trainers can start off from what the trainees have seen, heard, done, suffered during morning hours when they were doing their shopfloor training under the shopkeeper. The training can be made appealing and meaningful if it is done through experience sharing, case studies, role playing, practical exercise. . . .

A village entrepreneur who sets up a cycle shop for example must not only be motivated and familiar with basic principles of management. He must know all about cycles, cycle parts, cycle repairs, be skilled in handling his tools, and meet the needs of the village people for repairing their

cycles, and many things besides, such as flash lights, alarm clocks, etc.

The same applies to a tailor, a baker, a grocery store man. They could acquire these respective skills in the EDP training institution itself, provided the latter has the facilities, in technical institutions or craft centres in town, or in shops in town which are actually engaging in that trade.

Necessity is likely to force one to take the last alternative. Experience has shown that it has several advantages, as well as some disadvantages. It is the cheapest and often the only manner in which skill training can be imparted. The trainee learns to practise his skill in a situation very similar to the one in which he will work after the training. It enables him to get an over-all view of all the activities that go into his trade, and to see how management principles are applied to each activity, whether it be storekeeping, advertising, costing, etc. Finally, it is the traditional approach which has been used for over centuries for passing down skills from craft master to apprentice.

Michael Bodgaert, *Adult Education and Development,* 1983.

Village Polytechnics in Kenya

With 65 per cent of Kenya's population under age twenty-five and almost 50 per cent under fifteen, training and employment opportunities among the young are critical issues. At the same time, demand for affordable, good quality, locally produced goods and services has continued to grow within local communities — both urban and rural. Given Kenya's current foreign reserve problems, the need to substitute for imported consumer goods and materials has also become increasingly important. Hence, there exist economic reasons beyond employment generation for stimulating such local production.

The goals of the Village Polytechnic (VP) programme are various, and can be divided into three categories. At the level of the individual the major objective is to provide sound, informal training in marketable skill areas to primary-school leavers and other youths between the ages of sixteen and twenty-five. Beyond this, the programme is designed to assist VP leavers secure sustained employment, preferably in their home areas and in the form of self–employment in work groups.

As regards local communities, the goal is to provided needed, locally produced goods and services to communities throughout Kenya at a competitive price, while promoting self-sustaining and locally controlled economic development. Innovative thinking has particularly been shown in this area, as the architects of the programme saw that, by exploiting local markets, money would be retained in the given rural or peri-urban area, thus providing the basis for other local economic initiatives.

Finally, at the national level, the goals are to further the process of democratization and decentralization in the nation's development, while helping to shift the emphasis of Kenyan education toward non-formal training.

Trainees must be between sixteen and twenty-five years of age. The majority are between sixteen and eighteen upon entering the programme and between eighteen and twenty upon leaving. Figures for 1976 show that 75 per cent were male. Although specific data was not available, it is clear that most of the trainees, like the VPs themselves, are located in rural areas, as one of the basic goals of the program is to stem youth migration to the cities. The approximately 25 per cent of the trainees which are in the larger, urban-based VPs come in part from the urban slums, but there is evidence to suggest that most are rural youths who are temporarily in the cities looking for jobs. The exception may be the Mathare Valley VP in Nairobi, as it appears that most of the students there come from the surrounding slum area.

It is clear, however, that regardless of their place of origin, the vast majority of trainees are from the poorer segments of society. Although no figures on trainee incomes were available, most people interviewed maintained that trainees have hardly any income and were dependent on their families (or relatives, in the case of migrants) for support while training. It was also pointed out that it could be assumed that trainees were from poor families, since the 'well-connected youth' continue in school and find white-collar employmnt. On the other hand, VP youths are essentially primary-school leavers who have usually completed up to Standard VII (roughly 8th or 9th grade US), although there are also drop-outs from secondary schools. A quick interviewing of some twenty trainees yielded the information that only three of their fathers had gone beyond Standard VII education and only three had salaried jobs; average land ownership was 1.5 acres.

Training is offered in a wide variety of trade areas. In the 192 VP projects assisted by the government as of March 1978, there were thirteen trades which were taught in at least two VPs. They were: carpentry, masonry, tailoring/dressmaking, home economics, agriculture, metal work, leather work, motor mechanics, typing/book-keeping, plumbing, painting/signwriting, electrical and fitting/tanning.

The urban VPs are typically larger than their rural counter-parts and offer a broader range of courses. Many of the rural and peri-urban VPs teach agriculture to all the trainees, regardless of their speciality, in order to make them more self-reliant upon returning home. It is typical that the trainees are each given a small plot to work and that they get part of the return from the marketing of the produce they harvest. At one peri-urban VP, fish breeding will also be taught at a new project site.

As far as training methodology is concerned, the approach taken in VP programmes is termed 'work directed', with trainees taught skills through the production of marketable goods and services for which contracts have been secured. It is the responsibility of the local management committee and the manager, as well as the instructors, to secure such contracts from local sources for their VP. This is done for two basic reasons: first, to get trainees involved in income-producing, professional activities as quickly as possible and thereby instill discipline and a confident attitude toward self-employment; and, second, to raise money to defray the operating costs of the VP, which in many cases has become an important, low-cost producer of essentials within the community.

Unfortunately, in many cases VPs have been unable to effectively assist in the establishment of sound VP-leaver group enterprises, as evidenced by the thousands of VP leavers each year who must search for wage employment. The management obstacles to creating one's own enterprise are too great for most leavers, and their former instructors, overburdened with too many students, are unable to give the leavers the attention they require. Furthermore, even if a VP sets aside fair shares of contract profits for leaver tools and equipment, in some fields, like carpentry and masonry, this capital may represent as little as 10 per cent of that needed to get a successful enterprise off the ground.

The impact of the Village Polytechnic programme which is now in its fifteenth year of operation, must be considered in light of the original goals of the movement. To begin with, the 270 VP centres established by communities around the country are proof that, at least in the area of informal training, the development process has been significantly enhanced and decentralized in Kenya. Although greater public-sector involvement has led to a tighter structuring of the training programme itself, the programme has remained a significant innovation within the Kenyan educational system and has been accepted on equal footing with the country's formal education programme. It has positively influenced many communities to accept purely vocational or prevocational training as an integral part of the national development effort. In so doing, the individual VPs have in most cases established themselves as important parts of the communities in which they are situated.

Direct, community-level impact, however, is difficult to measure. While VPs and their leavers have successfully produced and marketed goods and services for local consumption, and local resources have been mobilized in support of the VPs, it is unclear to what extent local self-reliance and control have been enhanced. On the other hand, it is apparent that a majority of leavers (about two-thirds of those who find employment) remain in the local area and make a productive contribution to the broader community, and that at least some additional capital generated from these activities is retained locally for other community endeavours. VPs have also made a positive contribution toward increasing the awareness of communities as to alternative forms of local economic development.

Most important is the impact of the VP programme on the intended beneficiaries — the trainees and the leavers. In 1979, the programme turned out approximately 9,000 young people (at the cost of approximately US$150 per trainee per year in national and foreign funds) trained in a variety of important trades. This number is significant when it is taken into account that only 25 per cent of the 220,000 youths who sat for secondary-school entry exams in 1978 were accepted for further education. Among those who have taken technical qualifying exams for public licensing in various fields after completing their VP course, a high percentage have done very well, attesting to the quality of the skills training. This success in training has been accomplished at a relatively low cost and, for the most part, without screening out the country's poor from participation in the programme.

Fred M. O'Regan and Douglas A. Hellinger, *Assisting the Smaller Economic Activities of the Urban Poor*, 1980.

 # Training Pump Mechanics in India

The shop run by Sri Isamuddin in Kohir village is a simple, rather ramshackle stall, whose whole inventory of crumbling cigarettes, fleshy green betel leaves and silvery spices would fit inside a large paper bag. It is a stall like countless others throughout India, and its location next to a shady teashop on the main street is in no way remarkable. But Sri Isamuddin has set it up there for another reason. It keeps him near his other job: caretaker and overseer of Kohir village hand-pumps and water supply system. There are four hand-pumps to serve the 15,000 people in the village and the one near his shop is the most heavily used.

Sri Isamuddin is only twenty-one years old, has received a mere two days training for his part-time job and is unrewarded except by a tool kit and a certificate which confers on him a modest rise in status. But his services as a do-it-yourself maintenance man keep the village handpumps in working order and, when one breaks down, Sri Isamuddin acts as the link with the next tier in the maintenance system. He has a supply of already addressed and stamped postcards, one of which he puts in the mail to summon the block engineer. It then becomes the responsibility of the block engineer to call in the district mobile maintenance team if a major repair is necessary.

As well as his maintenance duties, Sri Isamuddin also functions as an informal public health official, trying to ensure that the water used by mothers and children in Kohir is kept as germ-free as possible. The area round the hand-pump near Sri Isamuddin's shop is immaculate. All dirt and refuse has been swept away and a small channel in the concrete area at the hand-pump's base is draining off the excess water.

Sri Isamuddin is a bright young man and is conscientious about his duties. He seems to have fully absorbed the various elements of his training which took place a few months ago at a two-day camp. Together with ninety-five others from the district, he was taught rudimentary preventive maintenance — greasing bolts, cleaning the pump head — and was made aware of the connection between contaminated or stagnant water and the spread of disease.

The village hand-pump caretaker training programme in Andhra Pradesh grew out of an earlier experimental programe in Tamil Nadu State, which had been set up with UNICEF assistance and encouragement. The intention was to try and design a maintenance structure for rural water supplies which could be applied all over the country. In an attempt to raise the health standards of mothers and children in rural India, UNICEF had for some years been heavily involved in programmes to provide over 150,000 'problem' villages with safe drinking water supplies. But in the early days, the drilling of boreholes and installation of pumps often seemed almost futile.

One of the main problems was the pump itself. Before 1974, the type of pump invariably installed was an old-fashioned cast-iron pump which, while it might have the apparent advantage of costing very little, broke down with monotonous regularity. This kind of pump, patterned on types used years ago in the rural Western world, was intended for use by a single family. Under the pressure of use by the two or three hundred families living in an Indian village, the pump's strength soon gave out. The Indian Government therefore invited UNICEF to help develop a heavy duty hand-pump which could withstand the strenuous requirement of providing a whole village population with a continuous supply of clean water. The result was the India Mark II Hand-pump, an all-steel pump which is produced under strict quality control and which, in the past seven years, has gained a reputation for technical reliability.

Properly installed, the Mark II hand-pump can function for considerable lengths of time without the need for major repairs. The design of the pump took account not only of the physical environment factors — typical depth of boreholes, ease of installation, heavy use — but also of the vagaries of the human environment. For example, the spout was bent so that village children could not poke sticks into it, and the handle reinforced so that the village children could swing on it without mishap. Above all, the pump was designed to be very easy to service and maintain.

The second major stumbling block to the successful provision of clean water in the 'problem' villages was the lack of a maintenance system and at the same time the lack of any sense of involvement by the village in the new pump's installation and upkeep.

Engineers are not trained as social workers; they tend to concentrate on the mechanics of their job and, once having installed a pump, replaced a pump, or repaired a pump, they get back into their landrover and drive away. Without a conscious effort to involve them, many village people thought of the pump as belonging, not to the community, but to the engineers. They were apathetic about its maintenance and its state of cleanliness, and if it broke down they simply went back to the supply source — usually an open well — which they had used before. The block or district engineer would only hear about the breakdown by chance.

In Andhra Pradesh, the village caretaker programme is still in its infancy. It is only a short time since Sri Isamuddin and his counterparts in other villages were trained. Not all of them bother as much as he does about taking off the pumphead and checking the chain, tightening the bolts and sweeping the concrete base. Not all of them would, like he did, take a twenty mile ride on the bus into town, paying his ticket and losing that afternoon's business, to fetch the block engineer when the postcard failed to bring him. Even he may lose his enthusiasm if he is not given a refresher course in a few months' time, or if his spanner is stolen and the water department fails to replace it. There are many details, major and minor, which could do with improvement. Sri Isamuddin is the first to admit failure so far in persuading all the local women to follow his example and keep the pump area spotless.

But the combination of Mark II hand-pump and village caretaker is having some definite effect. One survey in a nearby district carried out within the last twelve months discovered that over 95 per cent of the newly-installed Mark II hand-pumps were working. The equivalent figure of a few years ago at any one moment would have been around 30 per cent.

Maggie Black, *Development Forum*, 1982.

One Tier or Three?

The 'experts' have come up with the idea of a Caretaker who is normally an unpaid youth doing something else for a living and who, after being given two days' training, is responsible for keping the hand-pump clean, the bolts tightened and also doing some health education. For any major repair he corresponds with the Block Mechanic by a system of post cards.

It is immediately evident that this Three Tier System has been designed by engineers and economists who have never lived and worked in a village. They have never experienced what it is like to live without safe water for months because the hand-pump is out of order, when neither the Block Mechanic nor District Maintenance Units have shown the slightest interest in responding to repeated calls from the community.

In the State of Rajasthan, the Three Tier System had not worked: more than 50 per cent of the 20,000 hand-pumps were out of order. Already, disturbing reports were coming in that hand-pumps installed by private contractors through the PHED were poorly installed. Poor cement and adulterated materials had been used for foundations and washers and threads on the pumps' rods were beyond repair. The Indian Government had recently declared that the community would be given more responsibility in planning and decision-making through village elected bodies.

Both these factors contributed to the decision taken by the Rajasthan State Government to implement radical plans for a One Tier System. No other state government has so far put similar plans into practice. They are:

— The repair and maintenance of hand-pumps is the responsibility

Item	Three Tier System	One Tier System
Cost/pump/year to maintain	Rs.400-500	Rs.100 per pump Rs.50 for spares
Tools and equipment	Trucks, jeeps, heavy repair tools, special tools	Cycle, special tools
Educational qualifications	Mechanical degree holder: Diploma from Industrial Training Institute	Fourth standard pass: primary school level
Personnel	Superintending Engineer, Executive Engineer, Assistant Engineer, Block Mechanics, Caretakers	Handpump Mistri (HPM) at the village level
Training	No long-term training programme at any level. Only orientation time for Engineers and two days for Caretakers	Three months field training under TRYSEM, two months practical training on-site. Regular in-service training
Community participation	Marginal — at Caretaker level	HPM identified and selected by the village community: priority given to scheduled castes
Community accountability	None Answerable to the Government	The village has the right to recall the HPM. If he is not working satisfactorily he can be replaced
Institutional finance	No provision Tools are free to caretakers	HPMs take a loan from the nearest bank for Rs.2,500. There is a 50 per cent subsidy if the HPM is a member of a scheduled caste

of the community and not the government.

— A rural youth with some mechanical background (for instance cycle repair, blacksmithing, diesel pump repair) is selected by the community and is sent for three months' training to Tilonia. He is then officially called a 'Hand-pump Mistri' (HPM).

— Training is under a Government of India scheme called TRYSEM (Training of Rural Youth for Self-employment) and includes how to conduct major and minor repairs both above and below ground. The training is practical and on-the-spot. For two months out of the three he is under training, the HPM works on faulty pumps in the villages. The HPM is not to be a government employee, but is answerable to the community. The employment is part-time.

— After training the HPM looks after between thirty six and forty hand-pumps within a radius of 5–10km of the village where he is based.

— The State Government pays the HPM Rs.100/hand-pump/year under his charge and Rs.50/hand-pump/year for replacement of spare parts. After training the HPMs get a grant of Rs.250 from the Government, under the terms of the TRYSEM scheme, to buy tools. A set of special tools costs Rs.2,500. The government or training institution arranges for a loan from the bank so that the tools become the property of the HPM within a year.

— Tilonia trains trainers from Industrial Training Institutes from six other districts of Rajasthan so that the HPM system can be used all over Rajasthan for hand-pump maintenance.

— HPMs are selected mainly from the socially vulnerable groups, the 'scheduled' castes. Since March 1981, 133 HPMs have been trained at Tilonia to cover the districts of Ajmer and Bhilwara in Rajasthan. Trainers from the six Industrial Training Institutes are holding their own HPM training programmes.

Sanjit Roy, *Waterlines*, 1984.

 # Training for Self Reliance in Peru

We arrived at Culta in Peru, a rural community where more than 400 families live. it is typical of the altiplano community, where people depend on a subsistence economy, harvesting one crop a year, and where the social tie is basically the family. Culta, like other communities in the altiplano region, is organized around a communal assembly (composed of adult males only), which represents its decision-making body. The assembly elects a president, who represents the highest authority for the community. Any action undertaken within its boundary which ignores the president or the assembly can only meet with failure.

Immediately after arriving in Culta, we started talking with Pedro Camapasa, a local educational promoter who works a half-day with children in a small house provided by the community. Pedro is a young member of the community who has been selected by the local assembly to conduct an initial non-formal educational programe for children under five years old. He has received special training from the local chapter of the Ministry of Education.

Pedro's three-month training period generated a process of local mobilization in which parents and children are the main participants. The process started when parents themselves built the house for the programme and made some educational material for their children to play with. Mothers have organized themselves to prepare a daily lunch for the children who attend the programme.

Children are encouraged to understand their social, cultural and ecological environment by playing with things similar to those of their surroundings and those they have at home. In this way, children not only develop their ability to communicate by enriching their vocabulary but also develop their motor skills.

Pedro told us that he advises mothers to provide an appropriate meal for their children and encourages them to use the services of the local health promoter. The health promoter is a para-professional voluntary community worker. He, like Pedro, has been elected by the community and has also received special training from specialized technicians of the Ministry of Health.

The president of the community explains that prior to Pedro's training and activities, the children had to stay at home alone while their parents went to work on their plots. Older children, from six years and upwards, had to herd the sheep and llamas, while younger ones had to remain at home crying or sleeping. 'This is a waste of the best and most beautiful time in one's life' he says. 'Now children can learn and play at the Wawa-uta'. He has also observed that children who go to the Wawa-uta are more alert and less shy than those who do not go.

This educational programme, directed mainly toward children, allows the introduction of other activities such as health and nutrition. The Wawa-uta becomes the centre of community development with the active participation of the parents in improving the situation of children.

The health promoter provides vaccination and sanitary assistance to children attending the programme, and through them the promoter reaches their parents, giving them advice on how to protect the children from diseases and parasites. At the same time, the work of a nutrition promoter complements the efforts of the education and health promoters. Through nutrition activities, mothers learn new recipes to prepare nutritious and appetizing food, using local ingredients.

The parents participate in raising some small domestic animals provided by the local branch of the Ministry of Agriculture. Parents also learn appropriate technologies to improve their agricultural production. All these activities aim at improving the family protein intake and raising the family income.

Regarding health conditions, a doctor who has been working in Peru's rural areas for more than ten years, says that generally people do not use the local health services but prefer to go the 'curanderos' or the local pharmacist. Instead of being discouraged by this situation, the promoters of the programme believe that the training of traditional health practitioners can improve health services at the community level. These workers are especially important because doctors do not want to come to these communities. For this reason, mass training — including the use of closed-circuit television — is being encouraged by UNICEF so that people can learn to use health services more efficiently.

Unquestionably, Pedro's training has revealed the potential for development of his community. There are about 550 'Pedros' around the department of Puno, in southern Peru. This means that out of 1,500 communities, 550 are already providing their children with some basic services of health, education and nutrition. The overall impact on the quality of life in the communities is still to be measured. However, there is a significant accomplishment in the very fact that those communities which had been regarded as backward and resistant to change are now protecting more than 20,000 children from disease and death.

Talking to Pedro, parents reach a new understanding of the needs and nature of their children. This is very important in communities where children are not much valued before a certain age when they can start to

Saying yes to literacy (in Peru). (*UNA*)

work. The fact that so few of them survive creates feelings of resignation which too often are counter-balanced by a high birth rate.

Some ECLA (Economic Commission for Latin America) studies have shown evidence of the relationship between the mother's education level and infant mortality. Infant mortality is higher among women with less education. Although lower education of course is not a cause of infant mortality, it is in fact an intervening factor and training of more 'Pedros' should be expected to reverse this situation.

Yves J. Pellé and Orlando Lupo, *UNICEF News*, 1979.

Why We Need Female Extension Agents

There have been a number of occasions when project directors and planners have realized that the success of a project was being undermined by the fact that, although the main effort was aimed at male heads of households, much of the animal production work was performed by women. Consequently, without the direct involvement of women in project activities it was difficult for the project to achieve much success. In some cases it was possible to rectify this to some extent by involving women alongside men. However, this was not generally possible since women could not always both attend domonstrations and carry out their household duties. Furthermore, in highly segregated societies, it is not always acceptable for women to attend demonstrations alongside men.

Even if project authorities realize this problem they may find there is no easy solution since in many societies it is difficult for outsiders such as extension workers to approach female household members directly, or without the presence of the head of the household. This is well illus-trated by experience with an FAO poultry project in Pakistan: during a poultry vaccination campaign it was discovered that chickens kept by village women were being missed. It was culturally inappropriate for male veterinary assistants to visit these women, whose husbands were absent, to vaccinate their chickens. Many husbands has migrated. The realization that the unvaccinated birds were a potential source of infection for the whole village, and therefore an obstacle to the introduction of improved breeds, led to extensive training of female extension workers to improve contacts with village women.

This indicates the necessity for female extension staff, even though this may create further problems since it may be difficult for single women to travel alone in rural areas or for married women to travel without their husbands. Special solutions have, therefore, to be worked out, depending on the country, the region, the dominant religion and so on. In some places, it will be found appropriate to place local female extension workers within their home areas, or near to where their husbands work; in others, to train local women, either young unmarried women or married women, depending on local custom regarding the movement of women on their own. In areas where views on such matters are strictly held, female extension workers might work in pairs.

Problems may also arise when local custom prohibits women from meeting government officials or from attending meetings in public places. In a discussion of the practical problems of incorporating women in to the development process, it was suggested that such restrictions may be circumvented by holding a series of smaller meetings in the houses of the more prestigious families, or by persuading village leaders to set aside a certain place, such as the schoolroom after hours, for the use of women.

These difficulties have not meant that women have been completely excluded from animal projects for crop farmers. Indeed there have been some areas of success, notably in the dairy co–operatives in Anand, India, some of which were established by and are entirely managed by women. Dairying is, however, recognized more as a women's activity than other animal husbandry activities.

Clare Oxby, *World Animal Review*, 1983.

Time to Learn for Women

Following the failure of many formal education projects introduced to try to involve rural women in significant numbers, the non-formal Women's Education Project was launched in Upper Volta with two main objectives in mind: to gather data on the barriers preventing the full access of women to education, and to initiate experimental programmes to overcome these obstacles. As well as addressing the issue of women's excessive work-loads, the Project also tackled the problems of poor health and low standards of living: all were factors which preliminary sociological studies had pin-pointed as fundamental problems.

Three labour-saving technologies were introduced, with the idea that the time saved through using these could be devoted to training the women in improved agricultural methods, health and civic education, income-generating activities and literacy classes. Dynamic village women and traditional midwives were chosen by the villagers to attend special training courses in knowledge dissemination. Each village association organized by these women was given a mechanical grain-grinding mill and carts for the transportation of wood, water and crops; easily-accessible wells were dug, too. The plan was for the village women to utilize the equipment on their own behalf, but to have the opportunity to rent it out in order to earn collective revenues for the co-operative.

From 1976 to 1979 an evaluation of the Project was carried out to establish whether time did, in fact, constitute a significant barrier to educational activities; to determine the effectiveness of the appropriate technologies introduced and to assess how far the Project had increased the participation of women and girls in education programmes.

A vital contribution of the Project was the confirmation that though time is an obstacle, the careful choice of labour-saving appropriate technologies can release time to be spent on other activities. That the technologies introduced address three areas of work which are particularly time-consuming to Third World women — crop processing, water collection and fuelwood gathering — is also significant.

It reinforced the idea that educational programmes should be presented as a package. Thus, literacy classes should be integrated in a plan which introduces appropriate technologies and a dynamic functional education programme as the means to improve living standards. The commitment of the participants to the programme is the key to its acceptance in a broader sense as well. Prior to the inauguration of the Women's Education Project, missions were undertaken throughout Upper Volta to provide information on the Project to local authorities and to study the situation of women in various regions of the country. Villagers were kept abreast of the Project's plans on an ongoing basis. Local people were trained for managment functions at the central administrative level and for the conduct of the Project's multi-disciplinary operations at the local level. Villagers chose the women who they wanted to be trained as 'change agents'.

In the early stages of the Project, insufficient attention was given to the acquisition and management of communal equipment including training of millers and mechanics. Steps were taken to rectify this in later years; by 1978, villages wanting the technologies had to make requests and propose means of managing equipment as agreed beforehand at a village meeting. A village would thus commit itself to maintain, repair and make reimbursement for the equipment. The early problems experienced with this aspect of the Project serve to emphasize the importance of clarifying the management questions regarding who is to be responsible for the technologies on a daily basis, who is to own them, who is to repair them and on what terms others will be allowed to use them: all must be answered before technologies are introduced to a village.

Brenda McSweeney, *Appropriate Technology*, 1982.

The children of the mind are like the children of the body. Once born, they grow by a law of their own being, and, if their parents could forsee their future development, it would sometimes break their hearts. — R. H. Tawney

Producing Basic Literacy Materials

Kenya, like many Third World countries, is faced with the problem of high illiteracy. According to a survey carried out in 1976, slightly more than half of the adult population was found illiterate. In fact, when launching the National Literacy Campaign in December, 1978, the President of Kenya observed that it was estimated that 35 per cent of all male Kenyans above the age of fifteen, and 70 per cent of all female Kenyans in the same group, could not read and write. By declaring war on illiteracy, the President opened the way for all Kenyans to develop their abilities, enrich their lives, and participate more fully in the social, economic, cultural, and political life of this country.

That was four years ago and since then a lot of development has taken place. To realize the dream of a fully literate nation, a Department of Adult Education was created in 1979. By April the same year, the Department had recruited 3,000 full-time literacy teachers who were given a two-week course in methods of teaching adults reading, writing, and numeracy. On completing the introductory course they went back to their villages to start literacy classes.

Another 5,000 part-time teachers were recruited to support the programme and by 1981, the number of teachers involved in the programme had grown to about 15,000.

The next major problem which faced the Department after the recruitment of teachers was providing learning materials for the adults who had continued to enrol in large numbers. For a country with over forty two local languages, developing reading materials can be an enormous problem. The process of writing primers is a very slow one involving surveys of learners' needs, linguistic surveys, and organization of writers' workshops. If the primer being developed is to be functional, a period of pretesting must be allowed. It might take months or even a year before a primer finds its way into the literacy class. The bureaucratic tender system used to award printing contracts in the Third World countries does not make the production of primers any easier. . . .

Since 1979 primers in only fifteen languages have been produced. Furthermore, the primers so far produced represent large geographical and linguistic areas which do not necessarily have homogeneous learning problems. This makes it hard for the primers to focus on any specific learning needs. . . .

With the general economic recession in the Third World countries, progress is no longer a matter of seeking financial allocations to solve development problems, but of trying to find more cost-effective means of solving them. Kenya is no exception, and with the prevailing economic climate, the term 'low-cost' was bound to gain more prominence. In fact 'low-cost' has come to be equated with appropriate educational technology.

The Department seems to have recognized the problems with developing primers, and, in 1980, with the help of the British Council, it initiated an experimental project in low-cost print media. At that time, some teachers who had no primers had started to improvise in one way or another by producing their own teaching materials; those who could not produce materials were using books meant for children. . . .

What the Department wanted most was to provide teachers with skills to enable them to solve their teaching problems. The aim of the low-cost print project was therefore to test whether it was possible to reduce cost and time by producing materials locally instead of relying on national headquarters, and to improve teachers' skills in the design and production of learning materials.

To try the new idea, two pilot projects were begun in two districts of Kenya. Depending on their success, they would be expanded to cover the whole country. . . .

One year after the pilot project it was decided the projects should be expanded to cover the whole country. A national trainers' workshop was organized in Mombasa for two weeks in October 1981 to train trainers for the second phase of the project. During the two weeks, participants, who were drawn from all the administrative provinces of Kenya, were taken through the low-cost curriculum. As well as learning a variety of skills in low-cost materials production, they learned how to organize a low-cost training workshop so they could go back and present training workshops in their own districts. . . .

There is a lot of evidence which shows that relevant and cost-effective materials can be produced locally. Such materials, if well designed, can make the education being provided more relevant for individual and community development.

There are signs that the project is having a multiplier effect. Originally, the Department had planned to train only the full-time teachers, but observations in districts where training has already taken place show part-time and self-help teachers have also benefited. Most of them are now designing and producing word and syllable cards.

It is obvious that the success of the low-cost print project will depend very much on the availability of funds which will make it possible to train teachers in the remaining districts. Observations have also revealed the need to train officers who are in charge of the literacy programme at the district level to ensure efficient co-ordination.

Apart from the obvious benefit of training teachers in new approaches to materials' production, the project has taken learning closer to the community, since before they can design materials, teachers have to carry out a learners' survey to find out learners' interests.

There is no doubt that enormous economies can be achieved by simplifying the production of teaching materials. This low-cost project has shown that Third World countries do not have to be trapped in the web of modern education technology. There is need for continuous search for a technology which is appropriate both in cost and in meeting local educational needs. Low-cost print can be one way of achieving that technology.

Muriithi Kinyua, *Development Communication Report*, 1983.

Communication Technology

It was precisely the growing recognition of the need for better participation in development programmes by rural people, and for improved transfer of know-how to them, that led to the idea that communication media and techniques could be used to greater advantage than they had been. Extension systems, relying primarily on interpersonal communication, were (and still are) afflicted by many problems. Shortage of trained extensionists, of transport and fuel, and the sheer size of the task facing them made it necessary to redimension their efforts through the complementary use of mass media. And with illiteracy at such high levels, audiovisual training systems could be of prime importance.

Creative use of communication media could also provide a two-way channel for information flow between rural communities, and between these communities and the authorities. In this way, the information exchange necessary between all concerned in co-operative efforts could be assured.

This thinking is just as vaild today as it was when it began to be generally accepted about ten to fifteen years ago. Regrettably, systematic use of communication for rural development has not kept pace with acceptance of the theory.

Radio broadcasting for rural audiences is an example: there were excellent examples of how radio programmes had been able to support agriculture in industrialized countries. The Radio Farm Forums of Canada in the 1930s was one; and 'The Archers', the farmers' BBC radio series, begun in 1951 and still going today, was another. Both the Canadian Farm Forum experience and 'The Archers' in Britain were launched at a time when there was a serious need to improve agriculture and raise food production in the respective countries, a condition applying to developing countries today. 'The Archers' format, a dramatized serial about a farming family with 15 per cent instructional content and the rest entertainment had, and still has, an enormous following among town dwellers too. In fact, in 1953 one adult person in three of the whole population of Britain listened to it regularly, thus helping to keep the urban sector interested and informed about farmers and their

(*Mazingira, 1979*)

456

problems and enhancing the prestige of agriculture.

In the 1960s most countries had broadcasting services that covered the majority of the national territory, and today coverage is practically complete. True, radio receivers in the 1960s were relatively cumbersome and expensive, but this problem could be largely surmounted by organizing collective listening and providing each group with a receiver. Then as transistorized receivers became cheaper, smaller and lighter consumers of batteries, all of us involved in communication for rural development thought that the era of rural broadcasting in the Third World must be dawning. There were many technical assistance projects run by both bilateral and UN agencies such as UNESCO and FAO, and for a while some improvement took place in rural broadcasting in some developing countries.

Today, however, the general situation is abysmal. Rural broadcasting seldom, if ever, gets more than a fraction of total air time. A survey a few years ago showed that in most countries it got from 1-5 per cent of broadcasting time. An FAO consultant recently visited four African countries to study the possibility of media campaigns that should help reduce post-harvest food losses, estimated at 12 per cent of all grain grown in Africa. The consultant reported that in one country with well-developed broadcasting services and 90 per cent of the population living in rural areas, only 1.22 per cent of air time is given over to rural broadcasting. The average for the four countries was 17 per cent.

If such a situation were exceptional, the matter would not be so serious, but similar low priority for rural broadcasting is to be found almost everywhere. . . .

The missed opportunities and unrealized potential of rural broadcasting can best be illustrated by a couple of examples of what radio has achieved in some countries. In Ecuador, a one-minute advertising spot on prevention of goitre, repeated several times a day over one year, increased the proportion of households using·iodized salt from 5 per cent to 98 per cent.

In the late 1970s a disease threatened the vital cassava crop in the People's Republic of the Congo. An FAO-supported rural radio programme broadcast a warning that the crop should be harvested immediately. Harvesting cassava is exclusively women's work in central Africa, but the radio suggested that, in view of the crisis, the men should help. Reports soon began to arrive that a remarkable phenomenon was taking place in the villages: for the first time in memory, the men were helping the women to dig up cassava. They were doing so 'because the radio said to'.

Colin Fraser, *Ceres*, 1983.

At present, far from eradicating any of the problems, education seems mostly to bypass them in rural areas, and, in fact, to introduce new ones.

— Kusum Nair

The Barefoot Microchip

First came the idea of the agricultural extension worker, carrying the message of science and technology to improve agriculture, going barefoot to carry his message better and spread the green revolution wider. And we have also the barefoot doctors to ensure health for all. But the benefits of these marvels have reached only a tiny minority.

And now the barefoot microchip. Where could this microchip revolution and its marvels lead? Will it go the way of other whiz solutions that have found no answers to the problems of development, or could this sophisticated modern technology and its marvels provide the magic answer? . . .

Whether at global or village level, the dilemmas of the microchip involve transfer of technology, transfer of knowlede and issues of domination and co-operation. The new information technology and its trends, its constant adaptability to specific needs, the fantastic reductions in costs bringing it within reach of almost everyone, the diversification of all information media and the large range of technologies made possible by the microchip are all interlinked. However, the communciation needs of this technology and its priority in integrated national planning are not fully recognized.

At village and national levels it involves choice of priorities, and at the global level there are problems of standardization and the contradictions caused by competitve needs. The technology also throws up human problems — perception of man and his role, societal problems of the link between elites and people, between micro-societies and transnationals and states at the macro-level. Is there a technologically universal way to combine all the advantages of all the media and link them together?

The microchip revolution in electronic technology has been such that its full capabilities and problems are yet to be grasped even by its originators and promoters. The ability to store and process information in ever-smaller chips — whose final capacity may only be the finite number of electrons in a given size — continuous energy conservation and the attendant revolutions through satellites in communications technology, have reduced costs by factors of tens if not more, and the end is not in sight.

The marvels of this technology are already within the means and reach of the ordinary person in the industrialized world, but are still well beyond the means of the Third World. In future it could bring to the rural areas of the Third World access to a telephone and, through it, access to urban-centred national and international data bases, and ability to seek and obtain needed information. It can bring within reach of the village or comminity, access to modern medicine, can link up communities and teaching institutions, teachers and the taught, through satellite broadcasting, to receive and/or participate in conferences.

It can also work the other way,

458

rather less benignly. it can help those at the top of the pyramid to keep track of what is going on the bottom — and prevent too much movement, lateral or vertical, that may disturb the top, though it may well increase communication within communities, and horizontal communication among communities. But it would depend on those at the top, those who set up and control all these marvels. for it is they who will design or have designed what the system can or cannot do. it is they who will process the accumulated data, and decide how and how much of the processed data will be available and to whom.

In effect, those at the bottom can develop by increasing their capacity within remote communities to act and interact, but the means and processes, and the parameters, will be decided by those at the top. It is not merely an issue of technology in isolation or of costs in patricular but of the politics of development. It is not only an issue of Third World elites and and governments, and their lack of concern for their peoples as some may see it in the North. The whole question is also essentially linked to and governed by the realities of the international economic system. Rural-urban and people-elite dichotomies in the developing world are a replica of, and are tied to, North-South relations and the processes of the transnationalization of the global economy and assymetric interdependence.

The microchip revolution can enable the peasant quickly to know the prices that his products will command and perhaps plan his production. but he still sells to the middleman and the trade, and prices are determined elsewhere, not by him and, due to the transnationalization of the global economy, by forces beyond national frontiers and governments. The prices of coffee, cocoa or tea are not determined by the producer who cannot take advantage of the supply/demand equations. Long before he or his government (whether of Brazil, Ghana/Ivory Coast, India/Sri Lanka) even know the extent of any particular harvest or the effects of the weather, the information has long since been gathered through satellites in the sky and transmitted by the same microchip marvels to the giant transnational (TNC) conglomerates who process, assess and set prices. . . .

The basic problem facing the Third World countries trying to catch up with the industrial revolution that passed them by in the colonial era, is that they are now facing a major technological revolution in the industrial counties — in biotechnology, ocean technology, renewable energy, communications etc. the new technological structures that go with them and the ever widening gap between rural communities and urban centres within nations and among nations.

Could the microchip and its marvels help close this gap and break the dependence linkage or would they perpetuate it in an even greater way? Third World countries will no more be able to prevent the intrusion of all these 'marvels' than King Canute could stop the waves.

If they do use these technologies and the structures that go with them — for technology is not neutral and brings with it the value systems and contexts in which it was developed — they join the rat race and with it the

increasing problems of unemployment, poverty and income-gaps. If
they somehow decide not to join it,
they lose out on the world system,
and become even more peripheral.

Development Forum, 1983

Solar TV?

The problem is most certainly not the technology. Certainly, solar-powered television receivers, for example,
could be very useful in providing programmes to rural
areas with poor electrification. But if we had them, would
television programming for rural people improve? I
believe not, at least judging by rural radio programming,
which certainly has not improved in quantity or quality
since the arrival of the cheap and relatively widespread
transistorized receiver. In sum, all our experience leads
me to conclude that the mere availability of cheap and
appropriate communication technology does virtually
nothing to ensure its use for rural development.

Colin Fraser,
Ceres, 1983.

Video in the Village

It is not easy to talk to an Indian villager. It is not easy to have a talk
with him either. The professional
communicator from the city
flounders in the atmosphere of polite
indifference in the village that arises
out of a deep-rooted sense of distrust
after years of dealing with well-
meaning development schemes.
Theories and applications of communication for development, well-pro-
gnosticated in an urban environment,
find no sustenance in the parched
earth of rural India. The communicator must go through a long process
of unlearning before he can effectively communicate.

This has been the common experience of many small non-governmental agencies working among the poor
of rural India. For us the process of
unlearning began in 1974, when a
project for the Family Planning
Foundation took a team from our
Centre to Sultanpur, a small market
town in Saharanpur District of western Uttar Pradesh. We were outsiders who had come to assess the
impact of a package of centrally produced family planning films on a
rural audience. Disappointed with
the results of the survey and the lack
of response on the part of the village
people, we decided to carry out an
experiment on our own with a borrowed portapak. The results, however tentative they may have been,
taught us a great many things about

the people of Sultanpur, and about small-gauge video as a tool for communication. Initially, curiosity drew the villagers to the machine. They gathered to watch themselves on the little monitor screen. The children came first and remained our most loyal viewers right through the experiment. But the villagers did not remain mere spectators for long. Discussions, which turned into heated debates, were organized and taped. The uninhibited response of the villagers to what we thought was advanced technology helped to demystify the medium. They tinkered with the camera, did some of their own shooting, helped to tape more discussions, and watched themselves tirelessly on the monitor.

What did we learn from this experiment? We learned that machines are not the exclusive property of educated city dwellers; that video is a convivial tool that helps to turn a passive spectator into an active participant; that it lends a voice to those who have been silent for too long; that it finally opens up a dialogue between the communicator and the communicated to. Video, we discovered, had helped to make us less outsiders in the community. With one camera and a small videotape recorder, we did not disturb the flow of its existence, and the immediate playback facilities established our credibility; there was no mystery about what we were shooting in the village. A chance encounter grew into a long-term association. We came back to Sultanpur again and again. A health project initiated by us spread to more villages around Sultanpur. We continued to learn. The small experiment with which we

began led us to use video in a variety of situations.

Mirzapur, 40km from Saharanpur town, is one of a cluster of villages in western Uttar Pradesh where the landless peasants supplement their income by making ropes from wild grass from the forest. As the land around is largely infertile, rope making has become the only source of income for many families. But the individual producer is at the mercy of the forest contractor who sells the wild grass at a high price and the handful of buyers from the town who manipulate the rope prices to ensure maximum profits for themselves.

In Mirzapur a group of such landless peasants eventually formed a producers' co-operative and managed to obtain working capital as advance from the bank. In the effort to organize the co-operative, video was used extensively to document the problems faced by the rope makers, to initiate discussions, to mediate between individuals and groups and to create mutual understanding. Taped programmes produced with the help of the rope makers of Mirzapur are now being shown in the neighbouring villages, wherever rope makers are still struggling in isolation. Links are thus being established between different groups, and the co-operative is gaining in strength.

The intimate, participatory nature of video makes it an effective tool in organizing and raising the consciousness of a community. With increasing participation of the people in the use of video, the chance of manipulation by the medium are minimized. In the ideal circumstances, the community actively helps to make programmes for its own specific use. The Mandi experience is a case in point.

Mandi, a village 20km to the south of Delhi, was the scene of a struggle by Harijan landless peasants to obtain for themselves the right to till the land that had been officially allotted to them. This land was surrounded on all sides by the land of the rich and powerful Hindus who were preventing the Harijans from cultivating it. After a long and bitter fight they won, and even got the support of the administration and the police. A group of activists working in the area decided to document this struggle, with the help of the Delhi Dehat Mazdoor Union, a trade union of farm workers, landless peasants, stone quarry and brick kiln labourers who work in the immediate outskirts of Delhi.

The video documentation took the form of interviews with the peasants themselves. This and other pro-grammes taped with the help of this community were shown to a large audience at a fair organized by the labouring poor of the area. The programmes dealt with a variety of problems common to all poor communities — the double burden of the woman who works in the field and at home, or the problem of organizing resistance to exploitation. The tapes have also been shown in the villages near Saharanpur, and the Saharanpur experiences have been played back to the farmers and workers at Mandi. . . .

At Sukhomajri, a village north of Chandigarh, a curious use of video emerged. Sukhomajri is situated along the Himalayan foothills. Deforestation and unchecked cattle grazing had led to soil erosion and silting of the rivers of the area, but the building of a small dam and the provision of irrigation facilities to the villagers have brought about a regen-

New technologies: new audiences. (*WHO/UNESCO/S. Seraillier*)

462

eration of the ecosystem and given the villagers a vested interest in the maintenance of the catchment area. Video was used to document this process of change and to explore the difficulties of ensuring the people's participation in such programmes. During organization of the local people for this programme, the women were reluctant to participate in meetings with the men, although it is women who have the greater interest in the use and distribution of water resources. The problem was solved by a separate discussion for the women, which was taped on video and played back at the men's meeting. The machine became the mediator.

Anil Srivastava, *Ceres*, 1983.

They Can't See the Point

The Rendille are a nomadic people of northern Kenya who until recently, had almost no contact with the modern world. . . . They illustrate a problem of dealing with rural people that western experts have begun to perceive only recently. Educators, working in villages to introduce notions of health and hygiene, new farming techniques and improvements in livestock rearing, found that many of the films, charts and visual aids with which they sought to put across new ideas were greeted with blank incomprehension; villagers were unable to 'read' pictures. They were, in short, visually illiterate.

Visual literacy refers to the individual's capacity to extract information from a photo or illustration. Few are aware, that their capacity to understand many subjects comes mainly from constant exposure to television, films and book or magazine pictures. . . . In the homesteads and villages of Africa there are virtually no pictures to look at. . . . Due to lack of exposure to visuals and any form of education connected with interpreting them, people have difficulty in understanding pictures, and may even fail to realize that there is anything to understand. Rural audiences with a low visual literacy level are often unable to understand any messages in the teaching films they have been shown.

World Development Forum, 1983.

> Ill fares the land, to hastening ills a prey
> Where wealth accumulates, and men decay.
> — Oliver Goldsmith

People's Theatre: an Appropriate Medium

People's theatre has been used in Botswana for six years as a means of involving communities in expressing their problems, discussing them and taking action. It provides a good draw for people who are normally bored with development meetings; more important, the plays raise crucial problems which need to be faced by the community, and challenges the community to address these problems and take action on them. As a collective expression and a communal activity it creates the context for co-operative rather than individual thinking and action. As an oral medium in local languages it involves many people who are left out of development activities because of their illiteracy or lack of facility in English.

Community participation is both a major aim and an operational method of people's theatre. There is no specialist expertise required for people's theatre and local villagers are involved in all aspects of the work — identifying the problems to be presented, preparing and giving the performances, and working out the strategies for community action. From the perspective of the extension workers who organize the activity, this medium and the approach with which it is used helps to break down their traditional isolation from the people, and gives them an entry-point for working more closely with the villagers. It requires the extension workers to deal with local issues from a local viewpoint rather than bringing into the community, as they are often expected to do, a centrally-prescribed message of multi-media package. Since the actors and initiatives come from the community, there is a greater chance that the performance will lead to action.

This form of theatre is rough and improvized. There are no long rehearsals, scripts, or memorized lines. The actors agree on a scenario and improvize their words and actions within this basic structure. It works as theatre, as communication because people are playing their own roles, dealing with their own issues, and expressing their own ideas. They do not need to work up an artificial reaction to a certain situation — they have been in that situation and know just how it feels.

By using the people's media we are not only improving communication or harnessing a low-cost media, we are building on the cultural strength of the people and increasing their confidence and capacity in the process. By neglecting their own cultural expression, we inhibit people from active participation in the process of modernization, because an abrupt denigration of traditional forms of culture means denial of access to a kind of literacy to which they have been used. On the other hand, by using a people's theatre which builds on the villagers' own creative expression the creative forces that reside in the people are being brought to bear on the development process.

But mere expression of concerns and issues is not enough. The popular theatre programme is not designed simply to give people a chance to get their grievances and frustrations off their chest. It must lead to analysis and action! The theatre performance is merely the initial catalyst for an ongoing process of discussion, organization and action. At the end of each performance the community meets to discuss the problems presented, to work out solutions, and organize for action.

This notion of people's theatre is quite different from the conventional concept promoted by many development groups whose view is that since folk media works (because of its legitimacy and familiarity among the people) it should be used as a channel for development messages planned by development experts. In Botswana, on the other hand, people's theatre is used to express the people's own issues from their own perspective.

The drama provides an objective view of what is happening in the community which helps community members to stand back and look at it critically. Of course, other media have been used for similar purpose, e.g film and video. Drama has the same immediacy as video for 'playback' purposes but it has the added advantage of using the skills and resources in the community and avoiding the technical complexity and cost of video equipment. Video has been used in Botswana but its vulnerability to bumps, dust and other problems from rural use and the difficulties of arranging regular servicing makes it a liability rather than a useful tool.

The performance helps to make people more aware of their problems. However, in a one-day performance it is difficult to provide all the detailed information and advice necessary for people to change their way of doing things. A lot of questions remain to be answered. A follow-up programme is needed to help answer these questions and provide support for community action. The follow-up programme does not need to be elaborate. It may, for example, simply involve training extension workers to give factual talks, or helping villagers obtain the materials required for their community action projects. It is important, however, that the follow-up takes place as soon as possible after the performance.

People's theatre is not new. However, in many parts of the world it has been organized by professional theatre groups. The approach we have described, which is being used in Botswana, makes it possible for ordinary people to become involved and help themselves.

Martin L. Byram, *Appropriate Technology*, 1980.

A price has to be paid for anything worthwhile: to redirect technology so that it serves man instead of destroying him requires primarily an effort of the imagination and an abandonment of fear. — E. F. Schumacher

The Editor and the publishers acknowledge permission to reprint the extracts granted by the following (the numerals refer to the pages on which the quoted passage begins). The sources of illustrations are acknowledged below each reproduction.

Blond and Briggs Ltd.: 6 (Schumacher); 18 (Schumacher). *Volunteers in Asia, Inc.:* 8 (Darrow and Pam). *OECD:* 9 (Jéquier and Blanc); 327 (Jéquier); 330 (Jéquier); 429 (Jéquier). *James Robertson:* 11 (Robertson); 19 (Robertson). *The Courier ACP-EEC:* 12 (Nyerere); 165 (Muller); 266 (Mister); 299 (Traore). *Pergamon Press Ltd.:* 37 (Diwan and Livingston); 98 (Baldwin); 105 (Chambers); 245 (Foley); 322 (Reddy); 387 (Reddy). *New Scientist:* 65 (Salter); 145 (Harrison); 146 (Sattaur); 215 (Watt); 239 (Jones) 255 ('Japan's Water-mills'). *Fontana:* 29 (Dickson). *IDS Bulletin:* 33 (Green); 336 (Bell); 424 (Dore). *Institute of Economic Affairs, London:* 41 (Lal). *Third World Quarterly:* 43 (Griffin); 355 (Lall). *Westview Press Inc.:* 45 (Evans). *Martin, Secker and Warburg Ltd. and the Putnam Publishing Group:* 48 (Sale); 65 (Sale); 316 (Sale). *André Deutsch Ltd.:* 53 (Galbraith). *Penguin Books Ltd.:* 59 *(The Global 2000 Report). Impex India:* 63 (Schumacher). *Engineering Magazine:* 66 (Davis). *The Estate of Rachel Carson and Hamish Hamilton Ltd. and Houghton Mifflin Co.:* 68 (Carson). *New Internationalist:* 69 (Weller); 314 (Vogler); 422 ('Education for what?'); 423 ('Climbing Jacob's Ladder'). *Macmillan, London and Basingstoke, and Westfield Press, Inc.:* 82 (Stewart); 93 (Stewart); 112 (Stewart); 113 (Stewart). *UNIDO:* 114 (Garg); 265 (Spence); 278 (Barwell); 292 (Garg and Hoda); 412 (Garg). *The Financial Times, Ltd.:* 118 (Buxton). *Approtech:* 201 (Katz). *The Center for International Affairs, Harvard University:* 125 (Thomas); 132 (Wells); 384 (Timmer). *WIN News and Diplomatic World Bulletin:* 140 ('Food Situation Worsening in Africa'). *Oxfam:* 192 (Melrose); 297 (Melrose). *ADAB News:* 143 (Laumark). *Appropriate Technology Development Association:* 146 ('Amaranth Rediscovered'); 312 (Vogler). *Consortium on Rural Technology:* 155 (Gupta). *International Agricultural Development:* 159 ('Ethiopa's One-Ox Job;'); 255 ('World Bank Study cools sun pump rush'). *Overseas Development Administration:* 161 ('Farm Power for Farm Women'). *Earthscan:* 167 (Rao); 204 (Pearson); 205 (Agarwal); 206 (Madeley); 228 (Foley and Moss); 238 (Warigasundara). *UNICEF News:* 169 (Engel); 190 (Rifkin); 209 (Kamaluddin); 448 (Pelle and Lupo). *Food and Agriculture*

Organisation: 175 (Huss); 451 (Oxby). *German Foundation for International Development (Development and Co-operation):* 178 (D'Monte). *Development Forum:* 179 ('Progress — but for whom?'); 199 (Wilson); 200 (Trevedi); 227 (Madeley); 308 ('Small is expensive'); 339 (Chambers); 444 (Black). *World Development Forum:* 180 ('The Poor Man's Cow'); 361 ('What went wrong?'); 463 ('They can't see the point'). *The Ecologist:* 182 (Stiles); 236 (Shiva). *Marion Boyars Ltd.:* 195 (Illich); 285 (Illich). *The World Bank/Finance and Development:* 348 (Dahlman and Ewstphal); 359 (Streeten); 395 (Marsden). *World Water:* 218 ('Pour-flush Latrines in India'). *Canadian Hunger Foundation:* 250 (Greenwood and Perrett); 314 (d'Avanzo). *Ideas and Action — Freedom from Hunger Campaign (FAO):* 219 (Zimbabwean VIPs); 220 ('Community Participation as token jargon'). *African Business:* 225 (Harrington). *National Academy of Sciences:* 234 ('Barefoot Foresters'); 235 ('Fuel-efficient Stoves'). *Yayasan Dian Desa:* 232 (Soedjarwo). *GATE:* 237 ('The Tree Wall'). *International Institute for Environment and Development and Penguin Books Ltd.:* 240 (Ward). *Worldwatch Institute:* 247 (Flavin). *Development Communication Report:* 454 (Kinyua). *Appropriate Technology Development Institute:* 273 (Siegel). *German Foundation for International Development/Elizabeth Hoddy:* 283 (Hoddy). *International Development Research Centre:* 436 (Toomey). *Institute of Mining and Metallurgy:* 307 (Wels). *Commonwealth Science Council:* 309 (Woakes). *Delft University Press:* 372 (Carr). *Macmillan, London and Basingstoke, and St. Martin's Press Inc.:* 376 (Riskin). *Government of India Planning Commission:* 380 (India's Five Year Development Plan). *Government of Kenya:* 382 ('Kenya's Four Year Development Plan'). *SAREC (Swedish Agency for Research Co-operation with Developing Countries:* 390 (Edquist and Edqvist). *Development Digest:* 407 (French). *UNESCO Press:* 330 (Darrow); 428 (Darrow). *Science for the People:* 331 (Downs). Praeger Publishers for the International Institute for Environment and Development and the Overseas Development Council: 29 (De Sebastian); 97 (De Sebastian). *Adult Education and Development, Deutscher Volkshochschul-Verband:* 440 (Bodgaert). *Jonathan Cape, and Harper and Row:* 75 (McRobie). *ACCION International/AITEC:* 441 (O'Regan and Hellinger). *G. A. Natesan and Co., Madras:* 15 (Gandhi).

The International Labour Organisation, Geneva, *Technology for Basic Needs* by H. Singer, pp. 8-10; 25 (Singer); *Technology for Basic Needs* by H. Singer, p. 3; 32 (Singer); *ILO World Employment Programme Working Paper,* July 1981, pp. 52-58; 365 (Fluitman and White).

Ceres, the FAO Review on Agriculture and Development: 31 (McRobie); 148 (McRobie); 149 ('How Pumps Divide the Peasantry'); 150 (Clay); 176 ('Poultry Production'); 300 (Barwell); 333 (Biggs); 352 (Long); 425 (Higgs and Moore); 456 (Fraser); 460 (Fraser); 460 (Srivastava).

Entropy by Jeremy Rifkin, copyright © 1980 by Foundation on Economic Trends, reprinted by permission of Viking Penguin, Inc.: 51 (Rifkin); 147 (Rifkin); 256 (Rifkin).